Science Proficiency Review

PREPARING FOR YOUR EXIT LEVEL TEST

PAUL S. COHEN

Former Assistant Principal, Science
Franklin Delano Roosevelt High
Brooklyn, New York

JERRY DEUTSCH

Chemistry Teacher
Edward R. Murrow High School
Brooklyn, New York

ANTHONY V. SORRENTINO, D.Ed.

Director of Computer Services
Former Earth Science Teacher
Monroe-Woodbury Central School District
Central Valley, New York

AMSCO

Amsco School Publications, Inc.
315 Hudson Street / New York, N.Y. 10013

The publisher wishes to acknowledge the helpful contributions of the following consultants in the preparation of this book:

Michael J. Dick
Physical & Life Science Teacher
Gateway Middle School
Maumee, Ohio

Ned C. Owings
Science Coordinator
K-12 Curriculum, District One
Florence, South Carolina

Kaye Elsner-McCall
Educational Consultant
Secondary Science Teacher
Fulton & Cobb Counties, Georgia

Donna Popowich
Science Supervisor
Secondary Science Teacher
Clifton High School, New Jersey

Richard Lees
Chair, NJ Standards/Frameworks,
Earth Science Subcommittee, E.S. Teacher
Math, Science, & Technology Supervisor
Lyndhurst High School, New Jersey

Serena Troyan
Science Department Chair
Physical & Life Science Teacher
Gateway Middle School
Maumee, Ohio

Please visit our Web site at:

www.amscopub.com

Cover design: Merrill Haber

When ordering this book, please specify:
either **R 116 P** *or* SCIENCE PROFICIENCY REVIEW, EXIT LEVEL

ISBN-978-0-087720-045-1

PRINTED IN THE UNITED STATES OF AMERICA

12 13 10

To the Student

This book provides you with a comprehensive review of science to enable you to prepare for a Science Proficiency Exit Level Test.

The text contains twenty chapters. Topics in life, physical, and earth and space science are covered, as well as the scientific method, the history and nature of scientific inquiry, and the interactions of science, technology, and society.

The text presents the major ideas of each topic in a straightforward manner. The numerous illustrations that accompany the text clarify the concepts presented. Each chapter is divided into several topic sections, which are followed by Exercise sections that test and reinforce the main points covered. The questions are designed to test your abilities in *acquiring*, *processing*, and *extending* scientific knowledge.

In addition, special features called Process Skills appear at intervals throughout the text. The purpose of these features is to teach you a particular process-oriented skill or skills, such as interpreting graphs and diagrams. Each feature guides you through a process skill and then concludes with several follow-up questions that require you to apply the newly acquired skill on your own.

Throughout the book, important vocabulary terms are printed in ***bold italic type.*** These terms are defined in the text and appear with formal definitions in a glossary at the back of the book. Terms of secondary importance and words that may be unfamiliar to you are printed in *italic type*. These terms do not appear in the glossary, but are listed in the index.

Finally, the book includes three Practice Tests, which consist of 50 questions each. The tests focus on the major concepts, understandings, and process skills of the science curriculum. Again, the ability to *acquire*, *process*, and *extend* scientific knowledge is emphasized in these practice tests. We wish you success in your studies!

CONTENTS

To the Student **iii**

Chapter 1 **Classification** **1**

Classification • Classification of Living Things • *Process Skill: Classifying Data* • Classification of Rocks • The Periodic Table of Elements • *Process Skill: Interpreting a Table* • *Process Skill: Making a Data Table*

Chapter 2 **Thinking Like a Scientist** **11**

Observation and Inference • Making Valid Inferences • Organizing Observations • Using Line Graphs • Using Bar Graphs

Chapter 3 **Science Safety Procedures** **20**

Safety at Home • Preparing Lab Work • Working With Chemicals • Working With Heat • Safe Use of Electricity • Working With Sharp Tools • After the Experiment • *Process Skill: Using Laboratory Equipment Safely*

Chapter 4 **Scientific Measurements** **28**

Making Measurements • Tools for Measurement • *Process Skill: Making Precise Measurements* • Mass • Weight • Volume • *Process Skills: Making Measurements; Performing Calculations* • Density • Using Density to Identify a Metal • Why Do Objects Float? • Measuring Temperature

Chapter 5 **Earth's Changing Surface** **42**

Earth's Surface • Surface Changes • Surface Materials • Minerals • Rocks • *Process Skill: Analyzing Data in a Table* • The Rock Cycle • Forces That Change Earth's Surface • External Forces • *Process Skill: Predicting an Experimental Result* • Internal Forces • Plate Tectonics • Ocean Floor Features • *Process Skill: Organizing Data Into a Table*

Chapter 6 **Earth's Motions in Space** **60**

Earth: Member of the Solar System • Earth's Rotation • Daily Time • Earth's Revolution • Yearly Seasonal Time • Seasonal Changes • Proof of Earth's Revolution

Chapter 7 **Atmosphere, Hydrosphere, and Lithosphere** **71**

Interaction of Spheres • Energy Exchange • Composition • Interactive Cycles • The Atmosphere • Defining Weather • Weather Elements • The Sun's Energy • *Process Skills: Designing an Experiment; Predicting Results* • Climate • Large-Scale Weather Systems • Air Masses • High-Pressure and Low-Pressure Systems • Fronts • *Process Skills: Organizing Information Into a Table* • Weather Forecasting • Interaction of Humans and the Atmosphere • Weather Hazards • Storms • Pollution • *Process Skill: Interpreting Data in a Table* • Hydrosphere • Oceans of Earth • Motions of Ocean Water • Winds and Waves

Chapter 8 **Simple Machines** **100**

Work • Machines and Work • Simple Machines • Compound Machines • Efficiency of Machines • *Process Skills: Observing and Classifying*

Chapter 9 **The Laws of Motion** **110**

Force, Mass, and Acceleration • Mass • Acceleration • **Laws of Motion** • Newton's Laws of Motion • First Law of Motion • Second Law of Motion • Third Law of Motion • *Process Skill: Using Math to Analyze Data*

Chapter 10 **Forms of Energy** **121**

Energy Basics • Energy • States of Energy • Forms of Energy • **Nature of Energy Transformation** • Law of Conservation of Energy • Energy Transformations • Unwanted Energy

Chapter 11 **Sound and Light** **129**

Sound • Sound Energy • Sound Waves • The Speed of Sound • Sensation of Sound • *Process Skill: Interpreting a Graph* • Music and Noise • Sound Can Be Harmful • **Light** • Light Energy • Light Can Be Reflected, Absorbed, or Transmitted • Lenses • The Electromagnetic Spectrum • Sensation of Sight

Chapter 12 **Matter** **143**

Matter • Defining Matter • Elements • Atomic Structure • Compounds • Chemical Bonds • Mixtures • Physical Properties • Phases • Changes in Phase • *Process Skill: Interpreting Data in a Table* • **Changes in Matter** • Physical Changes • Chemical Changes • Properties and Chemical Changes • Conservation of Matter • *Process Skill: Predicting Results* • Energy and Chemical Changes • Reversible Reactions • Rate of Reactions • *Process Skills: Making Predictions; Determining a Quantitative Relationship; Graphing Data*

Chapter 13 **Organisms and the Environment** **159**

Organisms and Their Environment • Communities and Ecosystems • *Process Skills: Interpreting a Table; Analyzing Graphs* • Nutrition • Food Chains • Food Webs • Food Pyramids • Symbiosis • Upsetting the Balance of Nature

Chapter 14 **Living Things Carry Out Life Processes** **170**

Characteristics of Living Things • The Cell • Nutrition • Locomotion • Sexual Reproduction in Plants • *Process Skills: Interpreting and Analyzing Data in a Table* • Adapting to Changing Climates

Chapter 15 **Biological Diversity and Heredity** **177**

Diversity of Life • Reproduction • Life Cycles • *Process Skill: Interpreting an Experiment* • Cellular Reproduction • Sexual Reproduction • Inheritance of Traits • Mendelian Genetics • Using a Punnett Square • *Process Skill: Interpreting the Results of an Experiment* • Mutations

Chapter 16 **Living Systems: Human Body Systems** **189**

Organization, Support, and Movement of the Body • Human Body Systems Are Interdependent • Levels of Organization in the Human Body • The Skeletal System • The Muscular System • **Regulation, Digestion, and Circulation** • Regulation • The Digestive System • The Circulatory System • **Respiration, Excretion, and Reproduction** • The Respiratory System • *Process Skill: Interpreting a Diagram* • The Excretory System • The Reproductive System

Chapter 17 **Scientific Discovery** **205**

Science Defined • The Scientific Method • **Examples of Major Scientific Discoveries** • Plate Tectonics • Heliocentric Solar System • Germ Theory • Atomic Theory

Chapter 18 **Energy and Resources** **216**

Sources of Energy • Measuring Energy • Energy Consumption • Fossil Fuels • Other Uses of Fossil Fuels • *Process Skill: Interpreting a Graph* • Other Sources of Energy • **Problems With Energy Sources** • Problems With Fossil Fuels • Problems With Hydroelectric and Nuclear Energy • **Energy for the Future** • Energy Conservation • Reducing Energy Consumption • Reusing Materials • Recycling • It's Your Decision • Renewable and Nonrenewable Resources • Using Renewable Resources • Solar Energy

Chapter 19 **Science, Technology, and Society** **234**

Relationship of Science and Technology • Science and Technology • Science and Technology Advance Each Other • Technology as a System • Technology in Use • **Interaction of Science, Technology, and Society** • Effects of Science and Technology on Society • Effects of Society on Science and Technology • Global Effects of Technology • **Making Decisions About Technology** • Technology Increases Our Choices • Assessing Technology • Technology and Decision Making

Chapter 20 **Environmental Changes in an Ecosystem** **249**

Changing Ecosystems • Biological Balance in Nature • Succession • Competition • **Environmental Threats That Upset the Ecosystem** • Geologic Events • Climatic Changes • Floods and Droughts • Storms • Forest Fires • Human Interference • **Results of Environmental Change** • Biodiversity • Extinction

Practice Test 1 **265**

Practice Test 2 **275**

Practice Test 3 **285**

Glossary **297**

Index **303**

1 Classification

Points to Remember —

- Scientists categorize groups of objects and organisms based on their characteristics.
- Biologists classify organisms into large groups called kingdoms and into small groups called species.
- Geologists classify rocks as being igneous, sedimentary, or metamorphic.
- Chemists classify elements as metals, nonmetals, or semimetals.
- Chemists use the periodic table of the elements to organize the elements according to their properties.

Classification

What is science? This question is difficult to answer. We all know that there are many different branches of science, such as biology, chemistry, physics, and geology. Still, what do we mean when we refer to a field as a "science"? Many dictionaries describe science as a body of knowledge systematically obtained. The methods used—scientific methods—involve procedures that are common to all branches of science. These procedures include classification, observation, and experimentation. We are going to learn about these techniques and see how they are used in different branches of science.

If someone gave you a box of old baseball cards, what would you do first? Perhaps you would separate them by team or by year, or even by color. What you would be doing is called ***classification.*** A classification system for a set of objects groups them together by properties, which you choose. This is an important technique used by scientists in all areas of science.

If you were asked to classify the items in Figure 1-1 into two groups, you might separate them as living things and nonliving things. If you look closely at the group of living things, you might separate it further into two smaller groups, plants and animals.

Figure 1-1. All things are classified as either living or nonliving. Identify the living and nonliving things in this illustration.

Classification of Living Things

Looking at Figure 1-2, you might separate these animals into three types. This classification at first may seem obvious, but sometimes it can be very difficult. We are not always sure how to classify something.

The three groups you might have chosen are mammals, birds, and fish. Where did you put the dolphin? The dolphin is a mammal. Scientists must carefully define the groups they use so that there is agreement on how to classify something. If the three groups you chose were animals that walk, animals that swim, and animals that fly, how would these groups be different? In which group would you place a penguin?

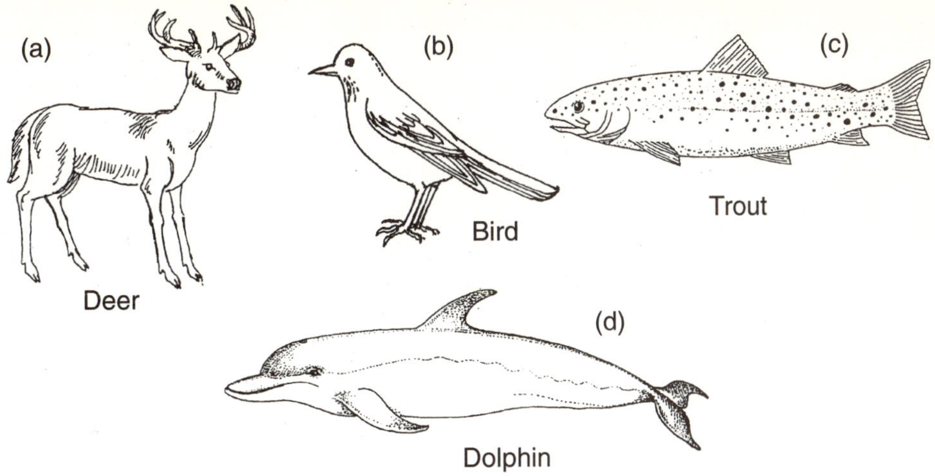

Figure 1-2. Scientists classify animals into different groups, such as mammals, birds, and fish. Where would the dolphin be placed?

In 1737, a Swedish scientist named Carolus Linnaeus devised a classification system for living things. He called the largest group a kingdom. Scientists now recognize five kingdoms of living things. The two kingdoms you are probably most familiar with are plants and animals. Each kingdom is broken down into subgroups called *phyla* (singular, *phylum*). Each phylum is broken down into classes, each class is broken down into orders, and each order is broken down into families. Each family is broken down into *genera* (singular, *genus*), and each genus is broken down into species. This may seem confusing at first, and you might wonder why we need so many different groups. However, think about how you address a letter. You include the country, state, city, zip code, street, and number. The country contains the largest number of locations. As you progress from state, to city, and eventually to street number, the areas get smaller and smaller. The system used by biologists works the same way. A kingdom contains an enormous number of different living things. As you progress down to species, the number of different kinds of living things in each group gets smaller and smaller.

Kingdom, phylum, class, order, family, genus, and species are assigned to every living thing (see examples in Table 1-1). We usually use only the last two names, the genus and species, to identify a living thing. The first letter of the genus is capitalized; the first letter of the species is not. Both are generally written in italics or are underlined.

Table 1-1. Classification of Living Things

Group	*House Cat*	*Red Maple*	*Lion*	*Human*	*Sugar Maple*
Kingdom	Animalia	Plantae	Animalia	Animalia	Plantae
Phylum	Chordata	Tracheophyta	Chordata	Chordata	Tracheophyta
Class	Mammalia	Angiosperm	Mammalia	Mammalia	Angiosperm
Order	Carnivora	Dicotyledonae	Carnivora	Primates	Dicotyledonae
Family	Felidae	Aceraceae	Felidae	Hominidae	Aceraceae
Genus	*Felis*	*Acer*	*Panthera*	*Homo*	*Acer*
Species	*catus*	*rubrum*	*leo*	*sapiens*	*saccharum*

The scientific name for a lion is ***Panthera leo,*** and for a house cat it is ***Felis catus.*** A lion and a house cat are both in the family Felidae, but they are different species. However, we say that they are both cats.

Process Skill

CLASSIFYING DATA

The table below gives the scientific classification for a variety of animals. Look at the table and answer the following questions.

Group	*Wolf*	*Dog*	*Horse*	*Grasshopper*	*Chimpanzee*
Kingdom	Animalia	Animalia	Animalia	Animalia	Animalia
Phylum	Chordata	Chordata	Chordata	Arthropoda	Chordata
Class	Mammalia	Mammalia	Mammalia	Insectae	Mammalia
Order	Carnivora	Carnivora	Ungulata	Orthoptera	Primates
Family	Canidae	Canidae	Equidae	Locustidae	Pongidae
Genus	*Canis*	*Canis*	*Equus*	*Schistocerca*	*Pan*
Species	*lupus*	*familiaris*	*caballus*	*americana*	*troglodytes*

1. Which animals are classified in the same genus but in different species?
   (1) dog and chimpanzee (2) horse and grasshopper (3) wolf and dog

2. Which animal is most closely related to humans?
   (1) wolf (2) chimpanzee (3) horse

3. What do all of these animals have in common?
   (1) They all belong to the same kingdom.
   (2) They all belong to the same phylum.
   (3) They all have a backbone.

Classification of Rocks

A geologist also uses classification. For example, a geologist classifies rocks into three groups—igneous, sedimentary, and metamorphic—depending on how they are formed.

1. ***Igneous rocks*** are produced by the cooling and hardening of hot, liquid rock. This melted rock material is called ***magma*** when underground and ***lava*** when it pours onto Earth's surface. Basalt is an igneous rock that forms by the rapid cooling of lava. Granite is an igneous rock that forms underground by the slow cooling of magma.

2. ***Sedimentary rocks*** form from particles called ***sediments*** that pile up in layers. These sediments may be small rock fragments or seashells. Sedimentary rocks usually form underwater when particles of sediment settle to the bottom of an ocean or lake and harden into rock over time. Sandstone, shale, and limestone are some common sedimentary rocks.

3. ***Metamorphic rocks*** are produced when igneous or sedimentary rocks undergo a change in form caused by heat, pressure, or both. These conditions alter the appearance and mineral composition of the rocks, changing them into metamorphic rocks. Marble and slate are metamorphic rocks formed from the sedimentary rocks limestone and shale. Gneiss (pronounced "nice") is a metamorphic rock that can be produced from granite, an igneous rock.

Classifying rocks by how they are formed enables geologists to learn about the history of a place by studying its rocks. Cozumel, an island off the Yucatán Peninsula in Mexico, is made up largely of a form of limestone called coquina, which is a sedimentary rock made up of seashells. Hawaii, an island in the Pacific Ocean, is made up of basalt, an igneous rock.

What can you conclude about the origins of these places? Hawaii has a very different origin from that of Cozumel. Hawaii was formed by active volcanoes that created new landmasses, while Cozumel was once part of the ocean floor.

The Periodic Table of the Elements

Chemists use classification in many ways. One way is to identify the ***elements.*** Elements are the building blocks of all matter. There are currently 110 known elements. A system of classification for the elements was first published in 1869 by Dimitri Mendeleev, a Russian scientist. Mendeleev knew about 70 of the elements. He arranged the elements in a table according to their weights and grouped them together by their properties (see Figure 1-3). If an element with the expected properties for a particular spot on the table was not known, a space was left for it. Mendeleev knew that more elements would be discovered.

Series	*Groups*						
	I	*II*	*III*	*IV*	*V*	*VI*	*VII*
1	H						
2	Li	Be	B	C	N	O	F
3	Na	Mg	Al	Si	P	S	Cl
4	K	Ca	□	□	As	Se	Br
	Cu	Zn	□	Ti	V	Cr	Mn
5	Rb	Sr	In	Sn	Sb	Te	I
	Ag	Cd	Y	Zr	Nb	Mo	□

□ represents an element that was predicted but had not yet been discovered.

Figure 1-3. The periodic table of Mendeleev.

The modern periodic table of the elements (see Figure 1-4) is arranged in a slightly different way, but it still uses the same basic concepts that Mendeleev used.

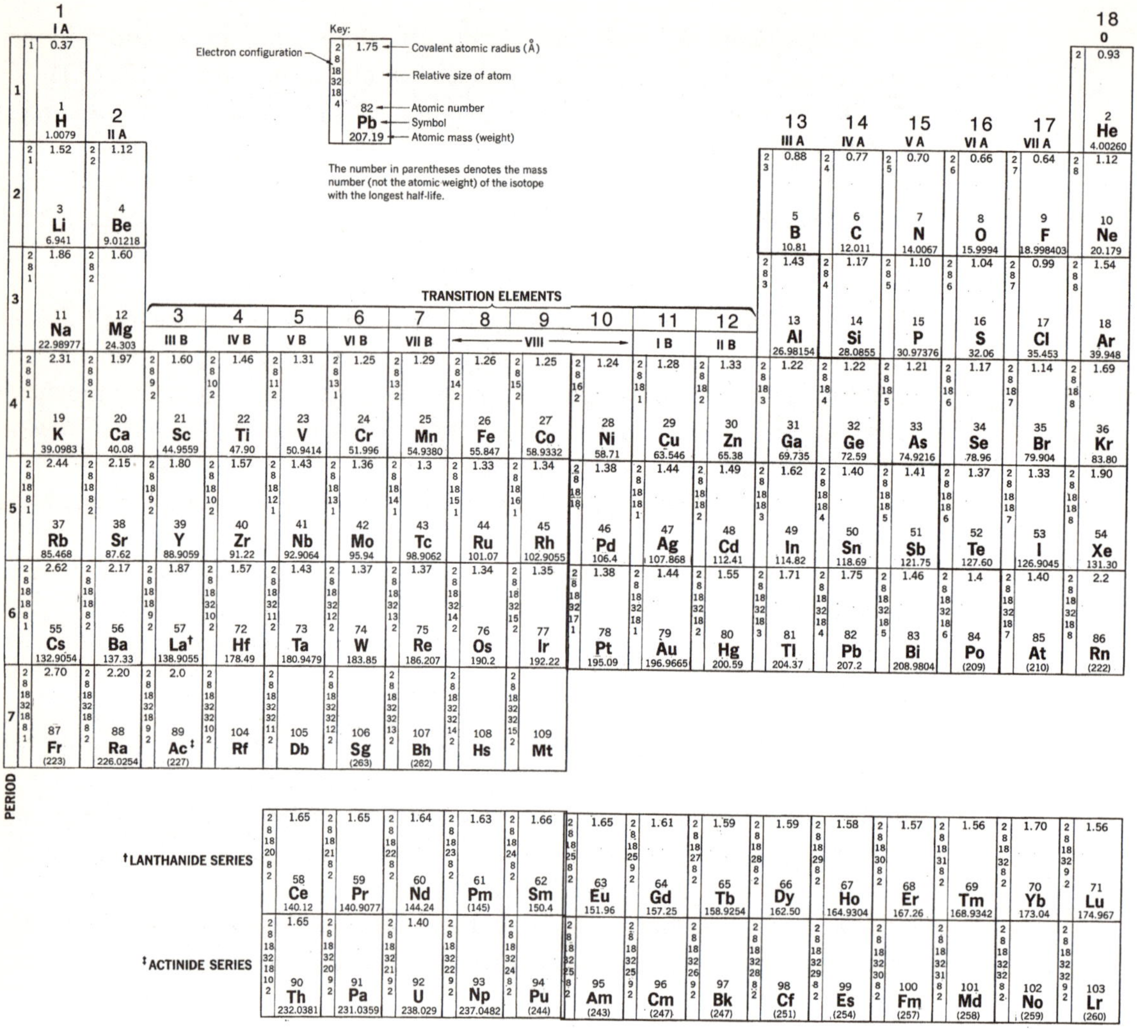

Figure 1-4. The modern periodic table of the elements.

EXERCISE

Base your answers to questions 1–5 on the following table and information, which describe the types of fertilization and development in different animals.

Many animals reproduce by a process called sexual reproduction. During sexual reproduction, sperm from the male joins with an egg from the female. This process, called fertilization, may take

Process Skill

INTERPRETING A TABLE

What do gold, silver, copper, iron, tin, and aluminum have in common? All of these elements are classified as metals. Oxygen, carbon, nitrogen, and helium are classified as nonmetals. The periodic table of the elements classifies elements as metals, nonmetals, and semimetals. In addition, elements with the most similar properties are placed in the same column, going up and down. These columns are called ***groups*** or ***families.*** Magnesium (Mg) and calcium (Ca) are very similar to each other. They have similar properties and therefore are found in the same group. Nitrogen (N) and phosphorous (P) have similar properties, and therefore they are found in the same group. Refer to Figure 1-4 to answer the following questions.

1. Which element would be most similar to chlorine (Cl)?
   (1) sulfur (S) (2) carbon (C) (3) fluorine (F) (4) sodium (Na)

2. Which element is classified as a metal?
   (1) sulfur (S) (2) carbon (C) (3) fluorine (F) (4) sodium (Na)

place either in the female's body (internal fertilization) or outside the female's body (external fertilization). After the egg is fertilized, it develops into an embryo. This process also may take place either inside the female (internal development) or outside the female (external development).

For example, a chicken lays an egg, which is already fertilized. The chicken then sits on the egg to keep it warm as the embryo in the egg develops. Chickens have internal fertilization and external development. The table below shows some animals, their class, their habitat, and the type of fertilization and development they undergo.

Animal	*Class*	*Habitat*	*Type of Fertilization*	*Type of Development*
Goldfish	Osteichthyes (bony fishes)	Water	External	External
Bluebird	Aves (birds)	Land	Internal	External
Bee	Insecta	Land	Internal	External
Dog	Mammalia	Land	Internal	Internal
Frog	Amphibia	Water and land	External	External
Lizard	Reptilia	Land	Internal	External
Whale	Mammalia	Water	Internal	Internal

1. Based on the table, which is required for external fertilization?
   (1) internal development
   (2) land habitat
   (3) water habitat
   (4) only fish can have external fertilization

2. Which is required for internal development?
   (1) land habitat
   (2) water habitat
   (3) external fertilization
   (4) internal fertilization

3. What generalizations can be made about mammals, based on this table?
   (1) They have internal fertilization and external development.
   (2) They have internal fertilization and internal development.
   (3) They have external fertilization and external development.
   (4) They have external fertilization and internal development.

4. A salamander belongs to the class Amphibia. What would you predict about salamanders?
   (1) They live both on land and in water.
   (2) They have internal development.
   (3) They live in water only.
   (4) They have internal fertilization.

5. Alligators have internal fertilization and external development. Which class might they belong to?
   (1) Amphibia (3) Osteichthyes
   (2) Reptilia (4) Mammalia

6. Which would contain the largest number of organisms?
   (1) kingdom (3) genus
   (2) family (4) species

Base your answers to questions 7 and 8 on the table below, which describes two types of chemical reactions.

Two Types of Chemical Reactions

Type of Reaction	*Definition*	*Example*
Synthesis	Two or more materials combine to form one new material.	sodium + chlorine → salt
Decomposition	One material breaks up to form two or more new materials.	water → hydrogen + oxygen

7. Which is a synthesis reaction?
   (1) salt → sodium + chlorine
   (2) hydrogen + oxygen → water
   (3) sugar + oxygen → carbon dioxide + water
   (4) ammonia → nitrogen + hydrogen

8. The reaction: chlorine + sodium iodide → iodine + sodium chloride is not classified as synthesis or as decomposition. This is because
   (1) there are two starting materials
   (2) there are two materials formed
   (3) there are the same number of materials formed as there were initially
   (4) no new materials are formed

For questions 9–11, use the information in the table below, which indicates the density, distance from the sun, diameter, and number of moons of each planet in our solar system.

Planet	*Density (grams/centimeter^3)*	*Average Distance From Sun (km)*	*Diameter (km)*	*Number of Moons*
Mercury	5.43	57,900,000	4,878	0
Venus	5.25	108,200,000	12,104	0
Earth	5.52	149,600,000	12,756	1
Mars	3.95	227,900,000	6,787	2
Jupiter	1.33	778,700,000	142,800	16
Saturn	0.69	1,427,000,000	120,660	18–22
Uranus		2,871,000,000	51,118	15
Neptune	1.64	4,497,000,000	49,528	8
Pluto	2.03	5,913,000,000	2,300	1

One way to classify planets is into the categories of high density and low density.

9. Of the planets with densities listed, how many are low density?
(1) 6 (2) 8 (3) 3 (4) 4

10. Uranus is similar in many ways to Jupiter, Saturn, and Neptune. The density of Uranus, in grams/centimeter^3, is probably closest to
(1) 5.3 (2) 0.01 (3) 1.3 (4) 8.2

11. The low-density planets are generally (excepting Pluto) located
(1) far from the sun and have large diameters
(2) far from the sun and have small diameters
(3) near the sun and have large diameters
(4) near the sun and have small diameters

Use the table to answer questions 12 and 13. The table lists some diseases, the type of disease, and the known cure for the disease.

Disease	*Type*	*Cure*
Pneumonia	Infectious	Antibiotics
Scurvy	Dietary	Vitamin C
AIDS	Infectious	None at this time
Sickle-cell anemia	Genetic	None at this time
Rickets	Dietary	Vitamin D
Strep throat	Infectious	Antibiotics
Athlete's foot	Infectious	Antibiotics (fungicide)
Tetanus	Infectious	None at this time; vaccine (preventive)

12. When Sally had an earache, her doctor prescribed penicillin, an antibiotic. Sally's doctor believed her ailment was
(1) dietary (3) infectious
(2) genetic (4) incurable

13. Iodine is often added to salt in order to prevent a disease called goiter. Goiter is probably
(1) dietary (3) infectious
(2) genetic (4) incurable

Process Skill

MAKING A DATA TABLE

The following data were collected by scientists studying the rocks of an island in the Pacific Ocean.

1: Very hard, green-and-brown speckled rock, which did not react with acid
2: Soft, solid white rock made up of small shells, which did react with acid
3: Hard light-and-dark-gray speckled rock, which did react with acid
4: Soft gray rock, which did not react with acid
5: Hard, shiny black rock, which did not react with acid
6: Medium-hardness, dark gray rock, which did react with acid.
7: Hard, gray-and-white streaked rock, which did not react with acid

To organize the data presented above into a table, we must identify the different types of information contained in the data. The description of each rock includes information on the rock's hardness, color, pattern, and whether or not it reacts with acid. These properties become the categories, or column headings, of our data table.

In a copy in your notebook, complete the table using the data presented above. The data for rock number 1 have already been filled in for you.

Rock Number	*Hardness*	*Color*	*Pattern*	*Reacts with Acid*
1	Very hard	Green and brown	Speckled	No
2				
3				
4				
5				
6				
7				

2 Thinking Like a Scientist

Points to Remember —

- Science is studied by using observations and inferences.
- Observations are made by the senses.
- Inferences are conclusions or predictions based on observations.
- Larger numbers of observations produce better inferences.
- Inferences may be tested through experiments.
- Limiting the number of variables in an experiment produces better inferences.
- Graphing observed data aids us in making inferences.

Observation and Inference

Science has many different branches, which involve the study of different things. All of the branches, however, have certain rules and procedures in common. These rules deal with the way we acquire scientific knowledge. The primary learning tools of the scientist are ***observation*** and ***inference.***

An observation is anything we notice through one of our five senses. An inference is a conclusion or prediction based on an observation or a series of observations. We are constantly making observations and inferences in everyday life, and it is important to understand the difference.

Suppose you walk into a room and notice a certain odor. You say, "It stinks in here! Something must have died!" You have just made an observation and an inference. You observed a foul odor, and you inferred that the cause of the odor was a dead animal. Your inference was based not on just one observation, but on several. If you had never smelled a dead animal before, you would not have been able to make the inference. The best inferences are those that are based on a large number of careful observations. Once an inference has been made, it is generally tested through further observations. In the case above, you would probably want to find the dead animal before you would be certain that your inference was correct.

EXERCISE 1

On the first warm day in March, Jennifer takes out her bicycle, which she has not ridden all winter. She finds it very difficult to pedal. She also is not able to ride as fast as she used to, even though the wind is behind her.

1. Jennifer *observes* that
   (1) she is out of shape
   (2) the bicycle needs oil
   (3) she is traveling unusually slowly
   (4) she needs to ride more often

2. One possible *inference* Jennifer can make is that
   (1) the wind is behind her
   (2) she finds it difficult to pedal
   (3) she cannot go as fast as she used to
   (4) she is out of shape

3. Which procedure would be *least* helpful to Jennifer if she wants to find out the cause of her difficulties?
   (1) trying to ride against the wind
   (2) oiling the bicycle and then riding the same route again
   (3) asking several of her friends to try riding her bicycle and listening to their observations
   (4) turning the bicycle upside down and turning the pedals with her arms, while checking to see if the wheels can turn freely

4. In one or two sentences, distinguish between an observation and an inference.

Making Valid Inferences

There are several procedures that scientists follow to make their inferences reliable. One of these is to make careful observations. In Exercise 1, Jennifer should have listened carefully to hear if her bicycle was making any unusual noises. The sound of the wheel rubbing against the brake, or an occasional squeaking sound would have helped her to locate the problem. While studying a culture of bacteria, the Scottish bacteriologist Alexander Fleming noticed that there were no bacteria growing in one region of a particular sample. By examining this sample more carefully, he discovered that a mold was growing in the region where there were no bacteria.

A second important procedure is to base an inference on a large number of observations. The greater the number of observations, the more likely the inference is to be correct. Fleming might have inferred that the mold kills bacteria, but before he was prepared to make this inference he performed hundreds of additional experiments. On the basis of only one experiment, he could not be sure that it was the mold, and not some other unknown factor, that had killed the bacteria. Fleming's careful observations eventually led to the discovery of penicillin, which has saved millions of lives.

A third procedure that scientists follow is to limit the number of variables. Variables are changeable conditions that can affect observations. Refer to Exercise 1. Jennifer's problem was probably caused by one or both of two variables: her condition and the condition of her bicycle. To make an accurate inference, she needed to eliminate one of the variables. If she tried several different well-maintained bicycles and found that she had difficulty riding all of them, then she could infer that the problem was with her and not with her bicycle. By eliminating a variable (the condition of the bicycle), she could make a valid inference. Fleming had to deal with such variables as food supply, light, temperature, and type of bacteria, any of which might affect bacterial growth. To limit these variables, he took two identical samples of bacteria, under identical conditions, and then added penicillin to one of them and not to the other. In this way, he could confidently infer that it was the penicillin and not some other variable which was affecting the bacteria.

An experiment that tests the effect of just one variable is called a controlled experiment. In Fleming's experiment, one sample of bacteria was treated with penicillin, while the other was not. All other conditions were kept the same in the two samples. The sample that was *not* treated is called the ***control.*** The sample that *was* treated is called the experimental sample. Only by comparing an experimental sample to a control can we make accurate inferences.

EXERCISE 2

Gordon wanted to find out the effect of adding lye to water on the temperature of the water. He filled a beaker with water and measured its temperature as 10°C. He then added some solid lye and stirred until the lye was completely dissolved. (**Warning:** Lye is a dangerous substance. Do *not* try this experiment yourself.) After 10 minutes, he read the final water temperature as 23°C. The air temperature in the room was 25°C. Gordon concluded that adding lye to water causes an increase in its temperature.

1. All of the following could be observed in this experiment *except*
   (1) the temperature went up 13°C during the experiment
   (2) stirring causes the water to get warmer
   (3) lye dissolves in water
   (4) the final temperature was below room temperature

2. Gordon repeated the experiment using the same amount of water but much more lye. This time the temperature increased from 10°C to 60°C. Once again he concluded that dissolving lye in water produces an increase in its temperature. Compared with his first experiment, this experiment produced an inference that is
   (1) equally valid, because he followed the same procedure
   (2) less valid, because large changes in temperature are not reliable
   (3) more valid, because the final temperature was too high to have been caused by heat from the surrounding air
   (4) less valid, because not all of the lye might have dissolved

3. We can make more reliable inferences if we
   (1) increase the number of observations and increase the number of variables
   (2) increase the number of observations and decrease the number of variables
   (3) decrease the number of observations and increase the number of variables
   (4) decrease the number of observations and decrease the number of variables

Organizing Observations

When we want to draw an inference based on a large number of observations, it is important to organize our observations in a logical way. Suppose that we are studying the rate of a chemical reaction at various temperatures. The common household chemical hydrogen peroxide, H_2O_2, breaks down to produce water and oxygen gas ($2H_2O_2 \rightarrow 2H_2O + O_2$). We measure how long it takes to produce 50 milliliters (mL) of oxygen from identical hydrogen peroxide solutions, at five different temperatures. In our first run, at 30°C, it takes 8.0 minutes. In our second run, at 50°C, it takes 2.1 minutes. In the third run, at 10°C, it takes 33.0 minutes. In the fourth run, at 20°C, it takes 16.0 minutes, and in the fifth run, at 40°C, it takes 4.1 minutes.

Before we begin to make inferences about the effect of temperature on this reaction, we need to organize our data in a more logical fashion. Here are two possible tables we could use to organize the data.

Time Needed to Collect 50 mL of Oxygen from Identical Solutions of Hydrogen Peroxide

A

Run	*Temp. (°C)*	*Time (minutes)*
1	30	8.0
2	50	2.1
3	10	33.0
4	20	16.0
5	40	4.1

B

Temp. (°C)	*Time (minutes)*
10	33.0
20	16.0
30	8.0
40	4.1
50	2.1

Which table is more helpful to us in making an inference about the effect of temperature on the time needed to produce a given amount of oxygen? Table A has more data, because it includes the order in which the observations were made. However, the extra data are neither helpful nor important! By listing the results in order of increasing temperature, as in Table B, we make it easier to observe a trend and easier to draw inferences. Table B shows a better way of organizing the data.

Using Table B, we can easily infer that as the temperature increases, the time needed to collect 50 mL of oxygen decreases. We can also infer that as the temperature increases, the chemical reaction proceeds more quickly. The speed at which a reaction proceeds is called the ***rate of reaction.*** In this case, as in most chemical reactions, an increase in temperature increases the rate of the reaction.

There are further inferences that can be drawn from Table B. Although we did not directly measure it, we can infer how long it would take to collect 50 mL of oxygen at 25°C. We infer that it would take less than the 16 minutes needed at 20°C but more than the 8 minutes needed at 30°C. We might even guess that it would take 12 minutes, which is half way between 8 and 16. However, we can draw a better inference by using an important visual tool for organizing data, called a ***line graph.***

Using Line Graphs

In Figure 2-1, we have graphed the data given in Table B. We chose to plot the temperature on the *x*-axis, and the time on the *y*-axis. We then entered our five measured points and connected the points. Note that the lines connecting the points are *not* straight lines. Our five points do not lie on a straight line, so we draw a smooth curve to connect our measured points.

Our five measured points show our five observations. All other points along the curve on the graph are inferences. We infer that our measurements establish a trend and that if we made other measurements they would follow the same trend. From our graph we can infer that, at a temperature of 25°C, it should take about 11 minutes to collect 50 mL of oxygen (see point *a* on the graph).

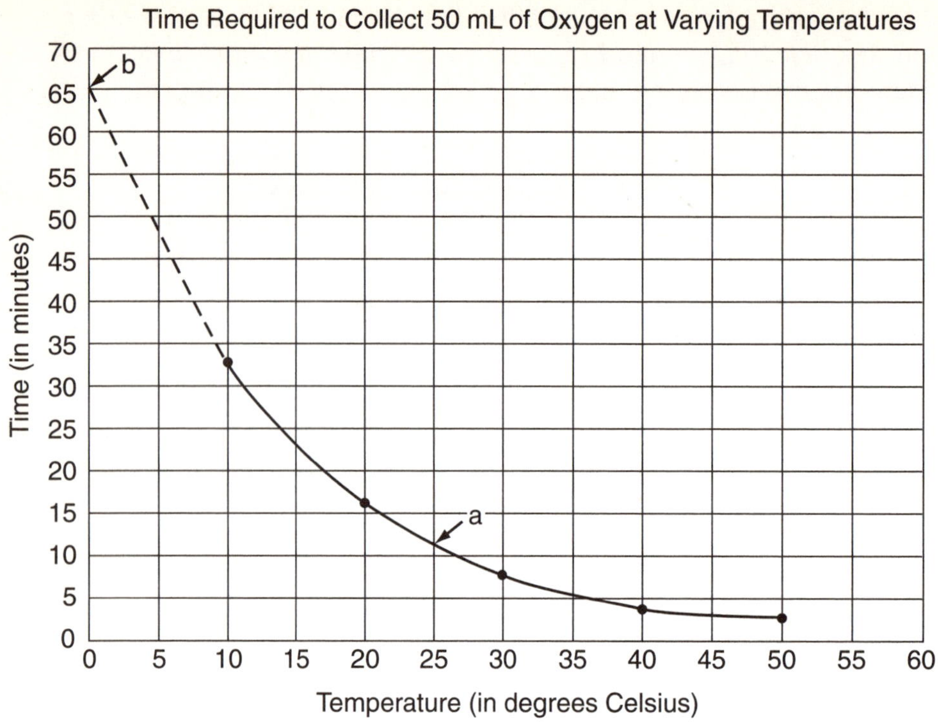

Figure 2-1. The line graph illustrates the effect of temperature on the rate of a reaction.

All of our measurements were made between 10°C and 50°C. On our graph, we extended the curve to show what might happen at a temperature of 0°C. When we extend a curve on a graph beyond our measured points, we are ***extrapolating.*** By extrapolating, we can infer that it would take about 64 minutes to collect 50 mL of oxygen at a temperature of 0°C (see point *b* on the graph).

By graphing our observations, we are able to infer values that lie between and beyond our measured points. The greater the number of observations, the more reliable the inferences.

EXERCISE 3

When gas is heated in a sealed container, the pressure of the gas increases. One unit commonly used to measure gas pressure is the torr, named after the Italian scientist Evangelista Torricelli, the inventor of the barometer. At sea level, 760 torr is considered standard air pressure.

Irina measured the pressure exerted by a gas in a sealed container at 10-degree intervals. Here are her data.

Temperature (°C)	*Pressure (torr)*
0	708
10	734
20	760
30	786
40	812
50	838
60	897
70	890
80	916
90	942

Irina graphed her data, as shown below.

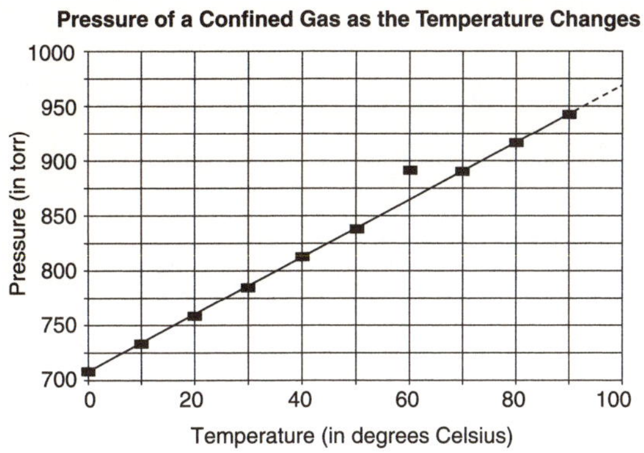

1. In her experiment, Irina *observed* all of the following *except*
   (1) as the temperature increased, the pressure increased
   (2) the pressure was 760 torr when the temperature was 20°C
   (3) the pressure at 60°C was greater than the pressure at 70°C
   (4) the pressure at 35°C was 800 torr

2. Which is the most valid statement about the pressure of the gas at 100°C?
   (1) Since Irina's measurements went up to only 90°C, no inference can be made about the pressure at 100°C.
   (2) Irina observed the pressure at 100°C to be about 970 torr.
   (3) Irina can infer by extrapolation that the pressure at 100°C would be about 970 torr.
   (4) The pressure at 100°C should be twice the pressure at 50°C.

3. Irina's graph does not go through the point she measured at 60°C. She infers that if she repeats the experiment under the same conditions,
   (1) she will get the same pressure at 60°C that she got the first time, 897 torr
   (2) she will get a different value for the pressure at 60°C, probably between 850 and 875 torr
   (3) she will get a completely different graph
   (4) she will get a different value for the pressure at 60°C, probably between 825 and 850 torr

4. This experiment will produce more reliable inferences if
   (1) the readings are made at 5-degree intervals instead of 10-degree intervals, over the same temperature range
   (2) the size of the container is varied during the experiment
   (3) the amount of gas is varied during the experiment
   (4) the readings are made at 10-degree intervals but only between 20° and 50°C

5. In one or two sentences, tell how a person can extrapolate information from data in a line graph.

Using Bar Graphs

Bar graphs are also used to help organize and illustrate observations. Suppose we wanted to compare the weather in Ohio to the weather in some other areas of the United States. One area of interest might be the amount of rainfall. On the following page is a table showing the average annual rainfall in eight cities. **Note:** Rainfall is expressed in inches in most weather reports.

City	Average Annual Rainfall (inches)
Columbus, Ohio	37.0
Toledo, Ohio	31.8
Cincinnati, Ohio	40.1
Cleveland, Ohio	35.4
Honolulu, Hawaii	23.5
Miami, Florida	57.5
Seattle, Washington	38.9
Los Angeles, California	14.8

A bar graph can be used to compare these eight observations, as shown in Figure 2-2. Using our bar graph, we can quickly observe that of the eight cities, Miami has the most rainfall each year, while Los Angeles has the least. Bar graphs are used to make clear and dramatic comparisons. Notice that, while in a line graph (Figure 2-1) both the *x*-axis and the *y*-axis show amounts, on a bar graph (Figure 2-2) only the *y*-axis shows amounts.

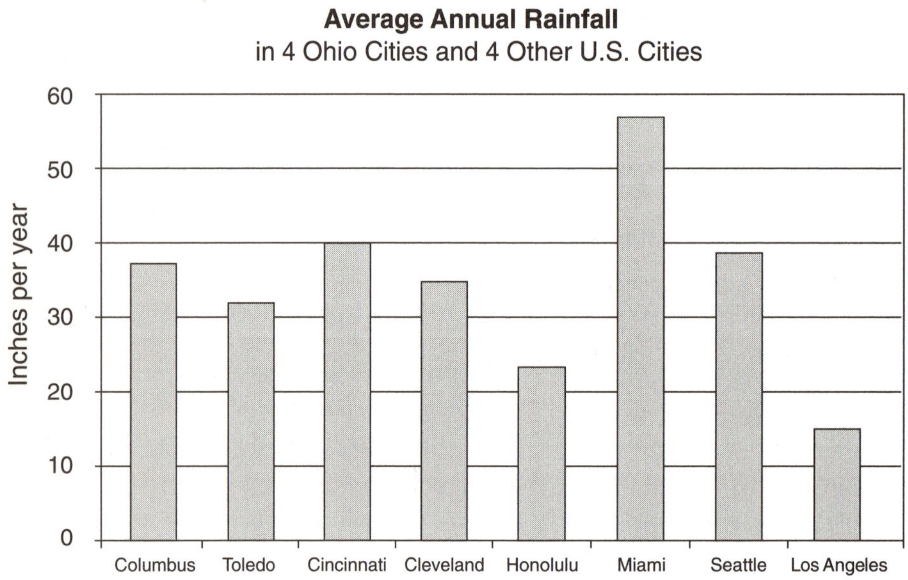

Figure 2-2. A bar graph makes it easy to compare observations.

EXERCISE 4

Use the bar graph in Figure 2-2 to answer the following questions.

1. The bar graph illustrates
   (1) only observations
   (2) both inferences and observations
   (3) only inferences
   (4) neither inferences nor observations

2. Which city in Ohio gets more annual rainfall than Seattle?
   (1) Columbus
   (2) Toledo
   (3) Cincinnati
   (4) Cleveland

3. Suppose we wanted to know which city has more rainy days, Seattle or Miami. Using our graph, we could
   (1) definitely observe that Miami has more rainy days
   (2) definitely infer that Miami has more rainy days
   (3) definitely infer that Seattle has more rainy days
   (4) make no definite inference about the number of rainy days

Hector is 1.8 meters tall. He measured the length of his shadow every Friday when the sun was at its highest point for a period of eight weeks. His shadow length decreased every week, eventually decreasing to .97 meters. He inferred that eight weeks later the shadow would be even shorter. When he measured it, however, he found it to be just over 1.3 meters.

4. Hector's final observation did not fit his inference. This was probably because
   (1) he should have measured his shadow more often than once a week
   (2) he should have made his measurements over a longer period of time
   (3) he probably measured his shadow inaccurately the last time
   (4) shadow lengths do not follow any regular pattern over time

5. Based on your knowledge of earth science, Hector's first measurement was probably made in the month of
   (1) June (3) March
   (2) September (4) January

6. When Hector looked at his shadow, he was
   (1) always facing north
   (2) always facing south
   (3) always facing west
   (4) facing a different direction each week

7. In one or two sentences, tell whether a line graph or a bar graph is better for comparing data from scientific observations.

3 Science Safety Procedures

Points to Remember —

- Many chemicals found at home and in the laboratory can be dangerous if not used properly.
- Dangerous chemicals include those that are toxic (poisonous), flammable (burn), and/or caustic (destroy skin).
- Never perform experiments without a supervising adult's permission.
- When working in the laboratory, observe proper safety precautions, including eye protection, safe clothing, and proper equipment.
- In case of an accident, inform the teacher ***immediately.***
- Never clean blood spills yourself. Report such spills to the teacher.

Safety at Home

Most of the chemicals we encounter are harmless. However, some chemicals used at home and on the job can be dangerous and must be handled carefully. Household substances that can be hazardous include bleach, ammonia, lye, paint thinner, some kinds of glue, and all medications.

Among the most dangerous household chemicals are those that give off ***vapors*** (fumes or gases). These vapors may be flammable, toxic, or both. A ***flammable*** substance is one that burns readily. Flammable vapors may explode from the touch of a spark. Gasoline and most paint thinners are highly flammable and give off flammable vapors.

Toxic substances are poisonous. Breathing toxic vapors can cause death or serious illness. Ammonia, paint thinners, and some kinds of glue produce toxic vapors. Chemicals that give off toxic or flammable vapors should be used only in well-ventilated areas (areas with a constant flow of fresh air) and away from sparks or flames.

Products containing dangerous chemicals are required by law to carry a warning label. A typical label might read, "**DANGER! Contents flammable. Do not use near flame.**" If the product is toxic, it may carry the skull and crossbones symbol, shown in Figure 3-1. Some substances are dangerous to touch. A ***caustic*** substance can destroy living tissue such as skin. A ***corrosive*** substance eats through skin, clothing, and most metals. Many oven cleaners and drain openers contain lye, an extremely caustic substance. Acid from car batteries is very corrosive. Keep these chemicals away from your skin and especially away from your eyes.

Figure 3-1. This symbol on a label indicates a toxic substance.

Some household chemicals become particularly dangerous when mixed with other chemicals. For instance, the cleansers bleach and ammonia (which produce toxic vapors) produce an extremely toxic gas when combined. Never mix chemicals together unless you are instructed to do so, since combining them may produce a hazardous substance, or even an explosion.

Many household substances are stored under high pressure in spray cans. Deodorants, hair sprays, spray paints, and even whipped cream are sold in pressurized containers. These cans may explode if punctured or heated. A typical label on a spray can might read, "**Warning! Contents under pressure. Do not puncture or incinerate [burn] can. Do not store at temperatures above 120°F.**" Always take such warnings seriously.

Even "safe" chemicals are safe only if used properly. For example, aspirin is usually a safe, beneficial drug, yet children have died from aspirin overdoses. **All drugs and chemicals should be stored out of the reach of children and pets, preferably in a locked cabinet.**

Before taking any medication or using any chemical product, always read the label carefully and follow the instructions. Remember that all chemicals can be dangerous if used improperly. Table 3–1 lists some common household chemicals and their dangers.

Table 3–1. Household Substances and Their Dangers

Substance	*Dangers*	*Special Precautions*
Ammonia	Toxic; gives off toxic vapors	Do not mix with other chemicals; use in well-ventilated area.
Chlorine bleach	Toxic; gives off toxic vapors	Do not mix with other chemicals; use in well-ventilated area.
Gasoline	Toxic and flammable; gives off toxic and highly explosive vapors	Keep away from flame; carry only in special containers.
Hair spray	Flammable vapors; pressurized can	Do not puncture or incinerate can; do not store at high temperature.
Lye (in oven cleaners and drain openers)	Toxic and caustic	Wear rubber gloves; protect eyes; do not mix with other chemicals.

Preparing Lab Work

Accidents in the laboratory can be avoided if certain precautions are taken ahead of time. Before doing *any* lab work, tie back long hair and wear protective clothing when necessary. Protective clothing includes aprons, lab coats, and safety goggles. Remove jewelry that could catch on something or interfere with lab work, such as large earrings or dangling bracelets and necklaces.

Don't attempt any laboratory work until you carefully read and understand all instructions. ***Never*** bring food and drink into the laboratory, and remove books and coats from the work area.

Listen carefully to all instructions given by the teacher. If there is something that you do not understand, ask your teacher to explain it to you *before* beginning the experiment. Be aware of the location of all safety equipment—this includes running water, fire extinguishers, fire blankets or sand, and eyewash or emergency shower. Know the proper fire drill procedure. If you need to leave the room in an emergency, be sure to turn off any burners or electrical equipment, if time permits.

Working With Chemicals

Special precautions must be observed in the laboratory when handling chemicals and laboratory glassware. There is always a danger of explosions, breakage, spills, and spattering. If you spill any chemical on your skin, clothing, or desk, immediately wash it away with plenty of cool water.

When testing any unknown substance, always assume that it may be dangerous. For example, to observe the odor of an unknown substance, fan the vapors toward your nose and sniff ***gently.*** Do ***not*** put your nose directly in the vapors or inhale deeply (see Figure 3-2). Also, ***never*** taste a chemical.

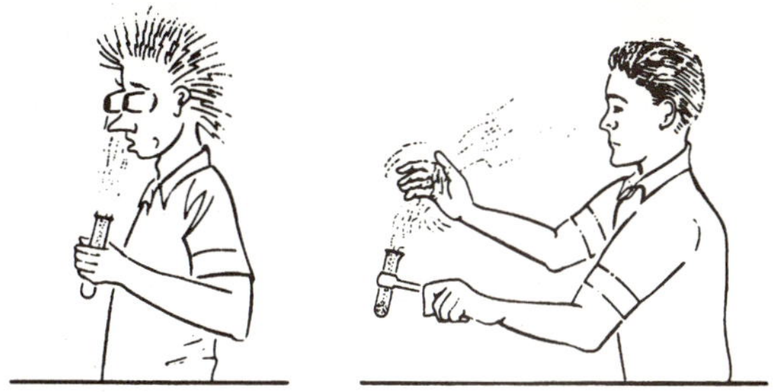

Figure 3-2. Use the safe method to test the odor of an unknown substance.

Working With Heat

Many experiments require that materials be heated. This can be done in various ways. Three common methods for heating are the hot plate, the alcohol burner, and the Bunsen burner (see Figure 3-3). Each method presents its own special hazards. A hot plate, for example, looks exactly the same whether it is hot or cold. Therefore, always assume that it is too hot to touch.

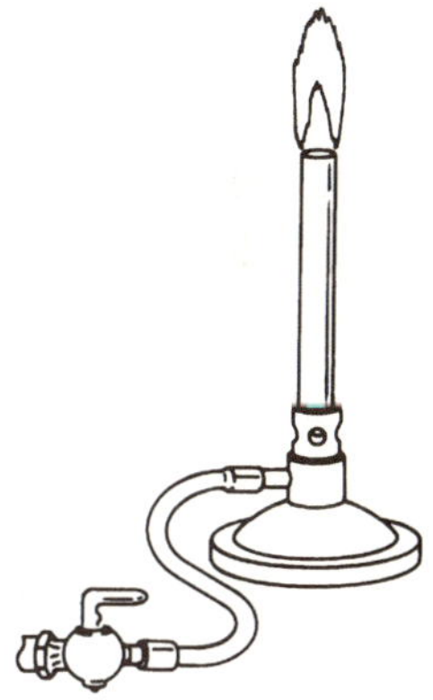

Figure 3-3. Be careful when working with any heating devices.

A test tube holder is designed to safely handle a test tube during and after heating. Special devices called tongs have been designed for handling other hot objects. These tongs are available in different sizes and shapes. (See Figure 3-4.) Similarly, objects that were recently heated can cause serious burns if they are touched. Report any burn, large or small, to the teacher immediately.

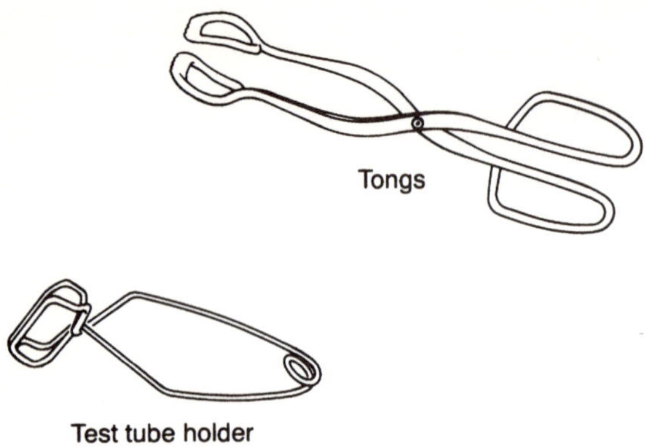

Figure 3-4. These tools are used to handle objects that have been heated.

Even water becomes dangerous when heated. Like other liquids, hot water may splatter when it reaches its boiling point. For this reason, wear safety goggles whenever liquids are heated. When heating a liquid in a test tube, make sure that the mouth of the test tube does not point toward you or anyone else. Heat liquids only in open containers. Figure 3-5 illustrates the right and wrong ways to heat a test tube.

Figure 3-5. Wear goggles and use caution when heating liquids.

Safe Use of Electricity

Electricity is dangerous and can injure or kill living things. Though it has become a common part of our lives, electricity should always be handled carefully. Observe commonsense safety rules whenever using electricity. Before working with any electrical device, check the wires for damage or breaks in the plastic or rubber insulating material. Electrical wires should not be used if they are broken or frayed, and plugs should be replaced if the wire is broken near the plug. If you find a wire that might be damaged, notify your teacher. Always unplug electrical devices before cleaning or repairing them. When removing a plug from an electrical outlet, grip the plug itself, not the cord attached to it. ***Never*** touch the metal prongs!

Water can act as a conductor of electricity, so ***never*** handle electrical appliances while you are in water or if you are wet. Keep all plugs, cords, and electrical appliances away from water. Also do not attempt to remove an electrical plug if your hands are wet.

Working With Sharp Tools

In a biology laboratory, students are sometimes asked to use sharp objects such as scalpels, probes, pins, and scissors. These tools are often called "sharps" by people who work with them. They require their own special precautions. ***Never*** touch the sharp edges of these tools. Even placing them on a table is a potential hazard. It is best to store sharps in drawers or cabinets when not in use.

If you or any of your classmates do get cut by a sharp, there are additional precautions you must take. Some types of hepatitis and some other contagious diseases are spread through contact with contaminated blood. A student may carry the disease and not show any of its symptoms. Therefore, the proper procedure is to assume that all blood is dangerous.

Report any cut, large or small, immediately to the teacher. Do not attempt to clean up any blood spills yourself. Do not touch the sharp instrument that caused the cut. Your teacher will know the proper way to remove the dangerous materials. These materials include broken glass, which is one of the most common causes of injury in the laboratory.

After the Experiment

When laboratory work has been completed, carefully clean and return all equipment to the appropriate place. Wash your hands, especially if you have been working with any chemicals or living things. Your lab station should be cleaned and made safe for the next students who will work there. Likewise, if you come into a lab and there is some unknown material on the desk, assume that it is dangerous and notify the teacher.

EXERCISE

1. Which is ***not*** a proper safety precaution?
   (1) wearing gloves when cleaning up spills
   (2) wearing goggles when heating liquids
   (3) tying back long hair when heating substances
   (4) picking up pieces of broken glass yourself

2. Which liquid produces flammable vapors?
   (1) water (3) ammonia
   (2) bleach (4) gasoline

3. A substance that is dangerous to touch is best described as
   (1) flammable (3) poisonous
   (2) caustic (4) toxic

4. If a warning label says "use only in a well-ventilated area," this probably indicates that the product
   (1) is corrosive
   (2) is under pressure
   (3) gives off dangerous vapors
   (4) has no dangers

5. When you heat water in a test tube, what is the most important safety precaution?
   (1) Use cold water.
   (2) Wear gloves.
   (3) Wear safety goggles.
   (4) Use a low flame.

6. All of the following are important pieces of safety equipment *except*
   (1) goggles
   (2) alcohol lamp
   (3) test tube holder
   (4) fire extinguisher

7. The student next to you has accidentally cut himself with a scalpel. What do you do first?
   (1) Attempt to stop the bleeding.
   (2) Help the student wash his wound.
   (3) Put away the scalpel.
   (4) Notify the teacher.

8. Which is the correct way to heat a liquid?
   (1) Student not wearing goggles holds a test tube pointing toward himself.
   (2) Student wearing goggles holds a test tube pointing toward herself.
   (3) Student not wearing goggles holds a test tube pointing away from himself.
   (4) Student wearing goggles holds a test tube pointing away from herself.

Process Skill

USING LABORATORY EQUIPMENT SAFELY

When doing laboratory work that requires heat, follow these safety precautions.

1. Always wear safety goggles when heating liquids, because they can splatter in your face and eyes.

2. Always wear insulated gloves when you are moving hot objects, unless the object has insulated handles.

3. Always be aware of people around you when you move anything hot.

4. Sources of heat can sometimes cause problems:
   a. Liquid fuels can spill and catch on fire; handle them carefully.
   b. Hot plates may not always look hot. Never leave a hot plate unattended while it is on and always turn it off immediately when you are finished using it.
   c. Gaseous fuels can cause explosions. Always wear safety goggles when using a gas jet and be sure to turn gas jets off when you are through using them.

5. A fire extinguisher should be nearby whenever you are working with flames.

What safety precautions should be taken when doing the activity shown in Diagram 1, page 27? When preparing hydrochloric acid, safety goggles and insulated gloves should be worn. The person performing the activity should be aware of other people in the area. Finally, there should be a fire extinguisher nearby.

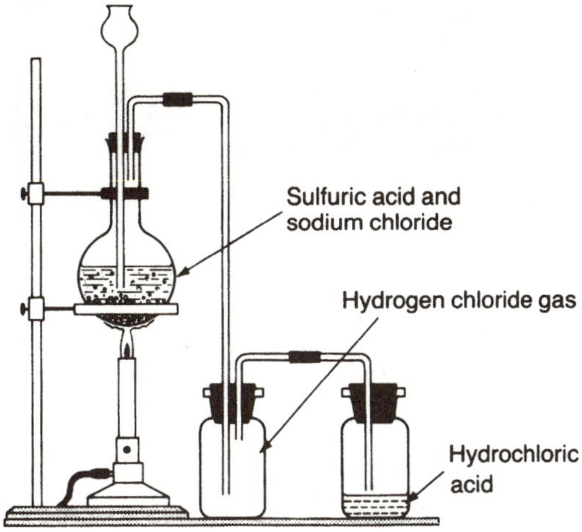

Diagram 1

Diagram 2 shows a setup used to determine what happens to a liquid's volume when it is heated. This experiment has been poorly set up, because too-rapid heating of the liquid may cause it to spurt out the top of the tubing. Examine the setup and use your common sense to answer the following questions.

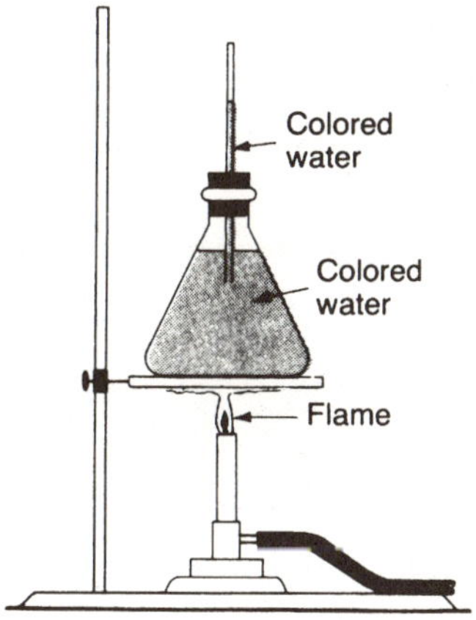

Diagram 2

1. A safer way to heat the flask is to
   (1) use a shorter piece of glass tubing
   (2) use more gas to make the flame bigger and hotter
   (3) use a greater amount of water in the flask
   (4) suspend the flask in a beaker of water, so that the flame is not directly on the flask

2. A safe way to quickly remove the flask from the heat source is to
   (1) remove it with your bare hands
   (2) remove it while wearing insulated gloves
   (3) remove it with a test tube holder
   (4) remove it by the glass tubing

4 Scientific Measurements

Points to Remember —

- Mass is a measure of the amount of material in an object. Mass may be measured in grams (g) or kilograms (kg) with a balance.
- Volume is the amount of space an object occupies. Volume may be measured in milliliters (mL) with a graduated cylinder.
- Length is the distance between two points. Length may be measured in meters (m) or centimeters (cm) with a ruler or tape measure.
- Temperature measures how hot or cold something is. Temperature is measured in degrees Celsius (°C) with a thermometer.
- In the metric system, a prefix is used to change the value of a unit:

  kilo = 1000 $\qquad$ milli = $\frac{1}{1000}$ $\qquad$ centi = $\frac{1}{100}$

- Density = $\frac{\text{mass}}{\text{volume}}$.
- An object floats if it is less dense than the liquid in which it is placed.

Making Measurements

How tall are you? If you can answer this question, it is because you have made a measurement. You probably used feet and inches to measure your height. In most other parts of the world, however, your height would be measured in meters.

Scientists make many different kinds of measurements. They all use the same set of units to express those measurements, called the ***metric system.*** Table 4-1 lists some of the many quantities that scientists measure and the units commonly used to express those quantities.

Table 4-1. The Metric System

Quantity	*Unit*	*Abbreviation*
Length	Meter	m
Mass	Kilogram	kg
Volume	Liter	L
Time	Second	s
Temperature	Degree Celsius	°C
Energy	Joule	J
Power	Watt	W
Force	Newton	N

Additional units are derived from these by using the prefixes listed in Table 4-2. For example, a kilowatt is equal to 1000 watts. A centimeter is $\frac{1}{100}$ of a meter, and a millisecond is $\frac{1}{1000}$ of a second.

Table 4-2. Prefixes Used in the Metric System

$\frac{1}{1,000,000}$	(0.000001)	Micro
$\frac{1}{1,000}$	(0.001)	Milli
$\frac{1}{100}$	(0.01)	Centi
$\frac{1}{10}$	(0.1)	Deci
1000		Kilo
1,000,000		Mega
1,000,000,000		Giga

Tools for Measurement

Your height is actually a measurement of length, the distance between two points. Length is commonly measured with a ruler. The standard unit of length is the ***meter,*** which is slightly longer than a yard. Smaller lengths are measured in centimeters (hundredths of a meter, cm) or millimeters (thousandths of a meter, mm), while long distances are measured in kilometers (thousands of meters, km). When using a ruler, one end of the object is placed at the zero (if it is clearly marked) and the ruler is read where it lines up with the other end of the object.

Process Skill

MAKING PRECISE MEASUREMENTS

Students were asked to measure the object illustrated in Figure 4-1. The following measurements were reported: (A) 2 cm; (B) 2.2 cm; (C) 2.22 cm; (D) 2.220 cm.

Only one of these measurements is scientifically correct. Scientists have agreed to follow certain standard procedures when reporting measurements. On the ruler illustrated in Figure 4-1, each small line represents $\frac{1}{10}$ of a centimeter. The object being measured is larger than 2 cm but smaller than 3 cm. If we report the length as 2 cm, we are not using our ruler to its fullest capability. Measurement A is therefore inadequate. The end of the object is between the second and third small lines past the 2-cm mark, as we can see in the enlargement of the ruler in Figure 4-2.

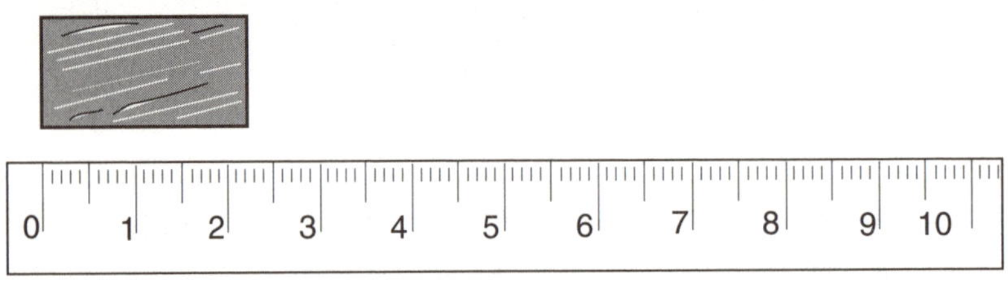

Figure 4-1

Since each small line represents 0.1 cm, we can express the measurement as 2.2 cm. However, we can be even more precise. Note that the object is slightly longer than 2.2 cm; try to estimate how much longer than 2.2 cm it is. To do this, imagine ten even smaller lines between 2.2 and 2.3 centimeters and estimate where the end of the object would be. One such estimate might be 2.22 cm. This is the correct expression of this measurement. Scientific measurements include one estimated place. Measurement D has two estimated places, the last 2 and the 0. Only one estimated place is permitted. In summary, when making measurements, always read between the lines.

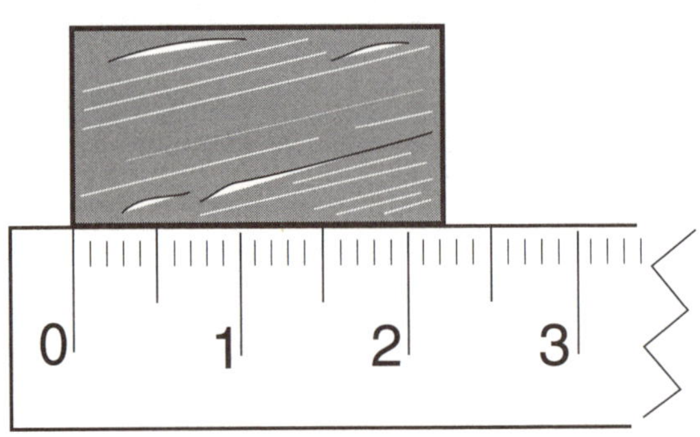

Figure 4-2

Mass

Mass is a quantity that measures the amount of material in an object. Scientists measure mass in units called kilograms. A kilogram (kg) is approximately equal to the mass of a quart of milk. Your mass is probably between 40 and 70 kilograms. In the laboratory, we are usually measuring small quantities and therefore need a smaller unit for mass. A gram is $\frac{1}{1000}$ of a kilogram and is about equal to the mass of a large paper clip.

Mass is measured with a balance, a device that compares the object being measured to an object of known mass. A triple-beam balance, like the one shown in Figure 4-3, can be used to measure masses to the nearest $\frac{1}{100}$ of a gram.

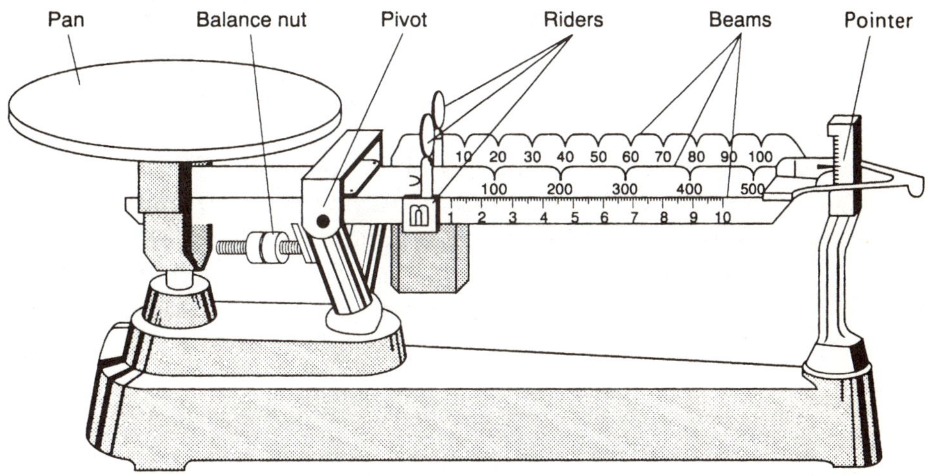

Figure 4-3. The triple-beam balance is used to measure mass.

Weight

Weight is a measure of the pull of gravity on an object. It depends on the mass of the object and on the strength of gravity. Since the strength of gravity differs from place to place, an object's weight may change when it is moved. Its mass, however, does not change. For example, your weight on the moon would be one-sixth of your weight on Earth, because the pull of gravity felt on the moon is one-sixth of that felt on Earth.

Weight is measured with a scale, like the ones in bathrooms and fruit stores, as illustrated in Figure 4-4. These scales contain springs, which are stretched or pushed by the weight of the object. In the United States, weight is generally measured in ounces, pounds, and tons. The metric unit of weight is the newton. Since an object's weight is not the same everywhere, scientists prefer to measure the mass of an object rather than its weight.

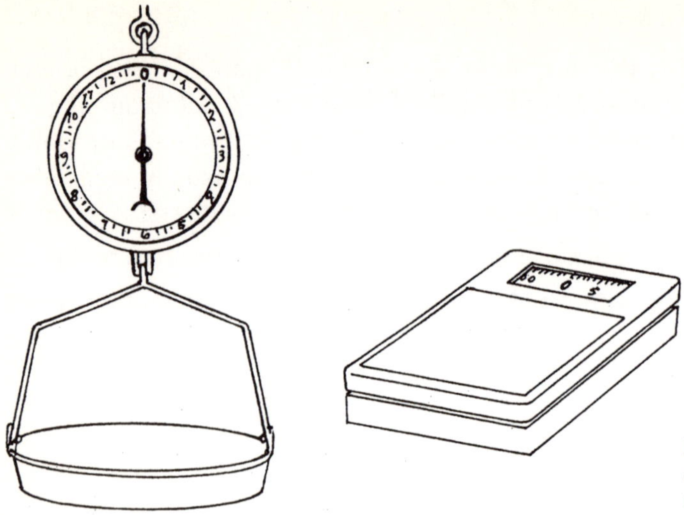

Figure 4-4. Scales such as these are used to measure weight.

Volume

Volume is the amount of space an object occupies. The volume of a rectangular object is determined by multiplying the measurement of its length times its width times its height (volume = length × width × height).

The box in Figure 4-5 is 2 cm long, 4 cm wide, and 3 cm tall. We determine its volume by multiplying 2 cm × 4 cm × 3 cm, which equals 24 cubic centimeters. The cubic centimeter, usually written as cm^3 or cc, is a unit of volume.

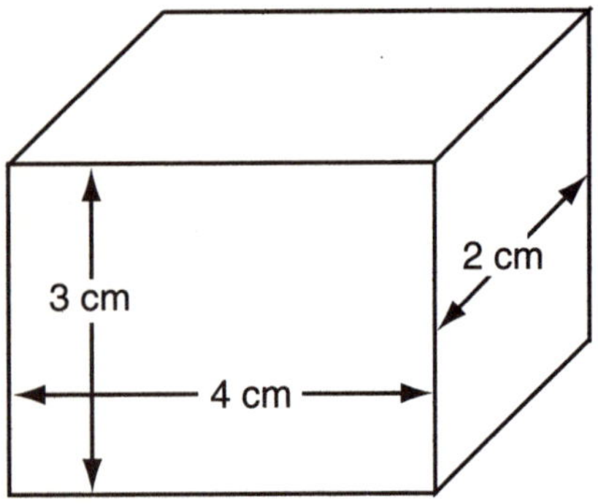

Figure 4-5. Multiply the measurements of length × width × height to find the volume of this box.

You are probably familiar with other units of volume, such as gallons, quarts, fluid ounces, tablespoons, and teaspoons. These units are commonly used to measure the volume of a liquid. The unit you use depends largely on how much liquid you are measuring. When measuring liquids and gases, a common unit of volume used by scientists is the liter. *One liter is equal to* 1000 cm^3. In the metric system, large volumes are measured in liters (L) and small volumes in milliliters (mL). Since a milliliter is equal to $\frac{1}{1000}$ of a liter, it is also equal to 1 cubic centimeter.

To measure the volume of a liquid, we pour the liquid into a container of known volume. One such container is called a graduated cylinder. A ***graduated cylinder,*** like those illustrated in Figure 4-6, has lines on it called

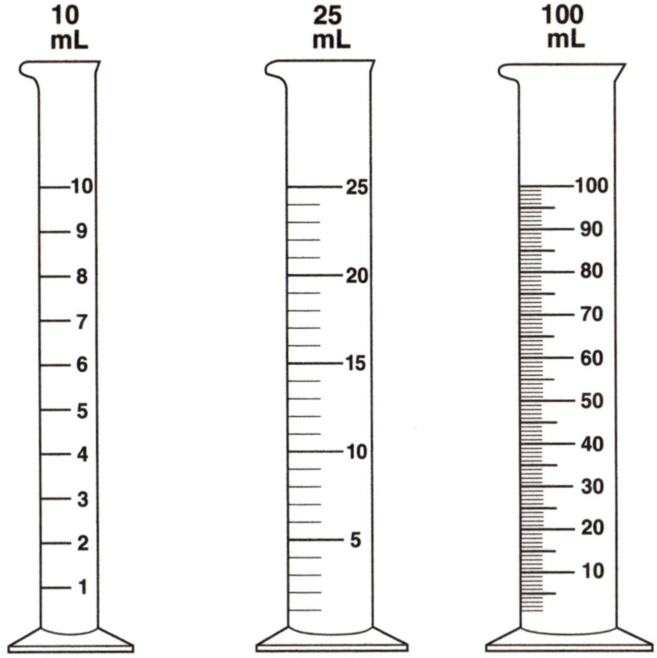

Figure 4-6. Graduated cylinders are used to measure the volume of a liquid.

graduations. Once the liquid has been poured into the graduated cylinder, the cylinder is placed on a level surface. As shown in Figure 4-7, we take the measurement by lining up our eyes with the liquid's surface. Water placed in any type of glassware forms a ***meniscus,*** a curve at the surface. We read the volume at the bottom of the curved water surface. In Figure 4-7, the volume should be read as 28.5 mL. Remember, the numeral in the last place is an estimate, between the 28 and 29 mL lines.

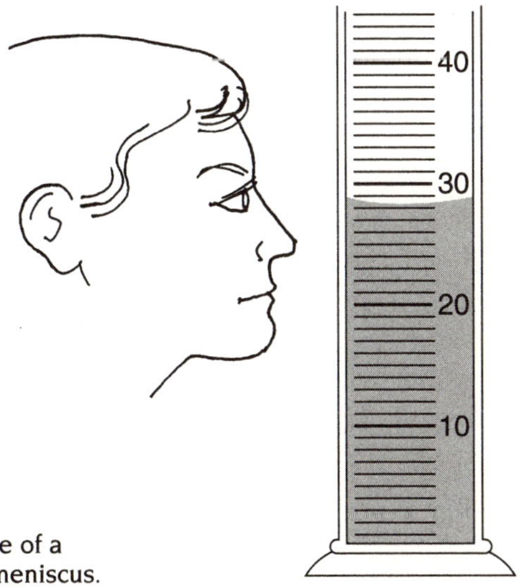

Figure 4-7. Read the volume of a liquid at the bottom of the meniscus.

MAKING MEASUREMENTS; PERFORMING CALCULATIONS

A graduated cylinder also can be used to determine the volume of small solid objects, as shown in the diagram. When an object such as a glass marble or small rock is placed into the water inside the cylinder, the water level rises. The change in the volume of the water is equal to the volume of the object.

For example, in the diagram, the graduated cylinder contains 20.0 milliliters (mL) of water. When the marble is placed in the cylinder, the water level rises to 25.0 mL. The change in the water's volume is 5.0 mL, so the volume of the marble must be 5.0 mL (or 5.0 cm^3). Study this procedure and answer the following questions.

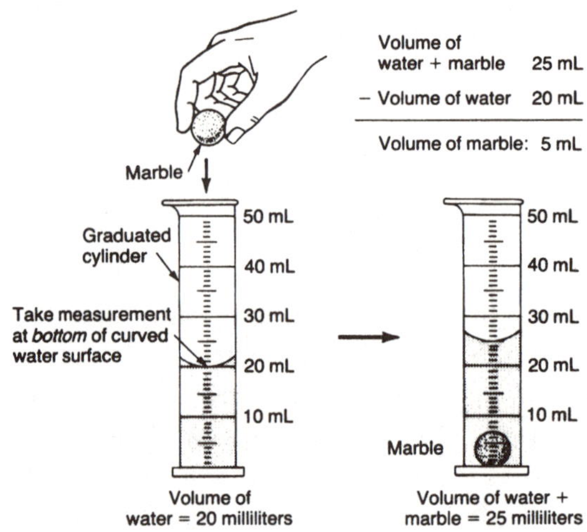

1. The diagram below indicates that the volume of the rock is
   (1) 30.0 mL (2) 20.0 mL (3) 10.0 mL (4) 5.0 mL

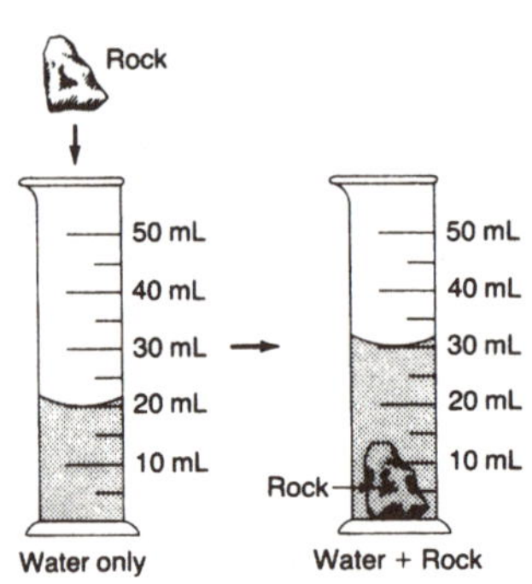

2. A graduated cylinder contains 20.0 mL of water. If a stone with a volume of 12.0 mL and another stone with a volume of 5.0 mL are both placed into this cylinder, the water level will rise to
   (1) 32.0 mL (2) 25.0 mL (3) 17.0 mL (4) 37.0 mL

Density

Why are airplanes made of aluminum, and fishing sinkers made of lead? You might answer that aluminum is a light metal, while lead is a heavy metal. Yet an aluminum airplane has a much larger mass than a lead fishing sinker. When we say that lead is heavier than aluminum, we really mean that if these two pieces of metal are the same size, the lead piece will be heavier. (If the pieces are not the same size, we need another way to compare them.) The quantity that compares the mass of an object to its size, or more specifically its volume, is called *density*.

Density is defined as the mass of an object divided by its volume (density = $\frac{\text{mass}}{\text{volume}}$). While the mass and volume of a piece of metal depend on the size of the piece, the density depends on only the nature of the metal and its temperature. (We discuss the effect of temperature later on.) Let's compare the densities of lead and aluminum. At room temperature, the density of aluminum is 2.7 grams per cubic centimeter, which we abbreviate as 2.7 g/cm^3. Since density is mass divided by volume, the unit of density contains a mass unit, grams, divided by a volume unit, cubic centimeters. When we say that aluminum has a density of 2.7 g/cm^3, this means that a piece of aluminum with a volume of 1 cubic centimeter has a mass of 2.7 grams. The density of lead at room temperature is 11.3 g/cm^3. Lead is about four times as dense as aluminum.

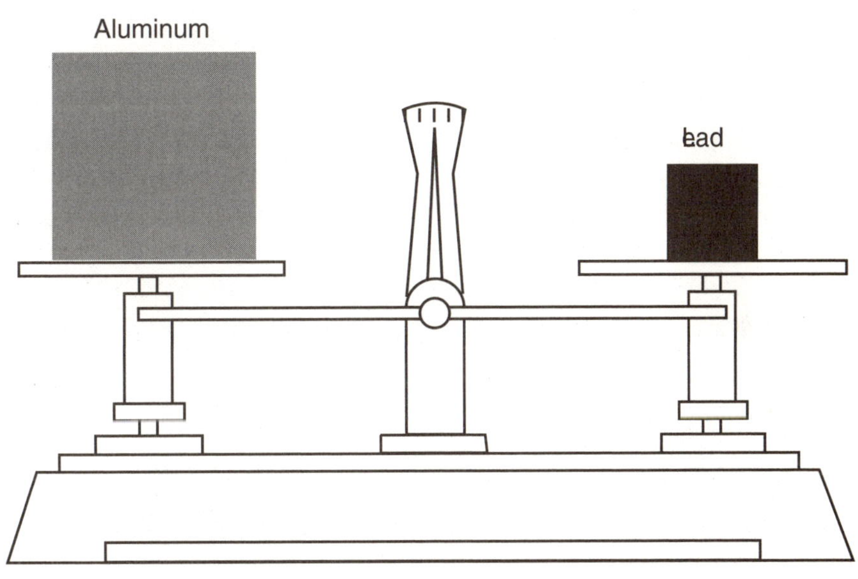

Figure 4-8. These metal cubes are of equal mass; the cube with less volume has greater density.

Which is heavier, 10 grams of lead or 10 grams of aluminum? Of course, this is a trick question. Since both are 10 grams, they are equally heavy. Which has a larger volume, 10 grams of lead or 10 grams of aluminum? This question requires some thinking. The density of aluminum is 2.7 g/cm^3, so 10 grams of aluminum has a volume *greater* than 1 cubic centimeter. Table 4-3 shows how the volume may be calculated. The density of lead is 11.3 g/cm^3, so a 10-gram piece of lead has a volume of *less* than 1 cubic centimeter. In comparing objects of equal mass, the *denser object* is the one with the *smaller volume*. Figure 4-8 compares pieces of lead and aluminum of the same mass.

Table 4-3. Performing Calculations with Density

Finding mass from volume and density:	$Density = \frac{Mass}{Volume}$
To get *mass* by itself, multiply both sides by *volume.*	$Volume \times Density = \frac{Mass}{Volume} \times Volume$
Volume cancels on the right side, so we get:	$Mass = Density \times Volume$

Finding volume from mass and density:	$Density = \frac{Mass}{Volume}$
To get *mass* by itself, multiply both sides by *volume.*	$Volume \times Density = \frac{Mass}{Volume} \times Volume$
Volume cancels on the right side, so:	$Volume \times Density = Mass$
To get *volume* by itself, divide both sides by *density.*	$Volume \times \frac{Density}{Density} = \frac{Mass}{Density}$
Density cancels on the left side, so we get:	$Volume = \frac{Mass}{Density}$

Which is heavier, 10 cm^3 of aluminum, or 10 cm^3 of lead? When comparing objects of equal volume, the object with the greater density has the greater mass. Thus, 10 cm^3 of lead weighs about four times as much as the same volume of aluminum, since lead is about four times denser than aluminum. Figure 4-9 compares pieces of lead and aluminum that have the same volume.

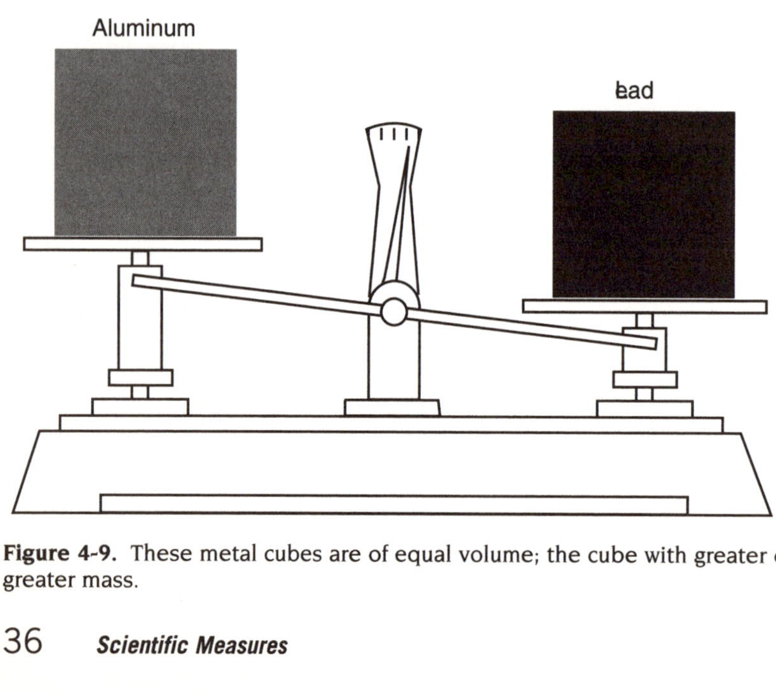

Figure 4-9. These metal cubes are of equal volume; the cube with greater density has greater mass.

The density of a 10-cm^3 piece of lead is 11.3 g/cm^3. What is the density of a 20-cm^3 piece of lead? The answer: still 11.3 g/cm^3! Table 4-4 gives the mass, volume, and density of several pieces of lead. As you can see, the density remains the same no matter what size the piece of metal is. The density of a material does not depend on the size of the object. This makes density a very useful property for identifying materials.

Table 4-4. Mass, Volume, and Densities of Several Pieces of Lead

Mass	*Volume*	*Density (Mass/Volume)*
113 g	10 cm^3	11.3 g/cm^3
226 g	20 cm^3	11.3 g/cm^3
1130 g	100 cm^3	11.3 g/cm^3

Using Density to Identify a Metal

Tin foil and aluminum foil look very much alike. Aluminum has a density of 2.7g/cm^3, while tin has a density of 7.3 g/cm^3. A chemist measures the volume of a sample of metal foil and finds it to be 5.0 cm^3. He then weighs it and finds its mass to be 36.5 g. Is it aluminum or tin? Remember, density is mass/volume. To find the density of the metal, divide the mass, 36.5 grams, by the volume, 5.0 cm^3. Thus, density = 36.5/50.0 = 7.3 g/cm^3. The metal was evidently tin.

Why Do Objects Float?

A wooden log floats on water, while an iron nail sinks. Why? The answer to this question lies in the densities of these materials. A material will float if it is less dense than the liquid in which it is placed. From this information, we can conclude that iron is denser than water, while wood is less dense than water. Water has a density of 1 g/cm^3. Therefore, any object with a density greater than 1 g/cm^3 will sink in water. The density of iron is 7.9 g/cm^3.

Some liquids do not mix with each other. For example, the oil and vinegar in salad dressing form separate layers, as shown in Figure 4-10. What can you conclude about the density of oil compared to the density of vinegar? Since the oil floats over the vinegar, the oil must be less dense.

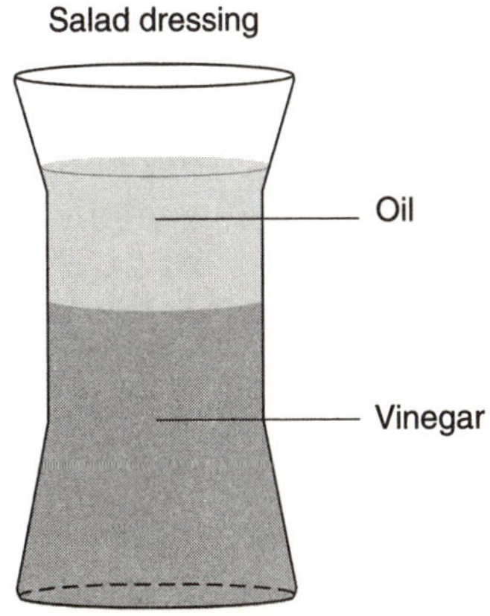

Figure 4-10. The less dense liquid (oil) floats on top of the denser liquid (vinegar).

Helium balloons float on air in much the same way that wood floats on water. The density of a helium balloon is much less than the density of air. Some balloons use hot air instead of helium. Since hot-air balloons float, hot air must be less dense than cold air. When most materials are heated, they expand. This means that their volumes increase while their masses stay the same. Since density is mass/volume, an increase in volume will cause a decrease in a material's density. In general, an increase in temperature causes a decrease in density.

Measuring Temperature

Temperature is a measure of how hot or cold something is. We measure temperature with a tool called a thermometer. The thermometer contains a liquid (usually alcohol or mercury) that expands when it is heated. As the liquid expands, it moves up a small tube. The height of the liquid tells us the temperature. As the liquid cools, it contracts and moves down the tube. Figure 4-11 shows a thermometer placed in hot water and a thermometer placed in cool water.

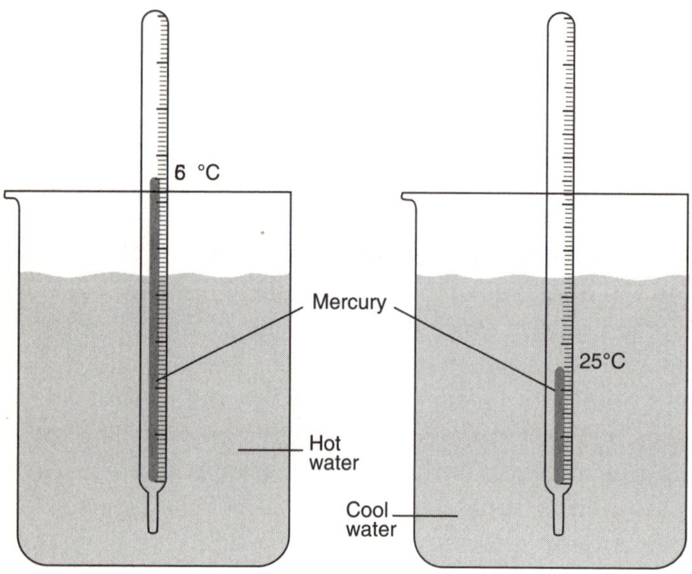

Figure 4-11. The height of the liquid in the thermometer tells the temperature of the liquid in the beaker.

There are two scales commonly used to measure temperature: Fahrenheit and Celsius. The unit for measuring temperature is either the degree Fahrenheit (°F) or degree Celsius (°C). Figure 4-12 shows a Fahrenheit thermometer and a Celsius thermometer with several important temperatures indicated. Two important temperatures are the freezing point and boiling point of water. Water freezes (turns to ice) at 0°C, which is the same as 32°F. Water boils (turns to steam) at 100°C, which is the same as 212°F. Two other important temperatures to know are standard room temperature (20°C or 68°F) and normal body temperature (37°C or 98.6°F). When reading a thermometer, as with all measuring devices, you estimate between the lines. Figure 4-13 shows a thermometer with a temperature reading of 25.5°C.

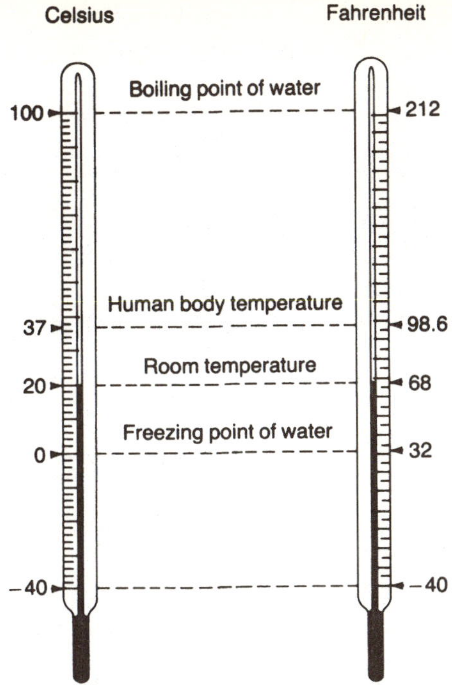

Figure 4-12. Some important readings on the Celsius and Fahrenheit thermometers.

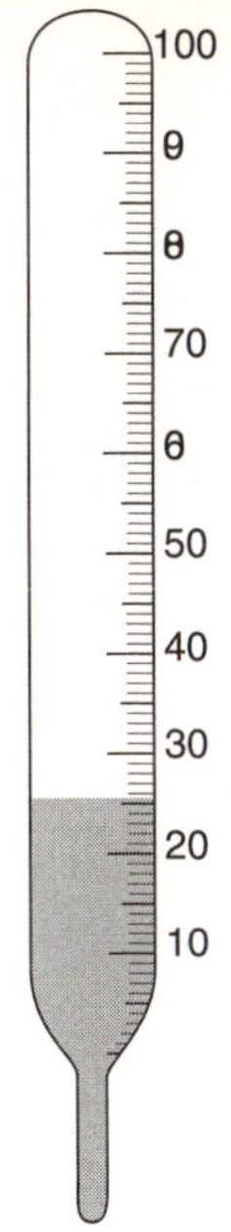

Figure 4-13. Reading between the lines: this thermometer shows a reading of 25.5° Celsius.

EXERCISE

A student measures the masses and volumes of four pieces of metal. The results are shown below.

Property	*Metal A*	*Metal B*	*Metal C*	*Metal D*
Mass Volume	10.0 g 2.0 cm^3	10.0 g 5.0 cm^3	30.0 g 5.0 cm^3	40.0 g 8.0 cm^3

1. Which metal is the most dense?
   (1) Metal A (3) Metal C
   (2) Metal B (4) Metal D

2. Which two pieces might be made of the same metal?
   (1) A and B (3) A and D
   (2) B and C (4) C and D

3. If we compared 10.0-gram samples of all four metals, which would have the largest volume?
   (1) Metal A (3) Metal C
   (2) Metal B (4) Metal D

4. If we compared samples of the metals with the same volumes, which would have the largest mass?
(1) Metal A (3) Metal C
(2) Metal B (4) Metal D

5. The density of a 100.0-gram sample of metal A should be
(1) 5.0 g/cm^3 (3) 50 g/cm^3
(2) 10 g/cm^3 (4) 100 g/cm^3

Questions 6 and 7 refer to the diagram below, which shows the relative densities of some liquids and solids.

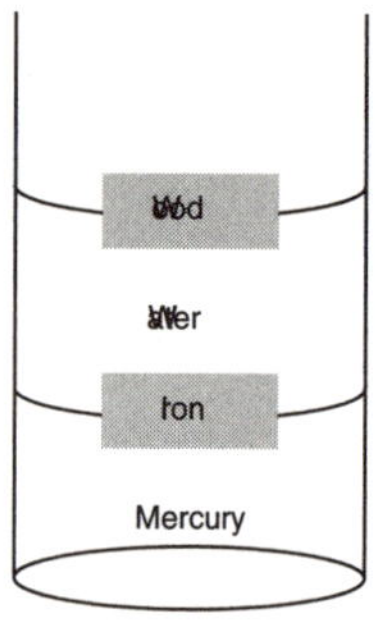

6. Which liquid in the diagram is the least dense?
(1) wood (3) iron
(2) water (4) mercury

7. Which solid in the diagram is the most dense?
(1) wood (3) iron
(3) iron (4) mercury

Questions 8–10 refer to the diagram below, which represents a beaker of ice water.

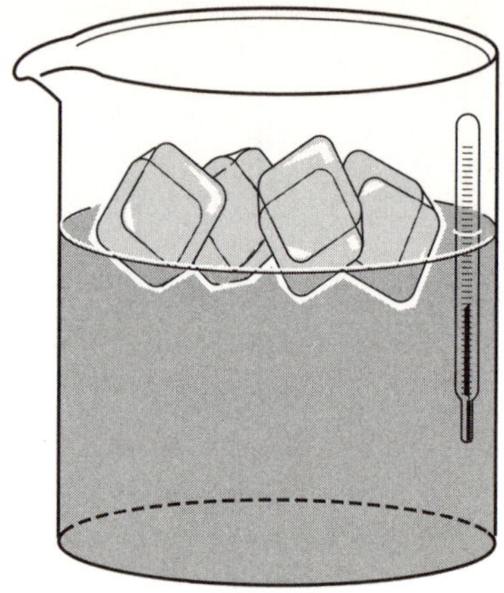

8. What is the temperature of the melting ice shown in the diagram?
(1) 0°C (2) 0°F (3) 32°C (4) 100°F

9. This diagram indicates that the ice is
(1) less dense than water
(2) more dense than water
(3) colder than water
(4) warmer than water

10. As water freezes to ice, its
(1) volume decreases
(2) mass decreases
(3) volume increases
(4) mass increases

11. Which two lines in the diagram below indicate the same dimensions?

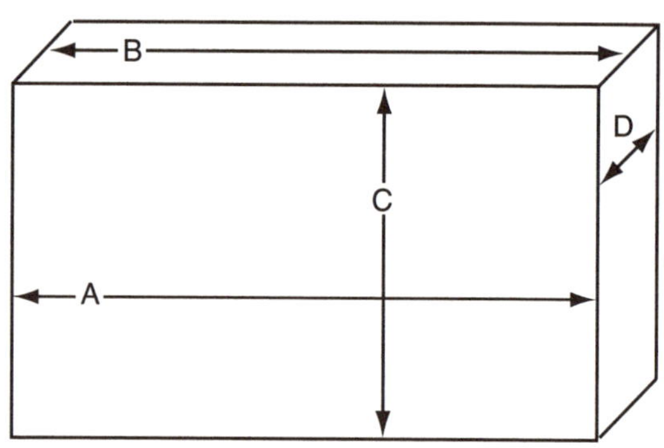

(1) A and B (3) B and C
(2) A and C (4) B and D

Questions 12–15 are based on the diagrams below.

A

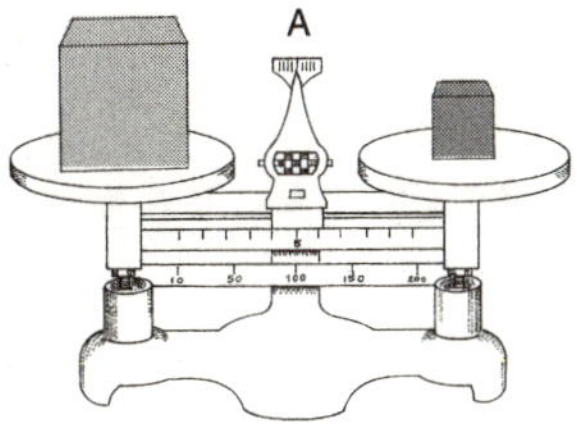

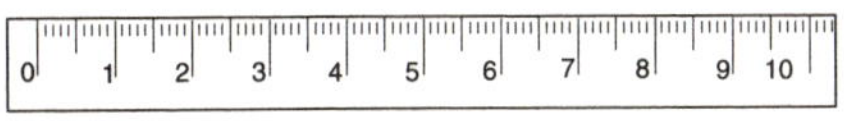

B

C

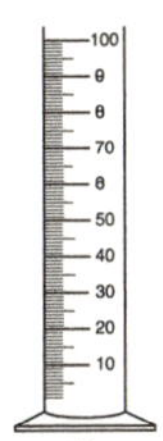

D

12. Which will give a reading in centimeters?
(1) A (2) B (3) C (4) D

13. Which tool can be used to determine whether water is cool enough to freeze?
(1) A (2) B (3) C (4) D

14. Which two tools could be used to determine the density of a liquid?
(1) A and B (3) B and D
(2) A and D (4) B and C

15. A student has pieces of lead and aluminum of equal volume. Which tool could help distinguish the lead from the aluminum?
(1) A (2) B (3) C (4) D

16. Volume can be measured in any of the following units *except*
(1) cubic centimeters (3) quarts
(2) liters (4) grams

17. The mass of the object shown on the balance is best reported as

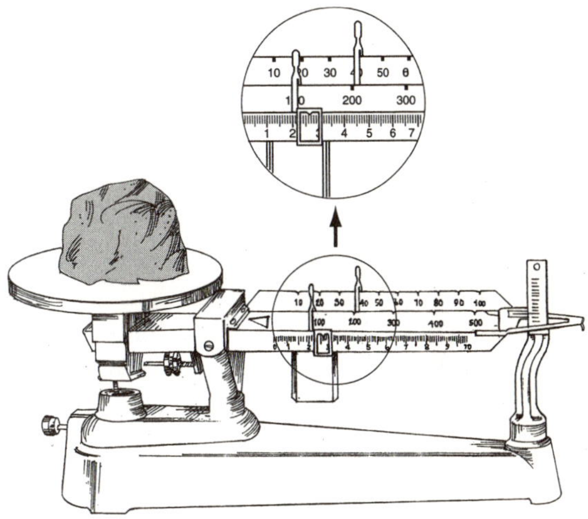

(1) 142.65 g (3) 100.65 g
(2) 142.65 mL (4) 100.65 mL

18. In one or two sentences, explain how you would find the mass and the volume of an irregularly shaped rock.

5 Earth's Changing Surface

Points to Remember —

- Earth's surface undergoes constant changes as a result of internal and external forces.
- Minerals are characterized by certain physical and chemical properties.
- Rocks, which make up Earth's crust, can be either igneous, metamorphic, or sedimentary.
- The three types of rocks are subject to processes that can change any one type into any other type; these processses make up the rock cycle.
- External forces that change Earth's surface include weathering and erosion.
- Internal forces that shape Earth's surface produce mountains, earthquakes, and volcanoes.

Earth's Surface

Surface Changes

The surface of planet Earth is constantly undergoing change. For example, new mountains rise as old mountains are worn away. Rivers carve valleys deep into the face of the land. Large rocks are broken down into smaller and smaller rocks. Two sets of forces, acting in opposite ways, shape Earth's surface. External forces wear the surface down, while internal forces build the surface up.

Surface Materials

Earth's rocky outer layer is called the ***crust.*** The surface of the crust consists of bedrock, rock fragments, and soil, as shown in Figure 5-1.

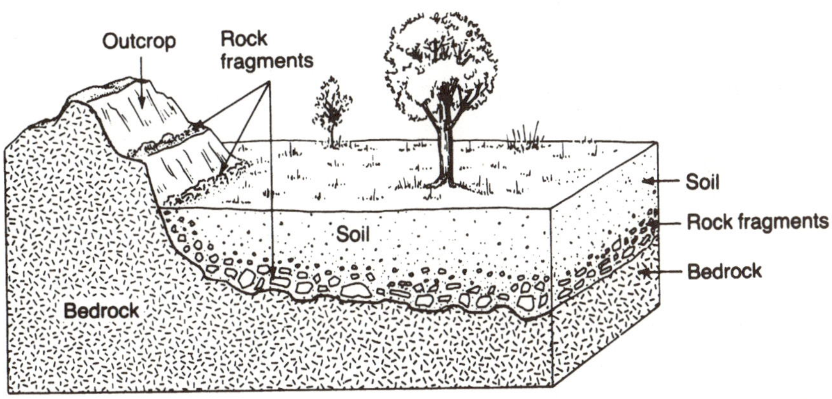

Figure 5-1. Earth's surface consists of bedrock, rock fragments, and soil.

Bedrock is the solid rock portion of the crust. Bedrock that is exposed at Earth's surface is called an *outcrop.* Rock fragments are pieces of broken-up bedrock. They can range in size from giant boulders to tiny grains of sand.

Soil is a mixture of small rock fragments and *organic matter* (materials produced by living things, such as decaying leaves and animal remains). Soil and fragments of rock make up most of Earth's surface, with the bedrock hidden underneath.

Minerals

Rocks are composed of *minerals,* which are naturally occurring solid substances made of inorganic (nonliving) material. Feldspar, quartz, mica, and calcite are some common minerals found in the rocks of Earth's surface. Minerals have certain *physical* and *chemical properties* by which they can be identified.

1. The ***physical properties*** of minerals include hardness, cleavage, and color.

Hardness is the resistance of a mineral to being scratched. Minerals are assigned a number between 1 and 10 to indicate their hardness, with 1 being the softest and 10 the hardest. The hardness scale shown in Table 5-1 lists the minerals used as reference points. A mineral can be scratched by another mineral only if the second mineral has a higher number on the hardness scale.

Table 5-1. Hardness Scale of Minerals

Mineral	*Hardness*	*Mineral*	*Hardness*
Talc	1	Feldspar	6
Gypsum	2	Quartz	7
Calcite	3	Topaz	8
Fluorite	4	Corundum	9
Apatite	5	Diamond	10

Cleavage is a mineral's tendency to break along smooth, flat surfaces. The number and direction of these surfaces are clues to a mineral's identity. Cleavage often causes a mineral to break into characteristic shapes, as shown in Figure 5-2. Not all minerals have cleavage; some fracture unevenly when broken.

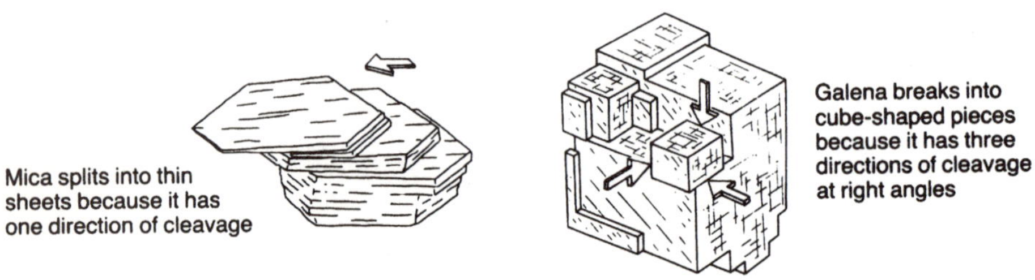

Figure 5-2. Cleavage in mica (left) and galena (right).

Color is not always a reliable guide to a mineral's identity. Various samples of the same mineral may have different colors. On the other hand, samples of different minerals all may have the same color. Color is best used together with other properties to identify a mineral.

2. Minerals also have ***chemical properties,*** such as how they react with an acid. Calcite, the chief mineral in limestone and marble, fizzes when hydrochloric acid is placed on it. The fizzing is caused by a chemical reaction between the calcite and the acid, in which bubbles of carbon dioxide gas are given off.

Rocks

The ***rocks*** that form Earth's crust are natural, stony materials composed of one or more minerals. Like minerals, rocks are identified by their physical and chemical properties. Rocks are classified into three groups—igneous, sedimentary, and metamorphic—depending on how they are formed.

Process Skill

ANALYZING DATA IN A TABLE

After reviewing the characteristics of each of the four unknown mineral specimens, use the chart below to try to identify each specimen and answer questions 1–4.

Specimen 1: Does not react to acid, white color, and can easily be scratched by specimen 3

Specimen 2: Reacts to acid, white color, and can easily be scratched by specimen 3

Specimen 3: Does not react to acid, white color, none of the other specimens will scratch it

Specimen 4: Does not react to acid, white color, can only be scratched by specimen 3

Mineral Identification Chart

Mineral Name	*Hardness*	*Acid Test*	*Common Color*
Quartz	7	No reaction	White
Calcite	3	Reaction	White
Gypsum	2	No reaction	White
Feldspar	6	No reaction	White

1. Specimen 1 is most likely
   (1) quartz (2) calcite (3) gypsum (4) feldspar

2. Specimen 2 is most likely
   (1) quartz (2) calcite (3) gypsum (4) feldspar

3. Specimen 3 is most likely
   (1) quartz (2) calcite (3) gypsum (4) feldspar

4. Specimen 4 is most likely
   (1) quartz (2) calcite (3) gypsum (4) feldspar

1. ***Igneous rocks*** are produced by the cooling and hardening of hot, liquid rock. This melted rock material is called *magma* when underground and *lava* when it pours onto Earth's surface. Different igneous rocks are generally identified by their color and by the size of the mineral grains (crystals) they contain.

Igneous rocks that form from rapid cooling of lava, called *volcanic* rocks, contain tiny crystals. Basalt is a dark-colored volcanic rock composed of crystals too small to be seen with the unaided eye.

Igneous rocks that form underground by slow cooling of magma develop large crystals. Granite is a light-colored igneous rock that contains large, easily visible mineral grains. Figure 5-3A on page 46 shows processes that produce igneous rocks.

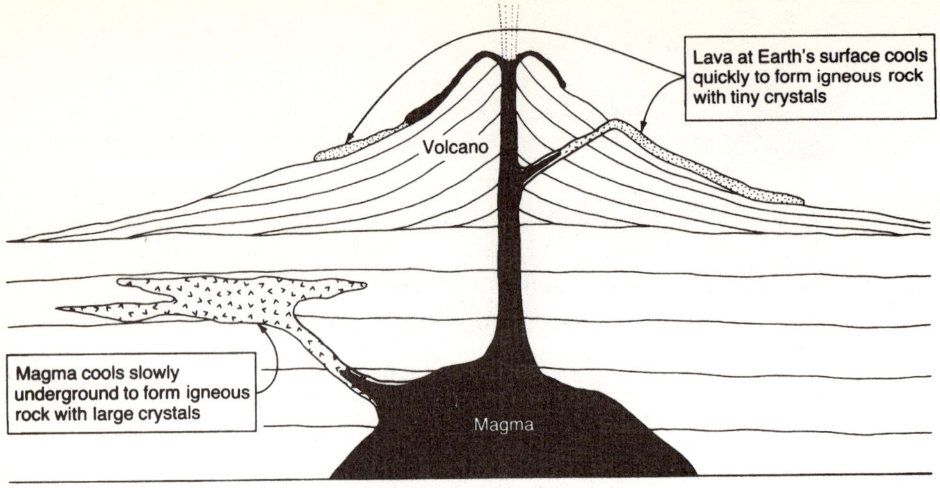

Figure 5-3A. Formation of igneous rocks occurs underground and at Earth's surface.

Magma cools at different rates depending on its depth below the Earth's surface. The closer the magma is to the surface, the smaller the grain (crystal) size of the minerals formed in the rock. In fact, lava on the Earth's surface may cool so rapidly that a glassy substance forms, not individual grains.

Carefully study Figure 5-3B, showing the relationships among common igneous rocks, grain size, rate of cooling, and environment in which magma solidified. Fill in the missing data in a copy of the table shown below.

Grain Size	*Environment*	*Cooling Rate*	*Letter in Diagram*	*Igneous Rock*
0.1 mm				
	Near surface		B	
		Slow		

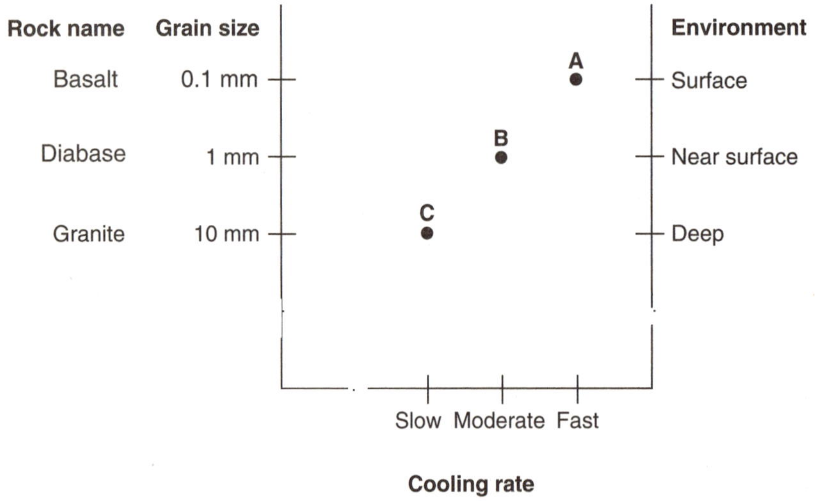

Figure 5-3B. Grain size of igneous rocks depends on the rate and depth at which magma cools.

2. ***Sedimentary rocks*** form from particles called ***sediments*** that pile up in layers. These sediments may be small rock fragments or seashells. Sedimentary rocks usually form underwater. For example, when a stream carrying particles of sediment empties into an ocean or lake, the particles settle to the bottom in layers. Eventually, these layers harden into sedimentary rock (see Figure 5-4A). Table 5-2 lists some common sedimentary rocks, what they are made of, and where they form.

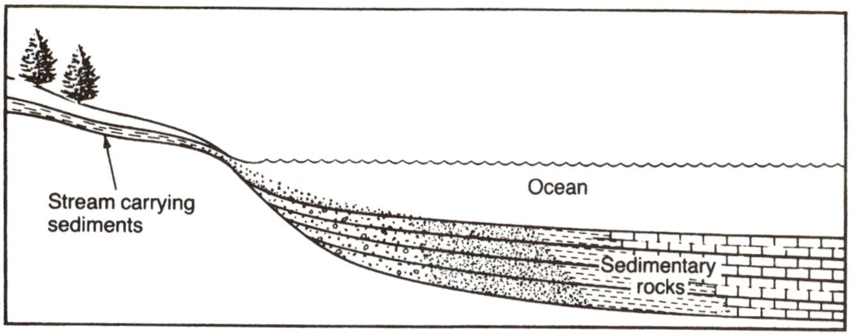

Figure 5-4A. Sedimentary rocks form in layers.

When entering the ocean, a river or stream loses energy and drops the sediments it is carrying. It drops the largest sediments first and, as the water flow continues to slow, it drops smaller and smaller particles. After sediments accumulate for millions of years, they become buried deep in the earth and harden into sedimentary rocks.

Figure 5-4B shows the accumulation of sediments as a river enters the ocean.

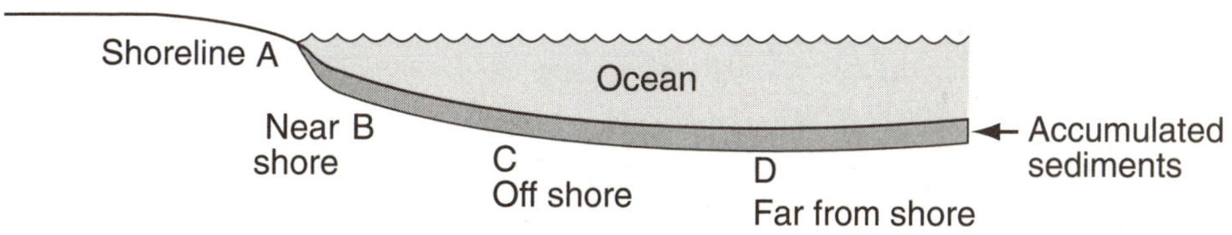

Figure 5-4B. Sediments accumulate on the ocean floor.

Sedimentary Rock Characteristics

Sedimentary Rock	*Description*	*Size of Particles*	*Formation Environment*
Sandstone	Cemented grains of sand	0.1-2.0 mm	
Conglomerate	Visible cemented rounded pebbles	2.0 mm	
Shale	Sheets of tightly compact clay particles	0.04 mm	
Siltstone	Powdery grains of cemented silt particles	0.01-0.06 mm	

Using Figure 5-4B, which shows depositing locations of sediment, select the letter that best shows the formation environment of each of the sedimentary rocks listed. Place the letter in the formation environment column in a copy of the table shown above.

3. ***Metamorphic rocks*** are produced when igneous or sedimentary rocks undergo a change in form caused by heat, pressure, or both. This can take place when magma heats rocks that it comes into contact with, or when forces deep underground squeeze rocks for a long period of time. These conditions alter the appearance and mineral composition of the rocks, changing them into metamorphic rock.

Marble and slate are metamorphic rocks formed, respectively, from the sedimentary rocks limestone and shale. Gneiss is a metamorphic rock that can be produced from granite, an igneous rock.

Table 5-2. Common Sedimentary Rocks

Rock Name	*Type of Sediment*	*Place of Formation*
Sandstone	Sand grains	Shallow waters near a shore pounded by waves
Shale	Clay particles	Deep, calm ocean waters; lake bottoms
Limestone	Tiny seashells	Warm, shallow seas

The Rock Cycle

The three types of rocks—***igneous, sedimentary,*** and ***metamorphic***—are subject to processes that can change any one type into another type. Igneous rocks can be changed into sedimentary or metamorphic rocks by various processes. Sedimentary rocks can be recycled into new sedimentary rocks, or changed into igneous or metamorphic rocks. Likewise, metamorphic rocks can become igneous or sedimentary rocks. All these changes and processes make up the *rock cycle,* shown in Figure 5-5 below.

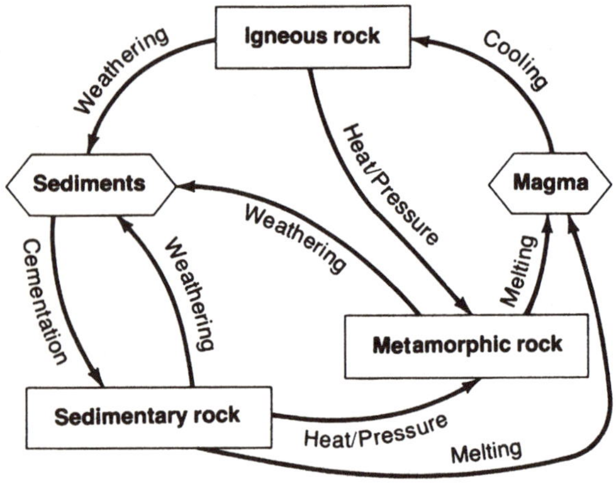

Figure 5-5. The rock cycle: igneous, metamorphic, and sedimentary rocks.

EXERCISE 1

Use Figure 5-5, which shows the rock cycle, to help you answer questions 1–4.

1. Which processes are necessary to change a metamorphic rock into a sedimentary rock?
   (1) weathering and cementation
   (2) melting and cooling
   (3) heat and pressure

2. Which statement is true?
   (1) Sediments melt to form sedimentary rock.
   (2) Igneous rock weathers to form sediments.
   (3) Magma cools to form metamorphic rock.

3. To become igneous rock, sedimentary rock must undergo
   (1) weathering and cementation
   (2) heat and pressure
   (3) melting and cooling

4. Schist is a metamorphic rock. This means it was formed by
   (1) cooling and hardening of magma
   (2) great heat or pressure, or both
   (3) buildup of sand grains
   (4) buildup of clay particles

5. Sandpaper is made of tiny grains of a hard mineral glued to paper and used to scrape softer substances. Based on the hardness scale given in Table 5-1, a good mineral to use in making sandpaper is
   (1) talc (3) gypsum
   (2) calcite (4) quartz

6. The mineral identification chart (on page 45) indicates the hardness of four minerals. Which mineral is hard enough to scratch gypsum but will not scratch feldspar?
   (1) mica (3) quartz
   (2) calcite (4) none

7. A *chemical property* that would help to identify a mineral is
   (1) luster (3) reaction to acid
   (2) hardness (4) cleavage

8. Rocks that form from layers of small particles are called
   (1) metamorphic rocks
   (2) sedimentary rocks
   (3) igneous rocks
   (4) volcanic rocks

9. Which process was involved in forming the mountain shown below?

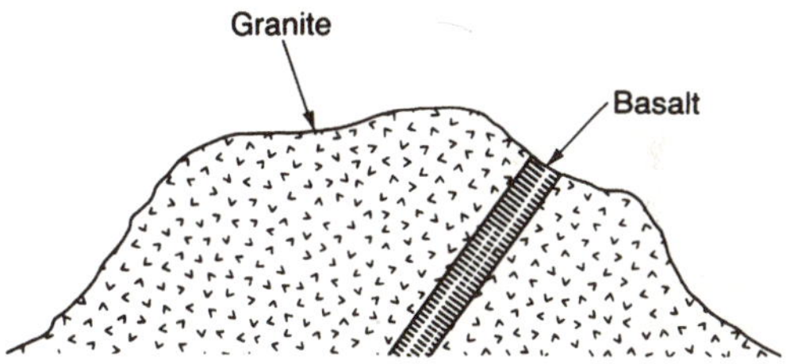

   (1) cooling of magma to form igneous rocks
   (2) buildup of sediments in shallow water
   (3) deep underground pressure
   (4) buildup of sediments in deep water

10. The illustration shows the mineral mica splitting into thin sheets. This is an example of

   (1) color (3) uneven fracture
   (2) hardness (4) cleavage

Forces That Change Earth's Surface

Various forces are constantly at work shaping and changing Earth's surface. Figure 5-6 illustrates these forces and their effects on Earth's surface features.

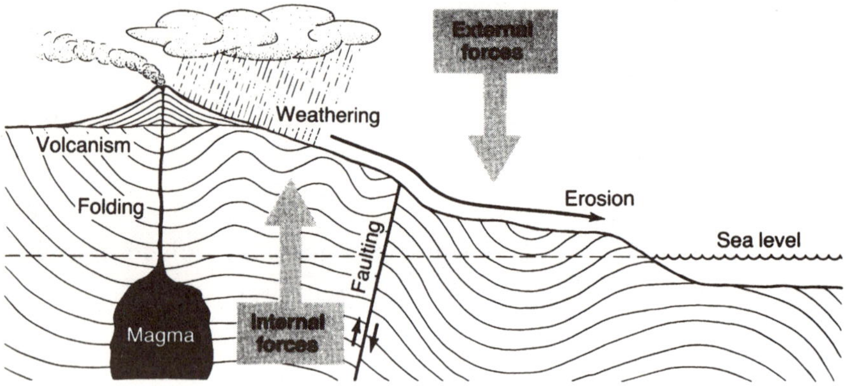

Figure 5-6. Earth's surface is shaped by the interaction of internal and external forces.

External Forces

External forces include the processes of weathering and erosion. Together, these processes wear down Earth's surface.

1. ***Weathering*** is the breaking down of rocks into smaller pieces. Both physical and chemical agents can cause weathering. In *physical weathering*, rock is broken into smaller fragments by physical agents. For example, when water seeps into cracks in a rock and freezes, the water expands, breaking the rock apart, as shown in Figure 5-7. The roots of plants growing in cracks can also force rocks apart.

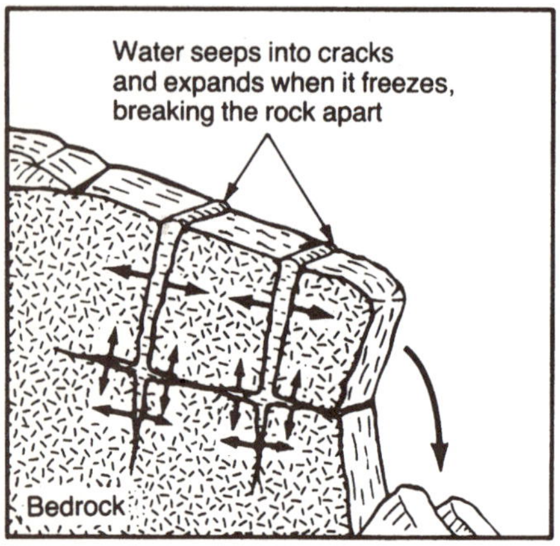

Figure 5-7. Physical weathering caused by water.

Chemical weathering is the breaking down of rocks through changes in their chemical makeup. These changes take place when rocks are exposed to air or water. For instance, when rainwater combines with carbon dioxide in the air, a weak acid is formed that dissolves certain minerals in rocks and causes the rocks to fall apart. Also, when oxygen and water react chemically with iron-bearing minerals in a rock, the iron is changed into rust, which crumbles away easily.

By breaking down rocks into smaller fragments, the processes of weathering assist in the formation of soil.

2. ***Erosion*** is the process by which rock material at Earth's surface is removed and carried away. Erosion requires a moving force, such as flowing water, that can carry along rock particles. This can be seen after a heavy rain, when streams turn a muddy brown from the rock material in the water.

Gravity and *water* play important roles in erosion. Gravity is the main force that moves water and rock downhill. Flowing water is very powerful; more rock material is eroded by running water than by all other forces of erosion combined. The Grand Canyon in Arizona is a spectacular example of erosion caused by running water (see Figure 5-8).

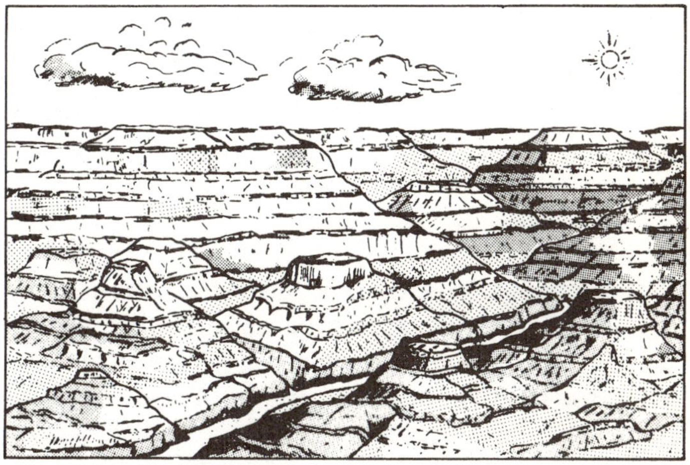

Figure 5-8. Erosion caused by running water carved the Grand Canyon.

Groundwater and glaciers also cause erosion. *Groundwater* forms from rain or snowmelt that filters into the soil. As groundwater seeps through cracks in the bedrock, the water dissolves rock material and carries it away. Eventually, this action may create large underground caves.

Glaciers are masses of ice that form in places where more snow falls in winter than melts in summer, such as in a high mountain valley. The snow that does not melt piles up over the years, and its increasing weight changes the bottom layers into ice. Gravity causes the ice to flow downhill, like a river in slow motion. As a glacier creeps along, it grinds up and removes rock material from the land surface.

Wind can also act as a force of erosion. In dry desert areas, sand grains blown along by the wind scrape and scour rock outcrops, slowly carving them into unusual shapes.

The forces of erosion are constantly at work, moving rock material from the continents into the ocean basins.

Process Skill

PREDICTING AN EXPERIMENTAL RESULT

Rocks in a stream constantly knock and scrape against each other and against the streambed as they are carried along by the flowing water. The longer the rocks are in the stream, the more they tumble about and strike one another. To simulate this action and study its effects, a student carried out the following experiment.

Twenty-five marble chips and a liter of water were placed in a large coffee can marked A. The can was then covered with a lid and shaken for 30 minutes. Then 25 marble chips and a liter of water were placed in a second can, marked B, and covered. This can was shaken for 120 minutes. The illustration below shows the materials used in the experiment. Keep in mind what you have learned about weathering to help you answer the following questions.

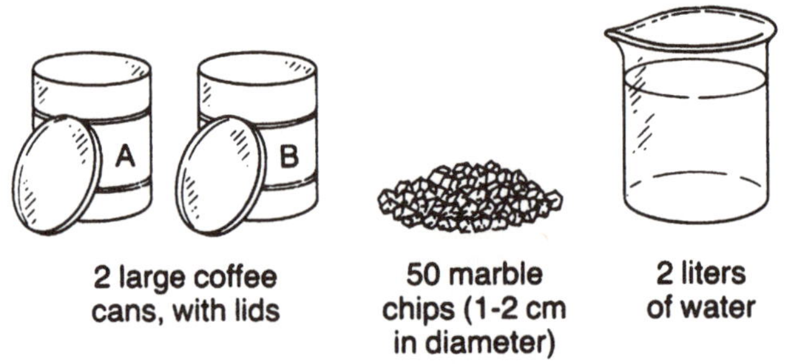

1. Which is the best prediction of the experiment's results?
   (1) The marble chips in can A will be smaller and rounder than the chips in can B.
   (2) The marble chips in can B will be smaller and rounder than the chips in can A.
   (3) There will be no difference between the marble chips in cans A and B.

2. Which graph best predicts what would happen over time to rocks in a fast-moving stream?

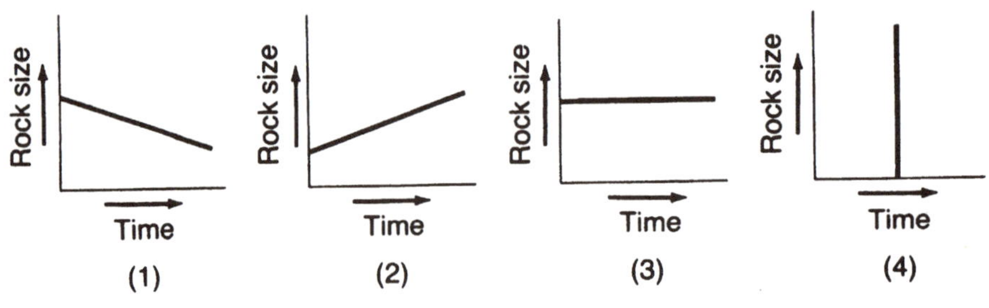

Internal Forces

Earth's internal forces also shape its surface. These forces produce *mountains, earthquakes*, and *volcanoes*, raising the land and building up Earth's surface.

1. ***Mountains*** are produced mainly by the processes of folding and faulting. ***Folding*** takes place when forces in Earth's crust press rocks together from the sides, bending the layers into folds. The land is squeezed into upfolds and downfolds, forming ridges and valleys (see Figure 5-9).

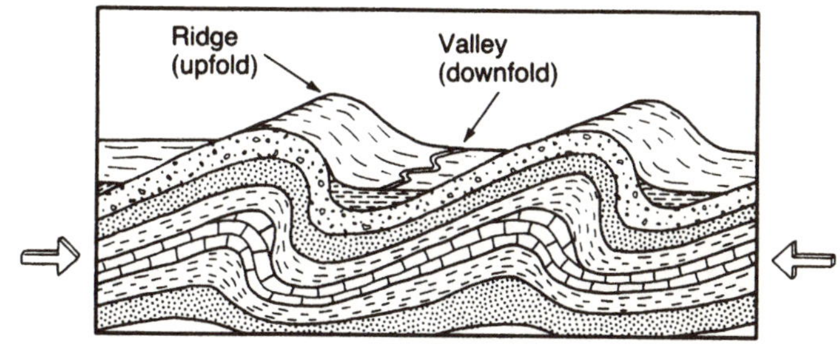

Figure 5-9. Folding: Forces in Earth's crust can squeeze rock layers into folds.

Faulting occurs when forces in the crust squeeze or pull rock beyond its capacity to bend or stretch. The rock then breaks and slides along a crack or fracture, called a *fault*, relieving the stress in the crust (Figure 5-10). Faulting can produce mountains in a number of ways, as shown in Figure 5-11.

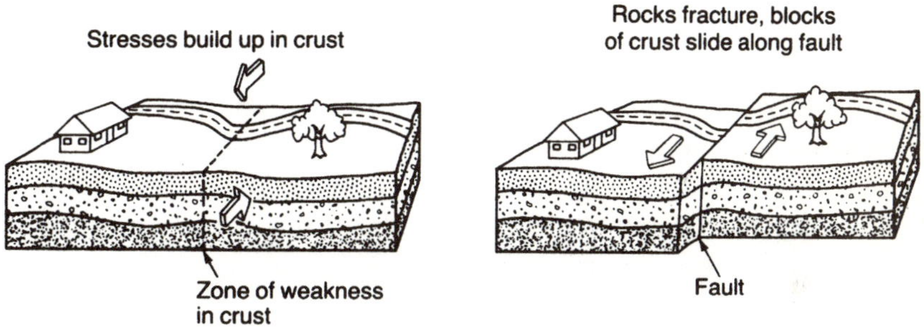

Figure 5-10. Faulting: When stresses in Earth's crust reach the breaking point, the crust fractures and slips.

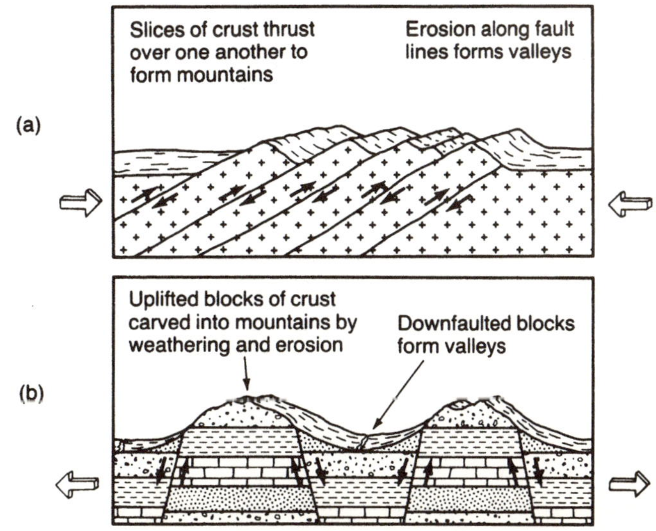

Figure 5-11. Mountains can be produced by (a) thrust faulting and (b) block faulting.

2. Sudden movements of rocks sliding along faults produce strong vibrations in the crust called ***earthquakes.*** Many earthquakes are associated with land uplift and mountain building.

3. Mountains can also be built by volcanoes. A ***volcano*** is a hole in Earth's crust through which lava flows from underground. During eruptions, the lava pours out onto the surface and cools to form solid rock, building upward in layers to produce a volcanic mountain, also called a volcano (refer to Figure 5-3). Mount St. Helens in Washington State is a volcanic mountain.

Besides mountains, other landforms that may result from uplift include plains and plateaus.

4. ***Plains*** are broad, flat regions found at low elevations. They are often made of layered sedimentary rocks that were formed underwater and slowly raised above sea level.

5. ***Plateaus*** are large areas of horizontally layered rocks with higher elevations than plains. They can form in several ways. A large block of crust may rise up along faults to create a plateau, or a plateau may be gradually uplifted without faulting. Plateaus can also be built up by lava flows.

Plate Tectonics

There is much evidence that forces at work inside Earth have raised the level of the land. For example, many mountaintops are made of sedimentary rock that was formed originally on the ocean floor. Folds and faults seen in many rock outcrops are also signs of crustal movements caused by internal forces. Scientists explain these forces and the movements they produce by the theory of ***plate tectonics.***

According to this theory, Earth's crust is broken up into a number of large pieces, or ***plates,*** that slowly move and interact in various ways. Some plates are spreading apart, some are sliding past each other, and some are colliding. These movements cause mountain building, volcanic activity, and earthquakes along the plate edges. Figure 5-12 shows Earth's major crustal plates.

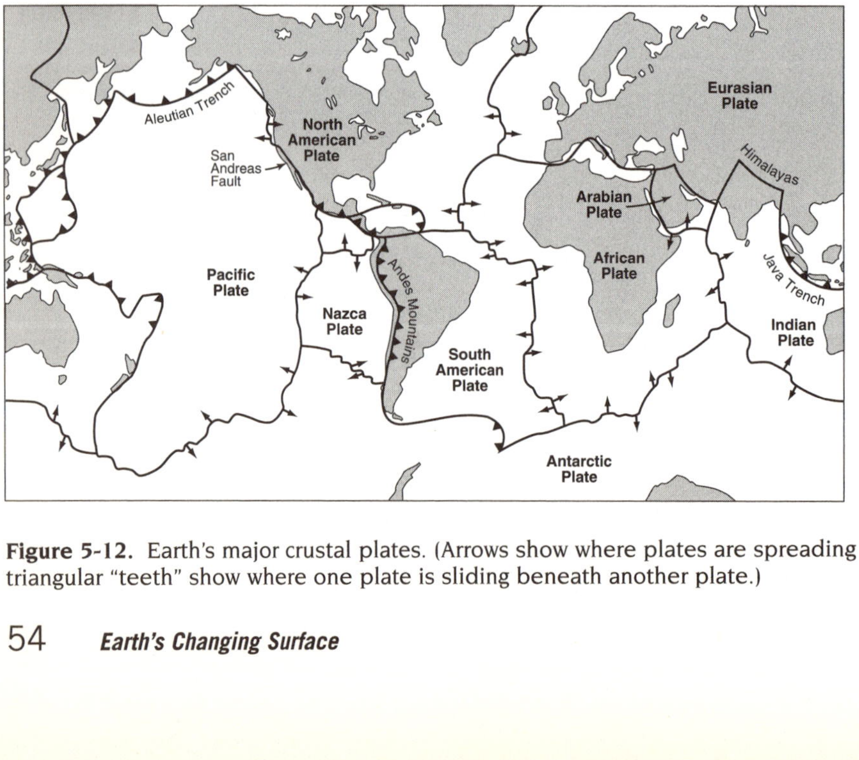

Figure 5-12. Earth's major crustal plates. (Arrows show where plates are spreading apart, triangular "teeth" show where one plate is sliding beneath another plate.)

Scientists believe that plate motions are caused by heat circulating in Earth's ***mantle***, the thick zone of rock beneath the crust. The heat softens mantle rock so that it flows very slowly, following the heat currents and carrying along overlying pieces of crust (see Figure 5-13).

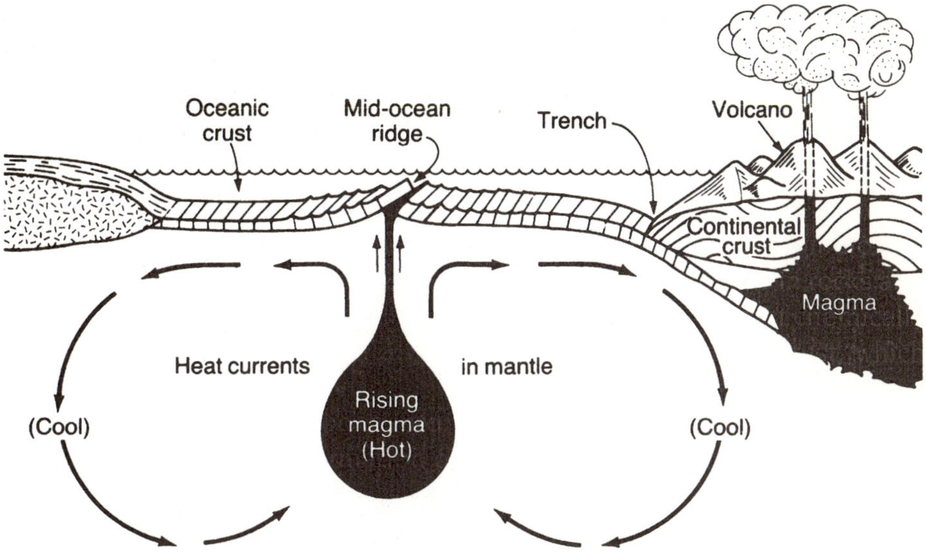

Figure 5-13. Plate tectonics: Heat currents in the mantle cause movements of Earth's crustal plates, producing many features on the seafloor and on the continents.

The processes of plate tectonics create many of Earth's surface features. The collision of two plates carrying continents produces great mountain ranges. The Himalayas were formed in this way.

When one plate slides sideways past another plate, a major fault and earthquake zone is produced. In California, the Pacific Plate is sliding past the North American Plate along the San Andreas Fault, sometimes causing severe earthquakes.

Where plates are spreading apart, ocean basins are formed. Large continents are broken into smaller landmasses that move away from each other in a process called ***continental drift***. This is taking place today where the Arabian Plate is splitting away from the African Plate, opening up the Red Sea.

Ocean Floor Features

Almost three-quarters of Earth's surface is covered by ocean water. The floor of the ocean is not all flat and featureless. Scientists have found that the ocean floor has mountains, valleys, plains, and plateaus. Many of these features, such as mid-ocean ridges and ocean trenches, are produced by the processes of plate tectonics.

1. A ***mid-ocean ridge*** is a long, underwater mountain chain where rising magma forms new ocean crust. The new crust is added to crustal plates that spread away from the ridge, as shown in Figure 5-13. This process is called *seafloor spreading*.

2. ***Trenches*** are underwater valleys that form the deepest part of the ocean floor. A trench is found where a plate of ocean crust collides with another plate and is forced to slide under it, back into Earth's mantle. This causes volcanic activity and mountain building along the edge of the upper plate (refer to Figure 5-13).

Process Skill

ORGANIZING DATA INTO A TABLE

After a heavy rain, Mary and Mike noticed how much faster the stream near their home was running. They also noticed that the color of the water was a dark, murky brown. They decided to keep notes on their observations of the changing stream over the next five days. No rain fell during the time of their observations.

Day 1: Water was dark brown, about 30 cm deep, and flowing very fast—about 1.5 meters/second.
Day 2: Water was brown, about 20 cm deep, and flowing rapidly—about 1 meter/second.
Day 3: Water was light brown, about 15 cm deep, and was flowing .6 meters/second.
Day 4: Water was almost clear, about 10 cm deep, and was flowing .3 meters/second.
Day 5: Water was clear, about 5 cm deep, and was flowing less than .3 meters/second.

Make a table of the five days of observations, showing the data collected.

Day	*Water Color*	*Water Depth*	*Water Speed*
1			
2			
3			
4			
5			

Based on the data collected, answer the following questions.

1. What is the relationship between water depth and water speed?
   (1) As the water depth increases, the water speed increases.
   (2) As the water depth increases, the water speed decreases.
   (3) As the water depth increases, the water speed remains the same.
   (4) As the water depth decreases, the water speed increases.

2. The changing color of the water was most likely the result of
   (1) the speed of the water
   (2) the depth of the water
   (3) the amount of sediments in the water
   (4) none of the above

Other ocean floor features include continental shelves, continental slopes, the deep ocean floor, and seamounts.

3. ***Continental shelves*** are areas of the seafloor that slope gently away from the coastlines of most continents. The angle of slope is so slight that if you could stand on a continental shelf, you would think you were on level ground.

4. ***Continental slopes*** drop away from the outer edges of continental shelves to the great depths of the ocean. These slopes are much steeper than continental shelves.

Continental slopes level off into the *deep ocean floor.* The deep ocean floor is not simply a flat plain; it also has ridges and valleys. Rising here and there from the ocean floor are tall underwater mountains called *seamounts.* Most seamounts were formed by volcanoes.

When the top of a seamount rises above the water's surface, an island is formed. The Hawaiian Islands are the tops of a chain of volcanic seamounts. Figure 5-14 shows the profile of an ocean floor that includes many of these features.

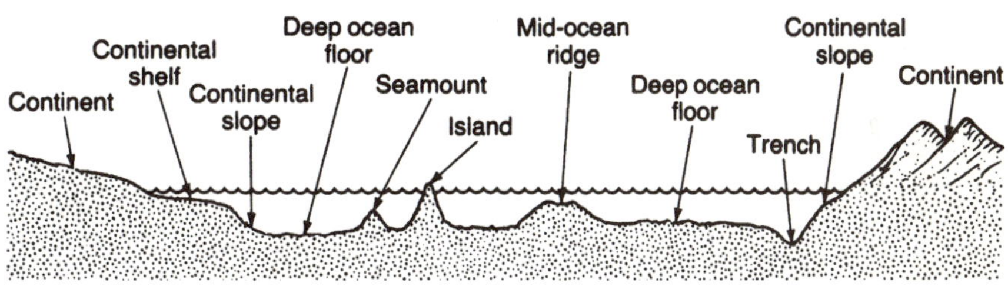

Figure 5-14. Major features of the ocean floor.

EXERCISE 2

1. When water freezes in cracks in a rock, the water expands, breaking the rock apart. This is a type of
   (1) glacial erosion
   (2) physical weathering
   (3) chemical weathering
   (4) groundwater erosion

2. The diagram below shows the mineral magnetite, which contains iron, changing into rust particles. This is an example of

   (1) physical weathering
   (2) chemical weathering
   (3) erosion by running water
   (4) the role of gravity in erosion

3. Erosion is the process by which rocks at Earth's surface
   (1) are removed and carried away
   (2) crumble and decay
   (3) turn into rust
   (4) melt to form magma

4. The diagram shows how an underground cave changed over time. Which process caused this?

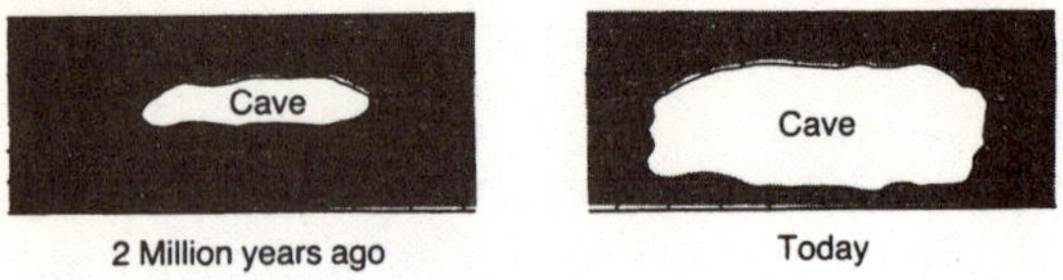

(1) erosion by glaciers
(2) physical weathering
(3) erosion by groundwater
(4) expansion of water changing to ice

5. A new road cut through a hillside in 1985 exposed a fresh outcrop of rocks. Looking at the rocks in the late 1990s, you see plants growing over the rocks, red-brown staining on the rocks, and broken rocks accumulated at the base of the outcrop. The evidence suggests that the outcrop is
(1) getting larger
(2) undergoing weathering
(3) remaining unchanged
(4) being affected by humans

6. The shape of the land in the series of diagrams below most likely changed because of

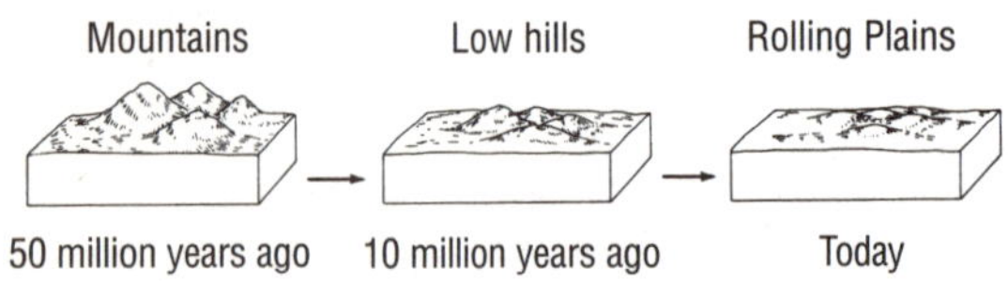

(1) weathering and erosion
(2) volcanoes and earthquakes
(3) folding and faulting
(4) melting

7. The rock layers in the diagram have been affected by

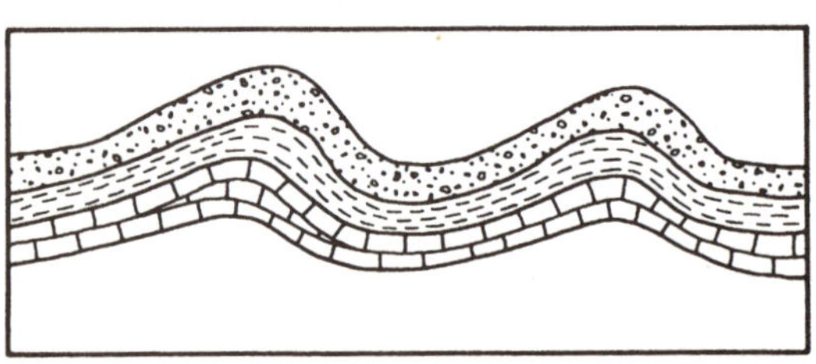

(1) volcanoes (3) groundwater erosion
(2) faulting (4) folding

8. Mountains can be produced by all of the following processes *except*
(1) volcanic eruptions
(2) folding
(3) weathering
(4) faulting

9. The San Andreas Fault cuts through California. The fault is a strike slip (horizontal slip) fault that produces major earthquake activity in California. The arrow showing the motion of the fault indicates that the shaded area of California is moving toward the

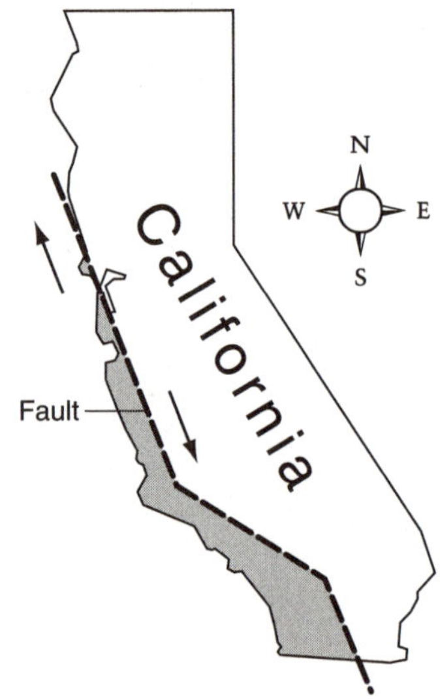

(1) northeast (3) northwest
(2) southeast (4) southwest

10. The theory that Earth's crust is broken up into large pieces that move and interact is called
(1) evolution
(2) mountain building
(3) the rock cycle
(4) plate tectonics

11. The ocean floor is best described as
(1) a flat, featureless plain
(2) having mountains, valleys, plains, and plateaus
(3) a flat plain with a deep valley in the center
(4) having plains and plateaus only

12. If crustal block A, to the left of the fault in the diagram, suddenly shifted downward several feet, what would most likely occur at location C?

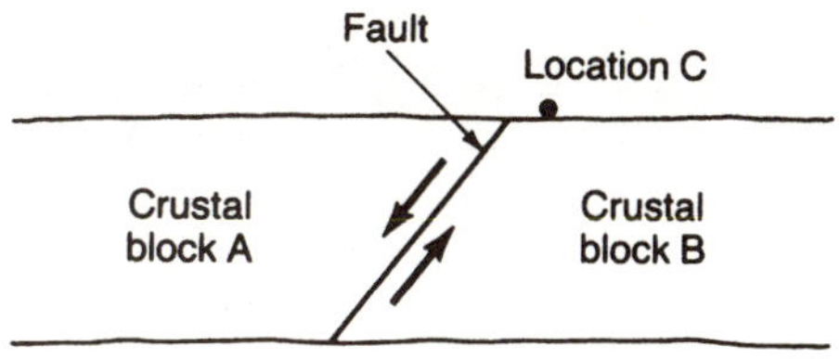

(1) An earthquake would occur.
(2) A volcanic eruption would occur.
(3) A mountain would form.
(4) An ocean would form.

13. The table below shows the depth of the ocean at various distances from a continent. At which distance from the continent is a deep trench located?
(1) 300 kilometers (3) 100 kilometers
(2) 200 kilometers (4) 500 kilometers

Distance From Continent	*Ocean Depth*
50 km	400 m
100 km	9000 m
150 km	1250 m
200 km	1100 m
250 km	200 m
300 km	950 m

14. Earth's crust is in motion. The Mid-Atlantic Ridge is a location of upwelling magma where new crust is being formed and pushed outward in two directions, east and west. If the crust is being produced at a regular and steady rate, then dating of rocks would give the same age at which positions?

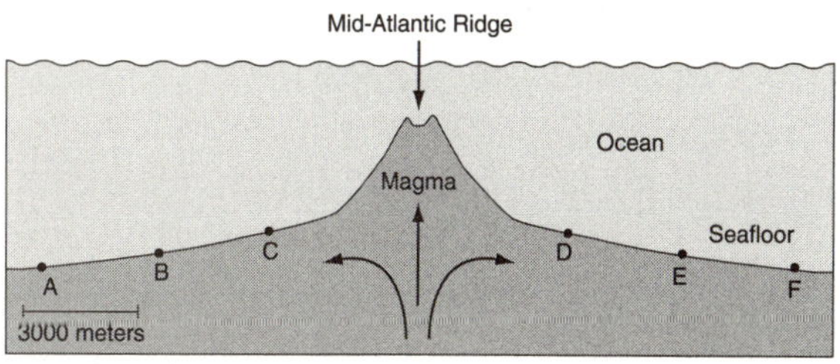

(1) positions A and B
(2) positions A and C
(3) positions B and D
(4) positions B and E

Base your answers to questions 15 and 16 on the following paragraph and the graph below.

The snow line is the elevation at which snow remains all year. That is, the accumulation of snow is greater than the amount of snow that melts during the year. The snow line is visible on mountains that exceed the snow line and have snow on their tops all year. The graph shows how the snow line changes with latitude.

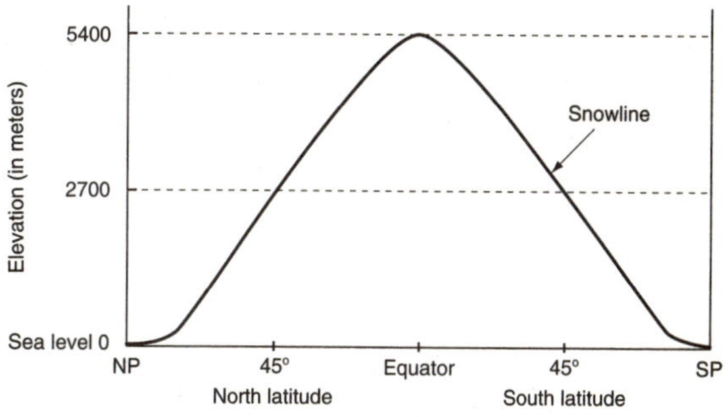

15. As you travel from the North Pole (NP) to the South Pole (SP), the elevation of the snow line
(1) increases
(2) decreases
(3) increases then decreases
(4) decreases then increases

16. A state is at 40° north latitude. How high would mountains have to be to have a snow line in this state?
(1) 305 meters (3) 1525 meters
(2) 610 meters (4) 3050 meters

17. In one or two sentences, tell why earthquakes are most likely to occur in a place where two crustal plates come together.

6 Earth's Motions in Space

Points to Remember —

- Rotation is the spinning of Earth around its axis.
- Earth's rotation causes day and night. It takes Earth 24 hours (1 day) to complete one rotation.
- Earth's axis is tilted 23.5°. The northern end of Earth's axis (the North Pole) points toward the North Star.
- Revolution is the motion of Earth traveling around the sun. It takes Earth 365.25 days (1 year) to complete one revolution around the sun.
- Earth's revolution combined with its tilt causes the sun's most direct rays to shift farthest north on June 21 and farthest south on December 21. Thus, the seasons on Earth are caused by Earth's revolution and its changing tilt relative to the sun's intense rays.

Earth: Member of the Solar System

The sun and the nine planets are the primary members of the solar system (see Figure 6-1). Earth is the third planet from the sun. Each of the planets revolves around the sun in a nearly circular orbit, and each planet rotates on its axis. It takes Earth 365 days to revolve around the sun and 24 hours to rotate on its axis.

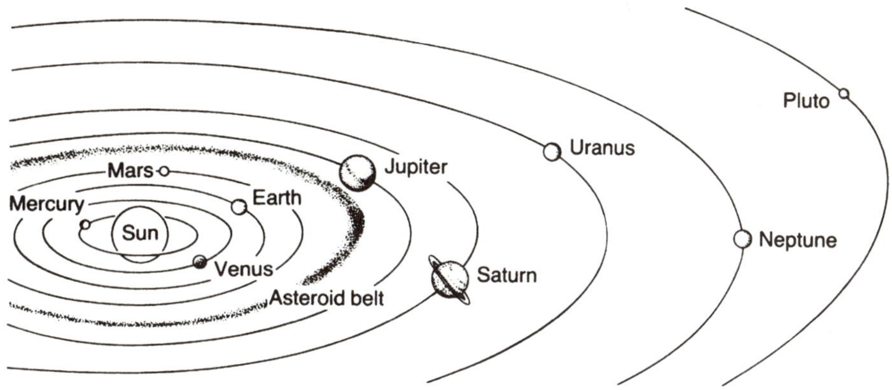

Figure 6-1. The solar system.

Earth's Rotation

People once believed that Earth stands still while the sun, moon, stars, and planets revolve around Earth each day. This seems reasonable, since we do not feel Earth moving, the sun does appear to move across the sky during the day, and the moon and stars appear to move at night. However, today we know that these apparent motions of the sun, moon, and stars are actually caused by Earth's own motion.

Earth spins like a top. This spinning motion is called ***rotation.*** Earth rotates from west to east—or, to put another way, looking down at the North Pole from space, we would see Earth spinning in a counterclockwise direction.

Extending through Earth between the North and South poles is an imaginary rod, or ***axis of rotation,*** around which Earth spins (see Figure 6-2). A basketball spinning on a fingertip gives a good idea of how Earth spins on its axis. The line from the fingertip through the basketball to the top of the ball is the axis of rotation.

The rotation of Earth has several results:

1. Earth's rotation causes the daily change from day to night. At any given time, half of Earth is in daylight, facing the sun, while half is in darkness, facing away from the sun. This is shown in Figure 6-3. Every day, all places on Earth, except the areas near the poles, experience this change from daylight to darkness. (Areas within the Arctic and Antarctic circles experience several weeks of continuous daylight or darkness at certain times of the year.)

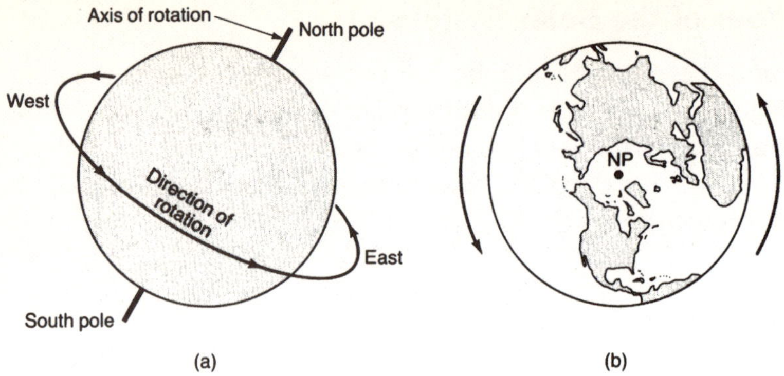

Figure 6-2. Rotation of Earth: (a) Earth rotates from west to east around its axis. (b) Viewed from above the North Pole (NP), Earth rotates in a counterclockwise direction.

2. The speed of Earth's rotation causes the length of one day to be about 24 hours. This is the amount of time Earth takes to rotate once on its axis.
3. The daily motions of the sun, moon, planets, and stars across the sky—as we perceive them—are also affected by Earth's rotation. These objects appear to rise in the eastern sky and set in the western sky because Earth rotates from west to east.

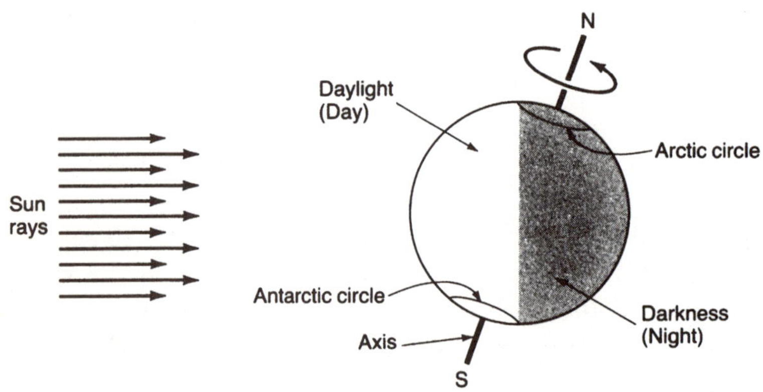

Figure 6-3. Earth's rotation causes the change from day to night.

Daily Time

The time of day at a given location on Earth is based on Earth's rotation. Earth completes one rotation each day. That is, each location on Earth rotates 360° around Earth's axis each day. There are 24 hours in a day. If we divide 360° by 24 hours, we find that Earth rotates 15° per hour. Table 6-1 and Figure 6-4 show how the time of day and the location of the sun in the sky are related to Earth's rotation.

Table 6-1. Rotation Positions on Earth

Line	*Rotation Angle*	*Time*	*Location of Sun*
A	Start	12:00 noon	High in the sky
B	45°	3:00 P.M.	In the western sky
C	90°	6:00 P.M.	Near western horizon
D	180°	12:00 midnight	Opposite side of Earth
E	270°	6:00 A.M.	Near eastern horizon

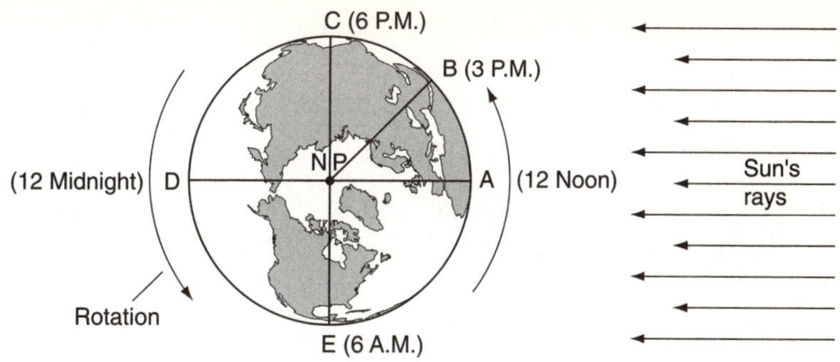

Figure 6-4. The time of the day and the sun's location in the sky are related to Earth's rotation.

EXERCISE 1

Viewed from above the North Pole, Earth rotates counterclockwise. Earth makes one complete rotation in one day. The side of Earth facing the sun is in daylight, and the side facing away from the sun is experiencing night. As Earth rotates every day, each city in Ohio goes through a period of daylight and darkness.

Questions 1 and 2 refer to the diagram below. Each lettered position in the diagram represents Ohio at a different time of day. For example, E represents Ohio at about 12:00 midnight.

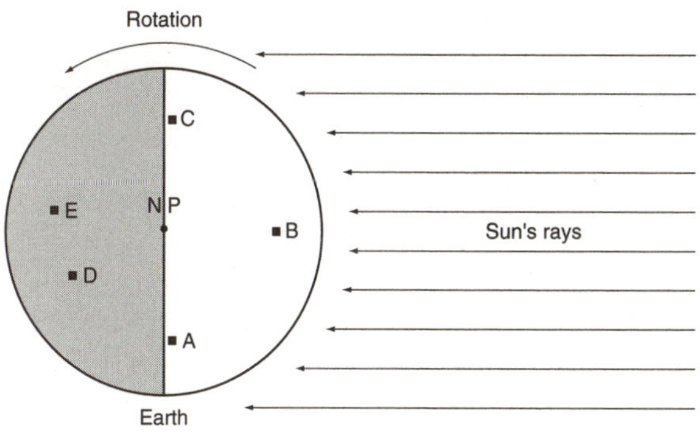

1. Which position represents Ohio at 12:00 noon, when the sun is most nearly overhead?
(1) A (2) B (3) C (4) D

2. Which position represents Ohio at 6:00 A.M., with the sun rising in the east?
(1) A (2) B (3) C (4) D

Questions 3 and 4 refer to the map below, which shows Ohio with about half the state in daylight and half the state in darkness (night). As Earth rotates from west to east, the portions of daylight and night in the state will change.

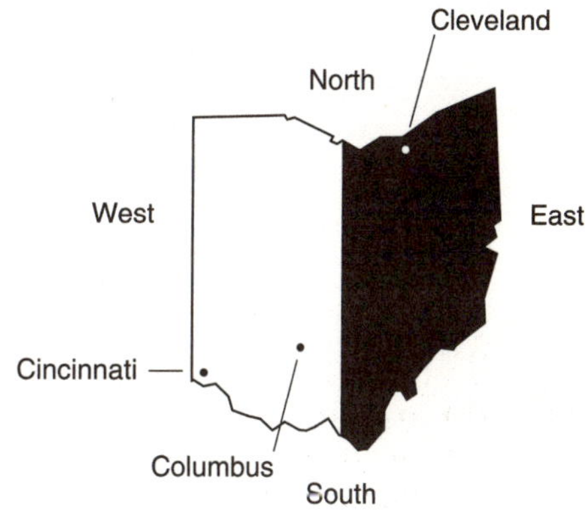

3. The time of day in Cleveland is most nearly
(1) noon (3) midnight
(2) dusk (sunset) (4) dawn (sunrise)

4. As Earth rotates from west to east over the next few hours, Ohio will
(1) move completely into night
(2) move completely into daylight
(3) not change its daylight or darkness
(4) change from completely night to completely daylight

5. The sun rises in the east and sets in the west because Earth rotates from
   (1) west to east (3) north to south
   (2) east to west (4) north to west

6. The daily change from daylight to darkness is caused by
   (1) Earth's revolution around the sun
   (2) the tilt of Earth's axis
   (3) Earth's rotation on its axis
   (4) the sun's light going out at night

7. Earth rotates once on its axis in one complete
   (1) year (2) week (3) month (4) day

8. The diagram below shows Earth as seen from above, looking down at the North Pole (NP). At location A, the time is 12:00 noon. What is the time at location C?

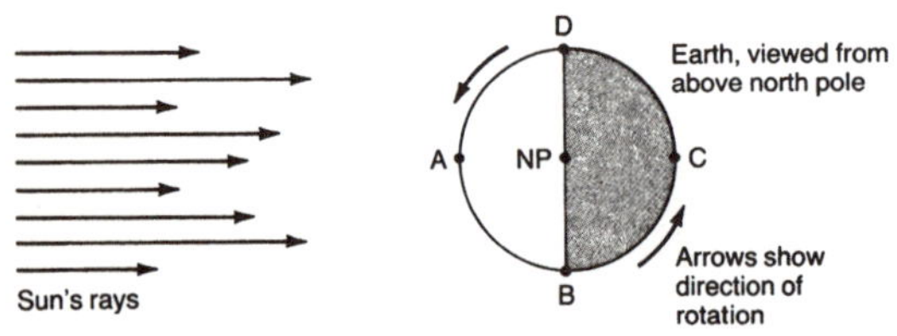

   (1) 3:00 P.M. (3) 12:00 midnight
   (2) 6:00 P.M. (4) 6:00 A.M.

9. Kareem saw the moon over a tree at 9:00 P.M. (position A in the diagram). An hour later, the moon had moved to position B. This change in position was caused by the

   (1) changing of Earth's tilt
   (2) rotation of Earth
   (3) revolution of Earth
   (4) rotation of the moon

10. In one or two sentences, explain why there is a day and a night at different places on Earth at the same time.

Earth's Revolution

Earth moves around the sun in a motion called ***revolution.*** The path Earth travels around the sun is called an ***orbit.*** Earth's orbit, although nearly circular, is actually slightly oval in shape (Figure 6-5).

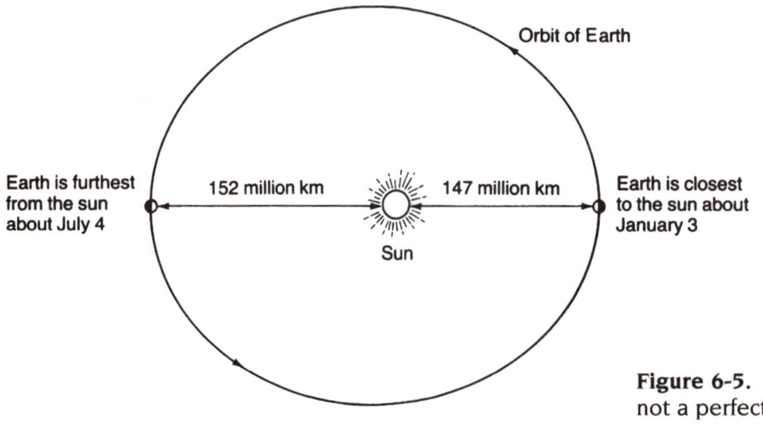

Figure 6-5. Earth's orbit around the sun is an oval, not a perfect circle.

Earth's revolution has two major effects. First, the time Earth takes to revolve once around the sun defines the length of a year. During that time, Earth rotates on its axis $365\frac{1}{4}$ times, so there are $365\frac{1}{4}$ days in a year. For convenience, the calendar year is 365 days long, and an extra day is added every fourth year (called a leap year) to make up for each leftover $\frac{1}{4}$ day.

Second, Earth's revolution around the sun, combined with the tilt of Earth's axis, causes the changing seasons on the planet. Earth's axis of rotation is not perpendicular (at a right angle) to the plane of its orbit; rather, it is tilted (Figure 6-6).

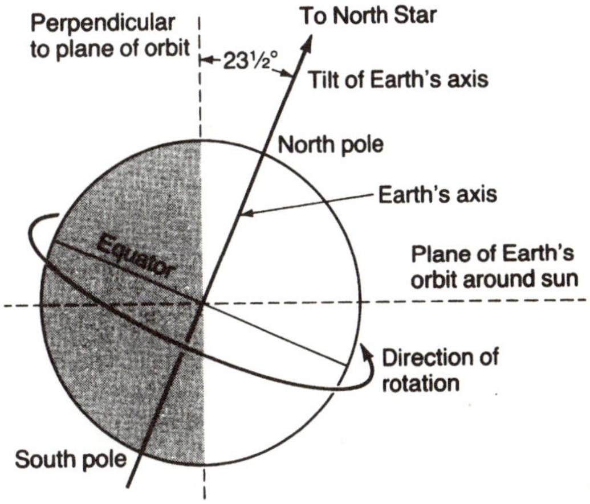

Figure 6-6. Earth's axis of rotation is tilted.

No matter where Earth is in its orbit, the axis is always tilted in the same direction in space, pointing toward the North Star. While all the other stars seem to move across the night sky, the North Star remains motionless because Earth's axis points to it.

Yearly Seasonal Time

The length of a year is based on Earth's revolution. Earth revolves completely around the sun once each year. That is, Earth revolves 360° during a 12-month period of time. If we divide 360° by 12 months, we find that Earth revolves about 30° per month. Table 6-2 and Figure 6-7 show how the date and the season on Earth are related to Earth's revolution around the sun.

Table 6-2. Revolution Positions of Earth

Points	*Revolution Angle*	*Date*	*Seasonal Information*
A	Start	December 21	Winter begins
B	30°	January 21	Winter
C	90°	March 21	Spring begins
D	180°	June 21	Summer begins
E	270°	September 23	Autumn begins

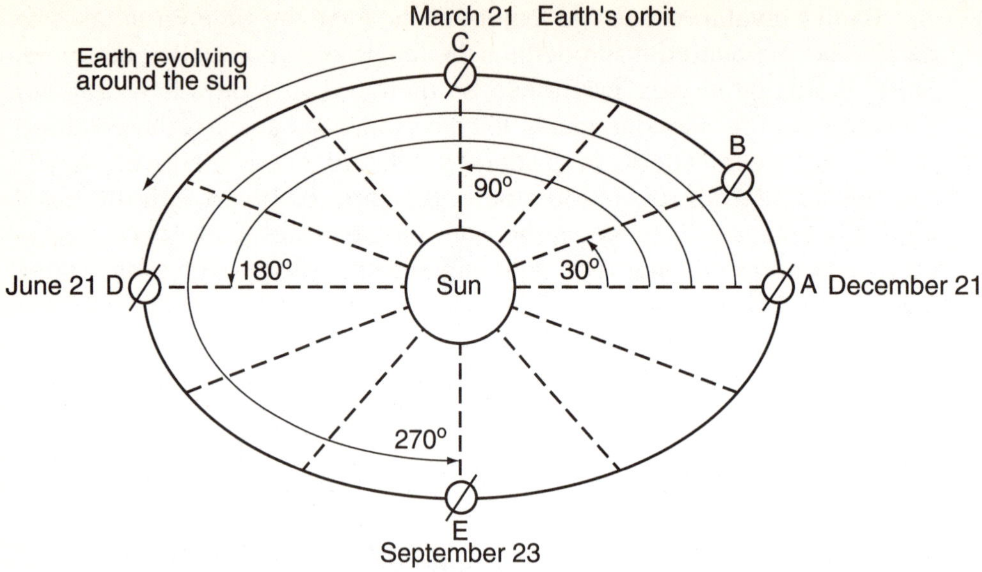

Figure 6-7. Earth revolves 360° around the sun in a 12-month period.

Seasonal Changes

Earth's orbit takes it closest to the sun in early January and farthest from the sun in early July (see Figure 6-5). This means that it is not Earth's changing distance from the sun that causes the changing seasons. The true cause is the tilt of Earth's axis as the planet revolves around the sun. Because the axis always points in the same direction while Earth orbits the sun, the northern hemisphere is tilted toward the sun part of the year and away from the sun part of the year (Figure 6-8). This causes changes in the length of daylight and in the angle at which the sun's rays strike Earth.

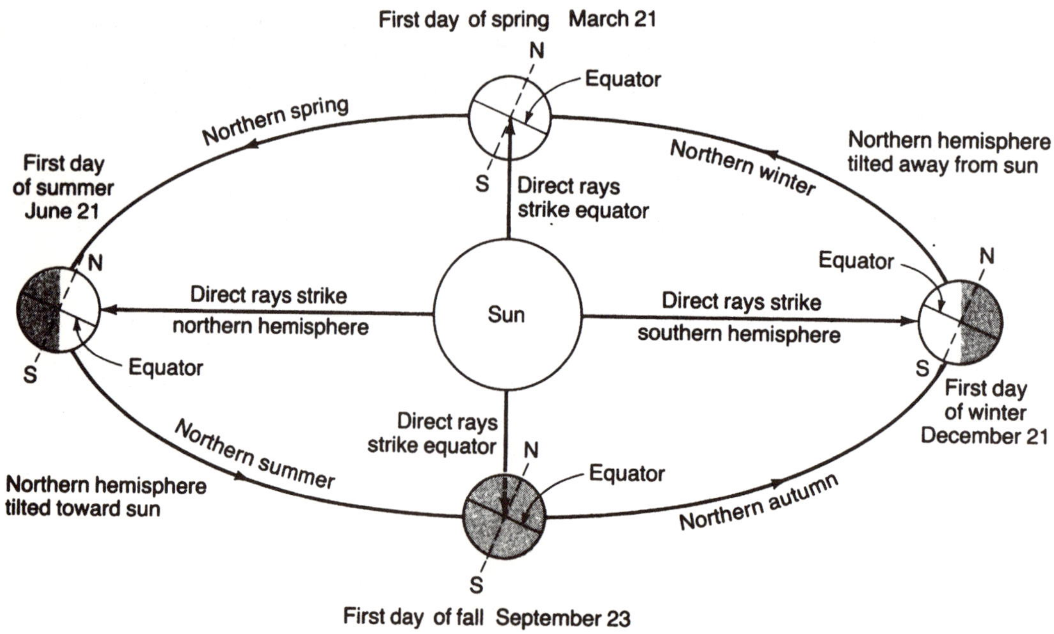

Figure 6-8. The seasons are caused by the tilt of Earth's axis and Earth's revolution around the sun.

In summer, the northern hemisphere is tilted toward the sun, and the sun follows a higher path across the sky. As a result, the sun's rays strike that part of Earth more directly and daylight lasts longer there (Figure 6-9). Areas north of the equator are therefore heated more effectively, producing the hot season. On the first day of summer in the northern hemisphere, the sun follows its highest path across the sky, and we have our longest period of daylight and shortest period of darkness (night).

In winter, the northern hemisphere is tilted away from the sun, so the sun follows a lower path across the sky. The sun's rays come in at a low angle in the north and the days are shorter there. Areas north of the equator are heated less effectively, and we experience our cold season. On the first day of winter, the sun travels its lowest path across the sky, and we have our shortest day and longest night.

In the southern hemisphere, the situation is reversed. During summer in the northern hemisphere, the southern hemisphere experiences winter. During winter in the northern hemisphere, the southern hemisphere has its summer.

During spring and autumn (fall), neither hemisphere is really tilted toward the sun. Both hemispheres experience moderate temperatures, between the extremes of summer and winter. On the first day of spring, and again on the first day of fall, the sun's rays shine directly on the equator, and day and night are of equal length. Table 6-3 summarizes information about seasonal dates in the northern hemisphere.

During any season, the sun is highest in the sky each day at noon. In the continental United States, the noon sun never reaches the point directly overhead but is always in the southern half of the sky (see Figure 6-9). As a result, shadows at noon always point in a northerly direction.

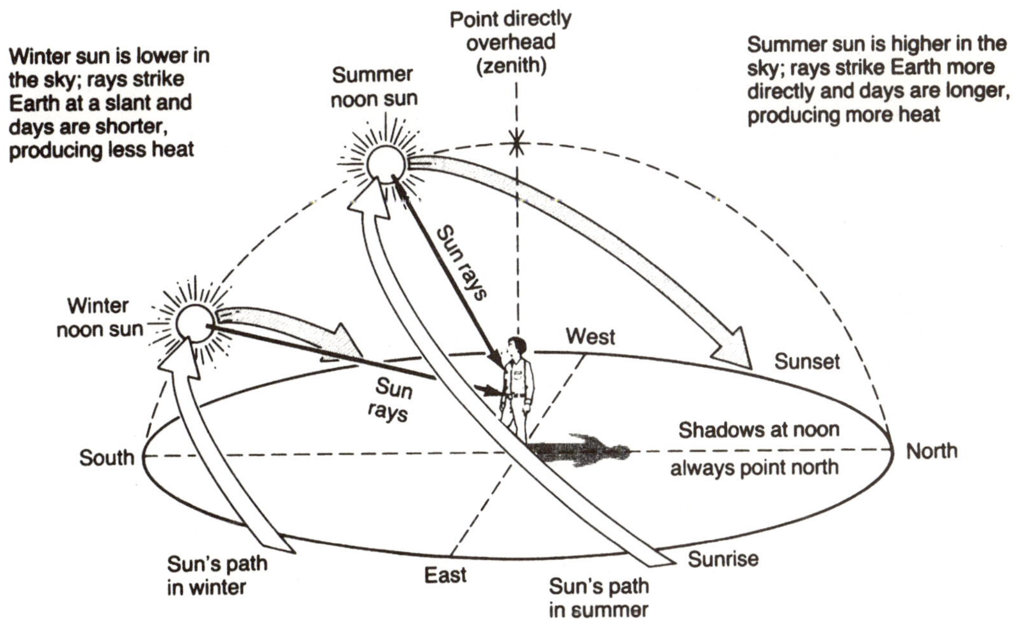

Figure 6-9. The sun's path across the sky changes with the seasons.

Proof of Earth's Revolution

Proof that Earth revolves around the sun comes from observations of stars. Stars in the night sky form patterns that have reminded people of animals or characters in myths. These patterns are called ***constellations.*** Two easily recognized constellations are Ursa Major (the Great Bear, which contains the Big Dipper) and Orion, the Hunter.

Table 6-3. Seasonal Information for the Northern Hemisphere

Date	*Season*	*Sun's Path Across the Sky*	*Length of Daylight and Darkness*
June 20, 21	First day of summer	Sun follows highest path across sky	Longest period of daylight; shortest period of darkness
September 22, 23	First day of fall	Sun follows path midway between summer and winter extremes	Daylight and darkness of equal length (12 hours each)
December 21, 22	First day of winter	Sun follows lowest path across sky	Shortest period of daylight; longest period of darkness
March 20, 21	First day of spring	Sun follows path midway between summer and winter extremes	Daylight and darkness of equal length (12 hours each)

During the course of the year, different constellations become visible at night. This suggests that Earth's night side faces different directions in space (Figure 6-10), so Earth's position in relation to the sun must be changing. In other words, Earth must be moving around the sun.

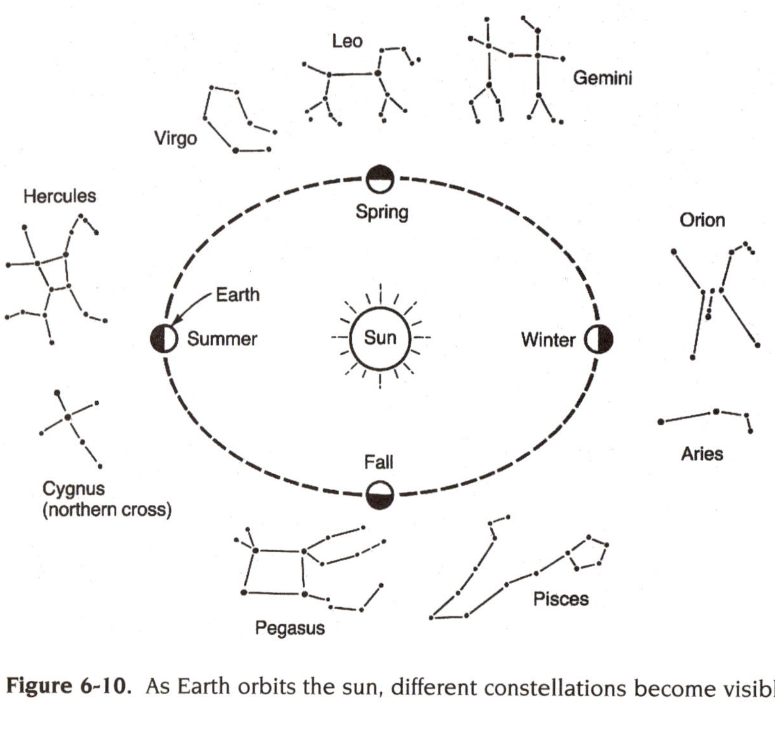

Figure 6-10. As Earth orbits the sun, different constellations become visible at night.

EXERCISE 2

As Earth revolves around the sun each year, the most direct rays of the sun move from 23.5° south of the equator to 23.5° north of the equator and back again. This is due to the combined effect of Earth's revolution around the sun and the tilt of Earth's axis. Questions 1–4 refer to the diagram.

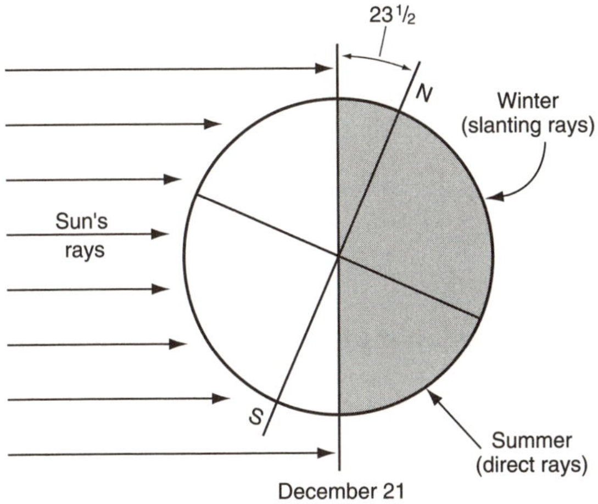

1. Philadelphia is located at 40° north latitude. The sun's most direct rays strike this city
   (1) between March 23 and June 21
   (2) between June 21 and September 23
   (3) on June 21
   (4) never

2. The most direct rays of the sun strike the equator on
   (1) June 21
   (2) December 21 and June 21
   (3) March 23 and September 23
   (4) December 21

3. During the course of summer months (June 21 to September 23), locations on Earth that receive the most direct rays of the sun are
   (1) farthest south
   (2) changing from north to south
   (3) changing from south to north
   (4) not changing

4. Earth is closer to the sun on December 21 than on June 21. We can conclude from this fact that the changing distance to the sun
   (1) has little to do with seasons on Earth
   (2) causes the seasons on Earth
   (3) causes summer in the northern hemisphere and winter in the southern hemisphere
   (4) causes winter in the northern hemisphere and summer in the southern hemisphere

5. According to the graph, the southern hemisphere of Earth has summer from

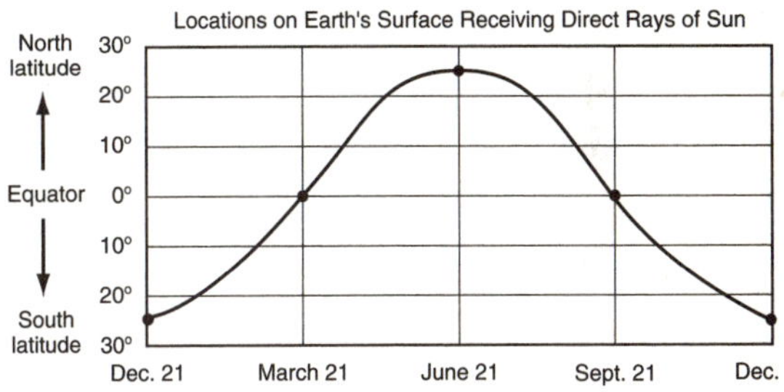

   (1) December 21 to March 21
   (2) March 21 to June 21
   (3) June 21 to September 23
   (4) September 23 to December 21

6. Earth makes one complete revolution around the sun in one
   (1) year (2) week (3) month (4) day

7. Yoshi read in the newspaper that there would be exactly 12 hours of daylight and 12 hours of darkness the next day. That next day's date could be
   (1) October 21 (3) December 21
   (2) June 21 (4) March 21

8. On July 10, the most direct rays of the sun strike Earth
   (1) in the northern hemisphere
   (2) in the southern hemisphere
   (3) at the equator
   (4) at the North Pole

Question 9 refers to the table below.

Length of Daylight for Three Cities in the U.S.

Date	*Atlanta 25°N*	*New York 40°N*	*Boston 45°N*
May 1	13 hr 35 min	13 hr 53 min	14 hr 15 min
June 1	14 hr 21 min	14 hr 49 min	15 hr 02 min
July 1	14 hr 23 min	14 hr 58 min	15 hr 14 min

9. In the month of June, as you travel north from Atlanta to Boston, daylight
   (1) increases
   (2) decreases
   (3) remains the same
   (4) first increases, then decreases

10. The diagram shows Earth at four positions in its orbit. At which position would the northern hemisphere be experiencing winter?

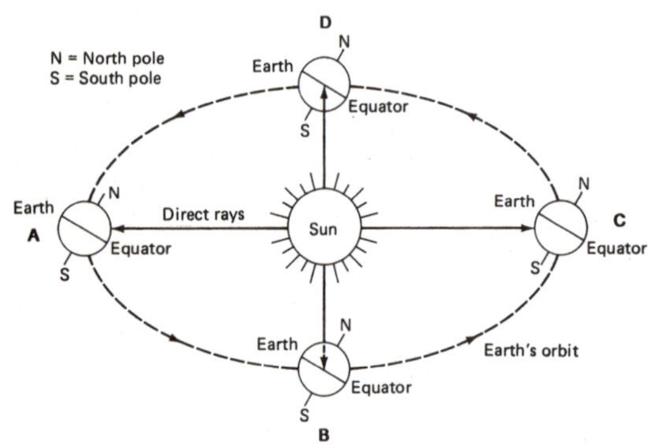

(1) A (2) B (3) C (4) D

Earth revolves around the sun in a nearly circular orbit. In one year, Earth makes one complete revolution around the sun. Therefore, during its 365.25-day trip around the sun, it completes a 360° orbit. That means Earth revolves at a rate of about 1° per day. Questions 11 and 12 refer to the diagram below.

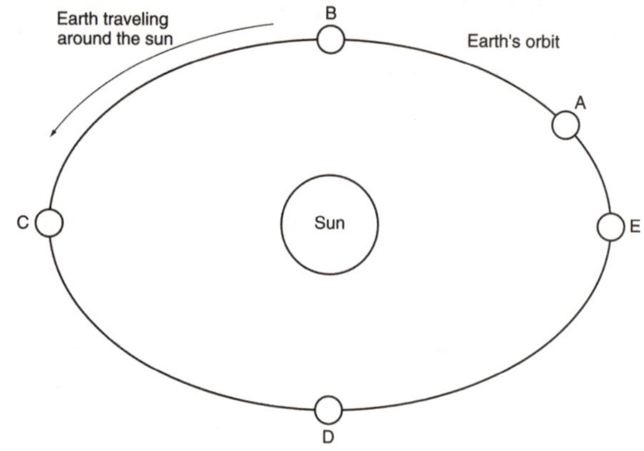

11. During a month (30 days), Earth revolves around the sun about
    (1) 7° (2) 30° (3) 180° (4) 15°

12. On January 1, Earth is located at position E in the diagram. Which point in Earth's orbit best indicates the position of the planet three months later?
    (1) A (2) B (3) C (4) D

---

13. Which constellation would be visible in the night sky from the position of Earth shown in the diagram?
    (1) Scorpio
    (2) Sagittarius
    (3) Aquarius
    (4) Gemini

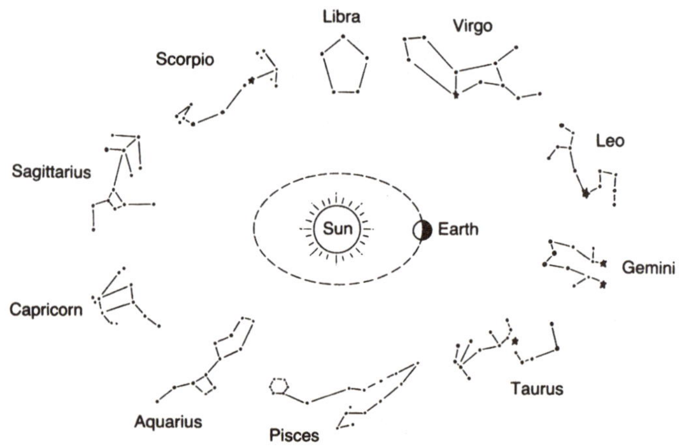

7 Atmosphere, Hydrosphere, and Lithosphere

Points to Remember —

- Earth's surface has three different "spheres": atmosphere, hydrosphere, and lithosphere. Among the three spheres, energy and matter interact.
- The sun is the primary source of energy affecting Earth's surface.
- Factors, or elements, that make up the state of weather are temperature, air pressure, humidity, wind, clouds, and precipitation.
- Climate is the average weather that prevails over a large area for a long period of time.
- Weather maps show the positions of air masses, fronts, and weather elements at many locations. Weather maps aid in weather forecasting.
- Humans are affected by the atmosphere in their daily lives. Major effects are caused by thunderstorms, hurricanes, tornadoes, and winter storms. Human pollution is changing the atmosphere.
- About 75 percent of Earth's surface is covered by water. The five oceans contain most of the liquid water on Earth's surface. They are in constant motion due to currents, waves, and tides.

Interaction of Spheres

Energy Exchange

Planet Earth can be thought of as consisting of three different spheres: a rock sphere, or ***lithosphere***; a water sphere, or ***hydrosphere***; and a gaseous sphere, or ***atmosphere*** (Figure 7-1). On Earth's surface, these three spheres come into contact with one another, thereby affecting one another through the interaction of energy and matter. The lithosphere was discussed in Chapter 5. This chapter focuses on the atmosphere, the hydrosphere, and the interaction of the three spheres.

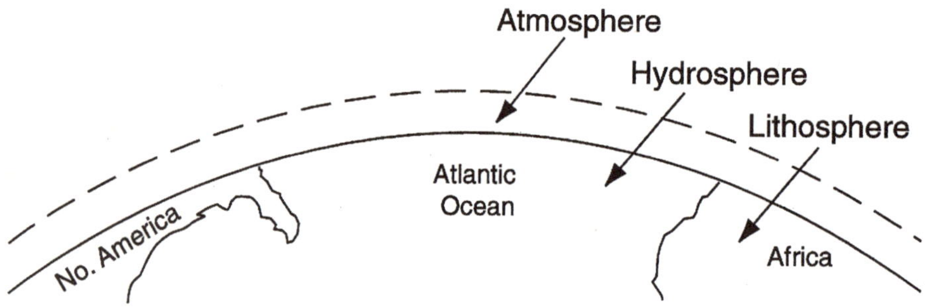

Figure 7-1. Earth consists of three "spheres": the lithosphere, hydrosphere, and atmosphere.

Energy is constantly being exchanged among the lithosphere, atmosphere, and hydrosphere. Energy effects can be observed in processes on Earth's surface. Table 7-1 gives examples of energy exchange between Earth's spheres.

Table 7-1. Examples of Energy Exchange Between Earth's Spheres

Atmosphere ⟶ Lithosphere Physical and chemical weathering by atmospheric gases and moisture cause rocks to crumble.
Atmosphere ⟶ Hydrosphere Waves are produced by wind blowing across water surfaces.The stronger the wind, the larger the waves.
Lithosphere ⟶ Hydrosphere Tsunami waves are produced by volcanic and earthquake activity on the ocean floor. Strong vibrations of the seafloor are transferred to the water, causing large waves to form.
Hydrosphere ⟶ Lithosphere Ocean waves breaking along beaches transport sand particles, causing coastline erosion.
Hydrosphere ⟶ Atmosphere Climate is affected by warm ocean currents traveling north and warming the land they contact, and by cold ocean currents traveling south and cooling the land they contact.
Lithosphere ⟶ Atmosphere Volcanic activity sends ash particles high into the atmosphere. These particles decrease the sun's radiation reaching Earth's surface, producing cooler temperatures.

Composition

Table 7-2 provides the average composition of Earth's atmosphere. The composition of the atmosphere is relatively constant. The interactions of the lithosphere, hydrosphere, and atmosphere assist in replenishing gases through cycles such as the water cycle, the nitrogen cycle, and the oxygen-carbon dioxide cycle.

Table 7-2. Composition of the Atmosphere

Nitrogen	78 %
Oxygen	21 %
Carbon dioxide	0.03 %
Other gases	0.17 %
Water vapor (variable)	1–3 %

Interactive Cycles

The sun's energy powers the ***water cycle.*** In this process, water moves between Earth's surface and the atmosphere, as shown in Figure 7-2. Heat from the sun changes liquid water into water vapor. This is called ***evaporation.*** For example, evaporation takes place when a puddle of rainwater

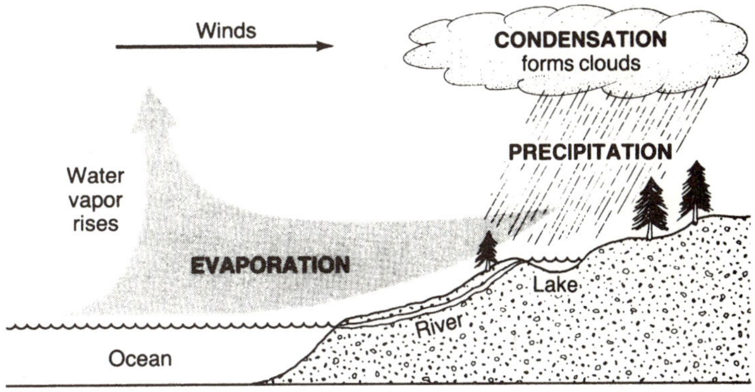

Figure 7-2. Water moves between surface and atmosphere in the water cycle.

shrinks and dries up on a hot, sunny day. Water enters the atmosphere by evaporation from oceans, lakes, and rivers, and by plants releasing water vapor through their leaves.

Warm air can hold more water vapor than cool air can. Therefore, when air cools, it cannot hold as much water vapor, and some of it changes back into droplets of liquid water. This process, called ***condensation,*** produces dew, fog, and clouds.

1. Dew is formed when water vapor condenses onto cool surfaces or objects. This takes place, for instance, when water droplets appear on the outside of a cold drink (see Figure 7-3 on page 74).
2. Condensation may also form tiny water droplets that remain suspended in the air. When this takes place near the ground, fog is produced.
3. When moist air rises high into the atmosphere and cools, condensation forms clouds. If enough water vapor condenses, the tiny water droplets may join together into larger, heavier drops that fall

to Earth as precipitation. Then the cycle of evaporation, condensation, and precipitation can begin all over again.

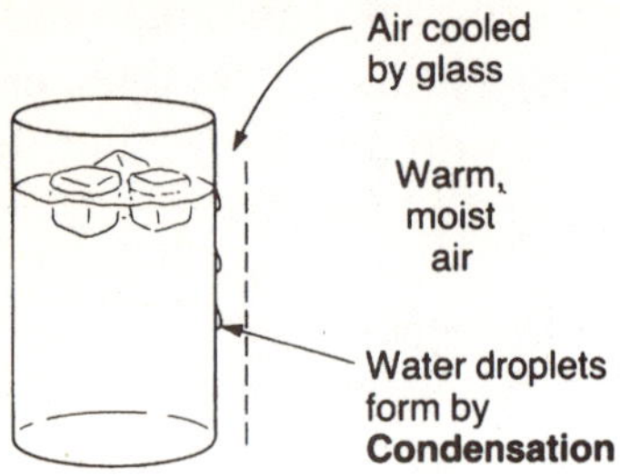

Figure 7-3. Condensation occurs on a cold glass.

Other cycles also affect the atmosphere by replenishing the primary gases in the atmosphere. In addition to water vapor, the most plentiful gases in the atmosphere are nitrogen, oxygen, and carbon dioxide. The exchange of these gases with the hydrosphere and lithosphere helps maintain a stable atmosphere that is capable of supporting life on Earth.

The ***nitrogen cycle*** maintains levels of gaseous nitrogen in the air, and at the same time provides nitrogen for protein production in plants and animals. Bacteria aid in the removal of nitrogen from decaying organic matter, and other types of bacteria return nitrogen back into a gaseous state. Lightning and nitrogen-fixing bacteria remove nitrogen from the air and place it back in the soil.

In the ***oxygen-carbon dioxide cycle,*** plants absorb carbon dioxide and produce oxygen in the process of photosynthesis. Animals breathe in oxygen and, through the process of respiration, return carbon dioxide to the air.

EXERCISE 1

1. Discuss energy interactions between Earth's spheres for each of the following statements. State which spheres are involved and how energy is exchanged.

Example Statement:

Waves along the beach wash sand grains in and out of the water.

Response:

Energy in ocean waves (hydrosphere) move sand grains (lithosphere).

A. Winds blow sand along the surface of a sand dune.
B. Hot air causes wavy heat lines to form above a parking lot pavement on a hot day.
C. Steam comes out from a crack in the side of a volcanic mountain.
D. Ocean waves pound a solid rock cliff and cause rocks to slide down the cliff.
E. Large ocean currents move across the ocean in the same direction as the global wind belts are blowing.

2. When it rained one morning, Willie saw a puddle of water form in the street. In the evening, the puddle was gone. The water probably disappeared by
   (1) evaporation
   (2) condensation
   (3) precipitation
   (4) expansion

Questions 3 and 4 refer to the diagram below, which represents the nitrogen cycle.

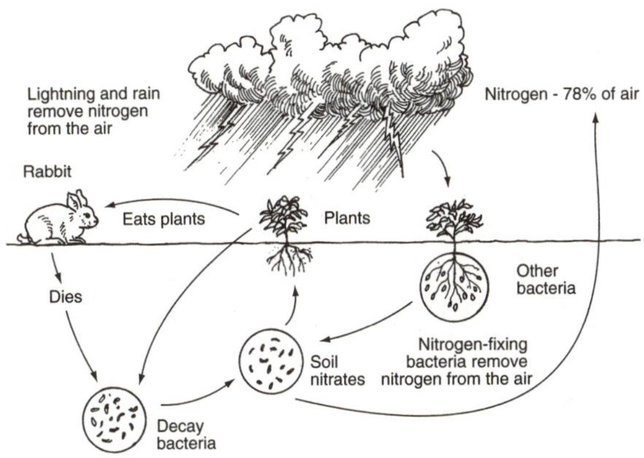

3. The process or factor that releases nitrogen gas into the atmosphere is
   (1) decaying bacteria
   (2) nitrogen-fixing bacteria
   (3) lightning and rain
   (4) other bacteria types

4. The process or factor that takes nitrogen out of the atmosphere is
   (1) animals eating plants
   (2) decaying remains
   (3) lightning and rain
   (4) absorption by plants

Question 5 refers to the diagram below, which represents the oxygen-carbon dioxide cycle.

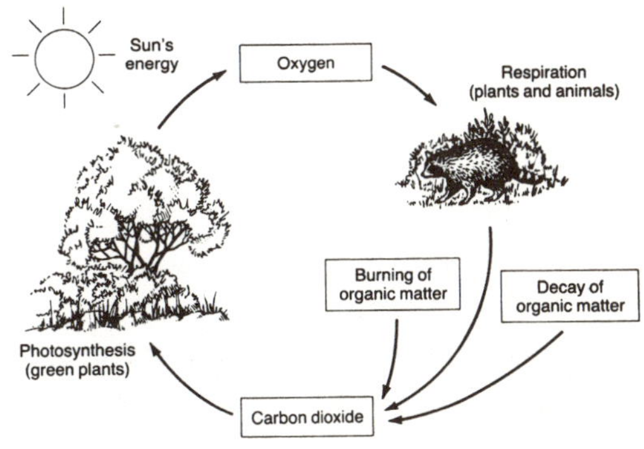

5. The destruction of trees in the rain forest causes
   (1) an increase in the production of oxygen
   (2) an increase in the production of carbon dioxide
   (3) an increase in carbon dioxide and a decrease in oxygen
   (4) a decrease in carbon dioxide and an increase in oxygen

The Atmosphere

Defining Weather

Surrounding Earth is a layer of gases called the atmosphere. These gases make up what is commonly known as air. ***Weather*** consists of the conditions of the atmosphere, such as heat, cold, sunshine, rain, snow, clouds, and wind. These conditions change from day to day and from place to place. Energy from the sun is the main cause of these changes.

Weather Elements

Weather is made up of a number of factors, or elements, including air temperature, air pressure, humidity, wind speed and direction, clouds and cloudiness, and precipitation. The weather at any given location can be described in terms of its elements.

1. ***Air temperature*** indicates the amount of heat in the atmosphere. It is measured with a thermometer.
2. Because air has weight, air presses down on Earth's surface with a force called ***air pressure.*** This force is measured with a barometer, in either inches of mercury or millibars. Variations in air pressure are caused mainly by temperature and altitude. Temperature affects air pressure because cool air weighs more than warm air. Altitude affects air pressure because places at high elevations, being farther up in the atmosphere, have less air weighing down on them.
3. ***Humidity*** is the amount of water vapor (water in the form of a gas) present in the air. Warm air can hold more water vapor than cool air can. Relative humidity is the ratio between the actual amount of water vapor in the air and the maximum amount of water vapor the air can hold at that temperature. For example, a relative humidity of 90 percent means that the air contains 90 percent of the water vapor it can hold at its current temperature. Relative humidity is measured with a hygrometer (sometimes called a wet-and-dry-bulb thermometer).
4. ***Wind*** is the movement of air over Earth's surface. Wind speed is a measure of how fast the air is moving, in miles or kilometers per hour. It is determined with an anemometer. Wind direction is the direction from which the wind is coming; it is *not* the direction the wind is blowing toward. For instance, a wind blowing from north to south is called a north wind. Wind direction is determined with a wind vane, which points in the direction from which the wind is blowing.
5. ***Clouds*** are masses of tiny water droplets or ice crystals suspended in the air. Cloudiness, the amount of sky covered by clouds, is observed directly and described with phrases like "partly cloudy" or "mostly cloudy." A sky completely covered by clouds is described as "overcast."
6. ***Precipitation*** is water, in any form, falling from clouds in the sky. Precipitation can fall as rain, snow, sleet, or hail. A rain gauge is used to measure the amount of precipitation in inches.

The Sun's Energy

The sun is the main source of energy in the atmosphere. As the sun heats Earth's surface, the surface radiates some heat energy back into the atmosphere. However, the sun does not heat Earth's surface evenly; consequently, the atmosphere is not heated evenly. This uneven distribution of heat energy in the atmosphere is the cause of weather.

The heating of Earth's surface depends to some extent on the nature of the surface, since some kinds of surfaces get hotter than others. For instance, pavement and sand get much hotter than do grass and water. On a larger scale, the surfaces of oceans, forests, and deserts are also affected differently by the sun. These surfaces, in turn, heat the air above them differently, producing variations in air temperature.

When air is heated, it becomes lighter (less dense) than the surrounding air. Therefore, warm air rises. Cool air is heavier (more dense), so it tends to sink. As air rises or falls, the surrounding air rushes in to replace it, causing air to circulate, as shown in Figure 7-4. This circulation, which can take place over a few kilometers or over thousands of kilometers, brings about changes in the weather. The circulation of air is referred to as ***convection***; the horizontal movement of air is known as wind.

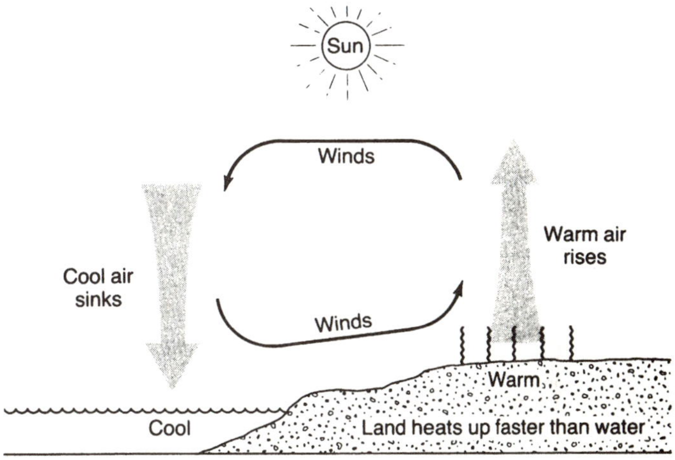

Figure 7-4. The air circulation pattern causes changes in weather.

The heating of Earth's surface also depends on the angle at which the sun's rays strike the surface (Figure 7-5). Near the equator, the sun's rays strike Earth most directly (Area A in Figure 7-5). This concentrates the sun's energy within a small area, heating the surface very effectively. However, since Earth's surface is curved, the sun's rays strike areas away from the

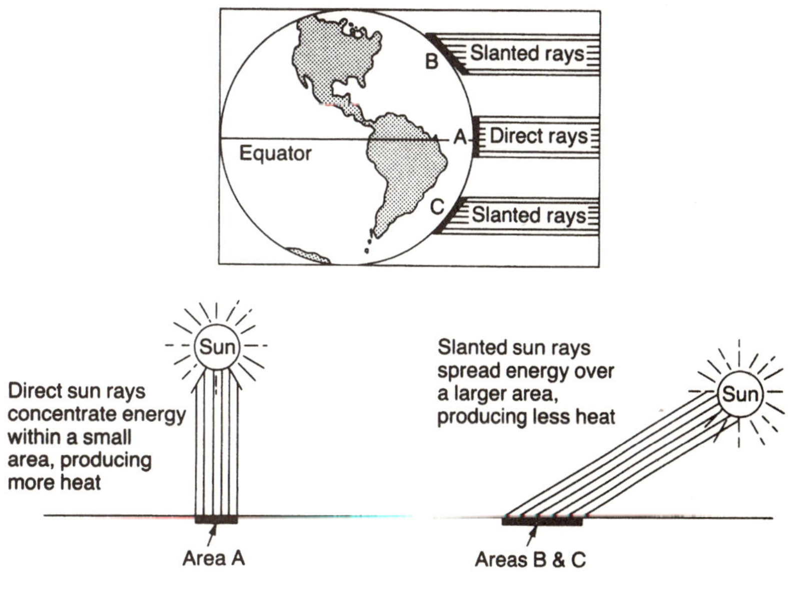

Figure 7-5. The angle of the sun's rays affects the heating of Earth's surface.

Process Skills

DESIGNING AN EXPERIMENT; PREDICTING RESULTS

Cold air is heavier and more dense than warm air. Therefore, cold air tends to sink and warm air tends to rise. To design an experiment that can demonstrate this, you would need cool and warm air, a way to control the flow of air, and several thermometers.

For instance, a small room with a window provides a means of controlling air flow (see diagram). On a day when the outside temperature is lower than the indoor temperature

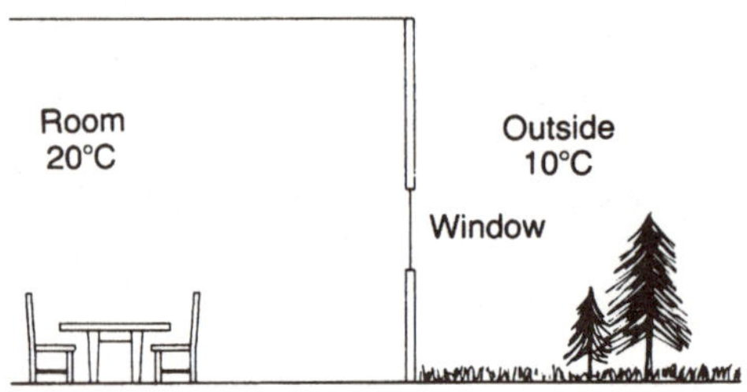

by at least 10°C, you could open the window to let in cold air and measure temperature changes in the room with thermometers. If it is true that cold air sinks and warm air rises, the lower levels of the room will get colder more quickly than the higher levels will. Answer the following questions, which are based on the diagrams below.

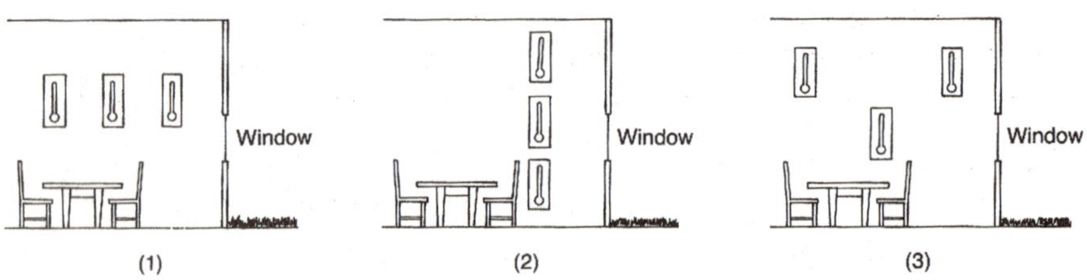

1. How should the thermometers be arranged in the room to show that cold air sinks and warm air rises?
2. If this experiment were performed on a summer day, when the outdoor temperature was 10°C higher than the indoor temperature, what would probably happen in the room after opening the window?
   (1) The temperature near the ceiling would increase faster than the temperature near the floor.
   (2) The temperature near the floor would increase faster than the temperature near the ceiling.
   (3) The temperature at all levels of the room would increase at the same rate.

equator at a slanting angle (Areas B and C in Figure 7-5). This spreads energy over a wider area, heating Earth's surface less effectively. The farther from the equator, the more slanted the sun's rays come in, and the less effectively they heat the surface.

The uneven heating of Earth's curved surface causes hotter air at the equator to rise and spread to the north and south, while cooler air near Earth's poles moves toward the equator to replace the rising air. Earth's rotation breaks up this simple circulation into complex global wind belts (shown in Figure 7-6), in which winds blow in different directions at different latitudes.

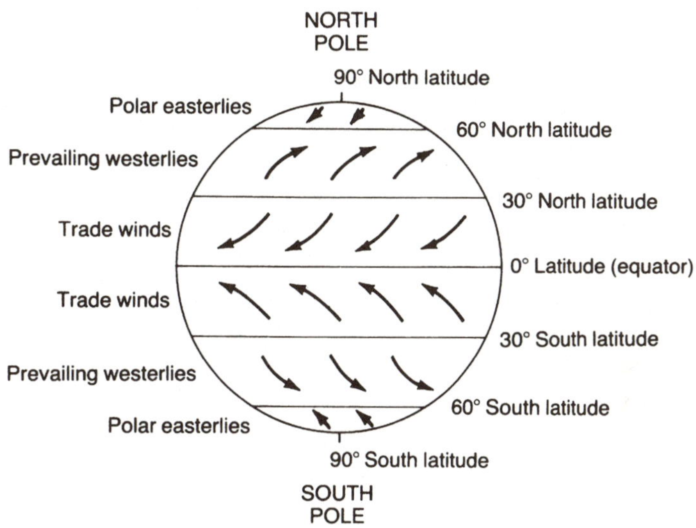

Figure 7-6. Earth's wind belts.

Latitude is the distance north or south of the equator. The lines drawn parallel to the equator on a globe or a map are lines of latitude. Winds that commonly blow in the same direction at a given latitude are called the prevailing winds. In the United States, the ***prevailing winds*** blow from west to east, so they are called westerlies.

Climate

Climate is the general character of the weather that prevails in an area from season to season and from year to year. It can be thought of as the average weather of a large area over a long period of time. A number of factors combine to produce different climates.

1. One factor is latitude, which is distance from the equator. Places at high latitudes, far from the equator, tend to have colder climates than places at lower latitudes. For instance, Canada is at a higher latitude than Mexico, so it has a colder climate.
2. Another factor is altitude, which is the height (elevation) above sea level of a place. Higher elevations are cooler than lower elevations. Just as a mountaintop is colder than its base, a city at a high altitude will have a colder climate than a city at the same latitude that is at a lower altitude.

3. Large bodies of water can affect climate. Land areas close to oceans or large lakes generally have more moderate climates (cooler summers and warmer winters) than areas far from water. Water absorbs and gives off heat more slowly than land does. Therefore, as the land heats up during summer, the water stays relatively cool, keeping coastal areas cooler in summer than places farther inland. In winter, the situation is the opposite. The water loses heat built up during summer more slowly than the land does, keeping coastal areas warmer in winter than areas farther inland.

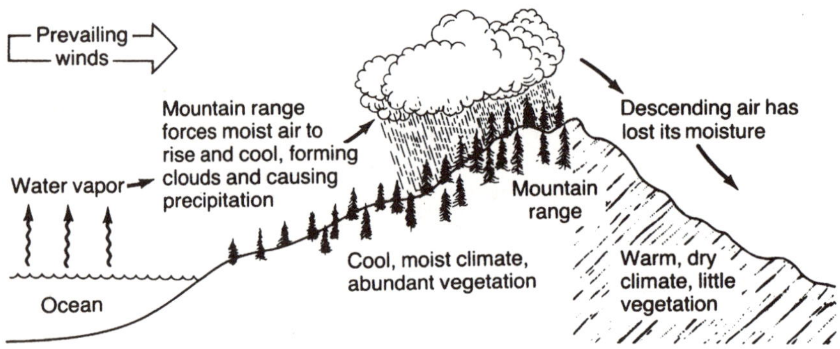

Figure 7-7. A mountain range can affect climate.

4. Mountain barriers can also influence climate. The side of a mountain range facing the prevailing winds tends to have a cool, moist climate, while the opposite side of the mountains has a warmer, drier climate. This is illustrated in Figure 7-7.

EXERCISE 2

1. The two temperature scales commonly used for thermometers are the Celsius scale (°C) and the Fahrenheit scale (°F), shown and compared in the diagram at the right. Both scales are divided into units called degrees. The two thermometers are aligned so that you can easily convert one scale to the other scale by using a ruler or the edge of a sheet of paper.

   Using the thermometer diagram, convert each of the following temperature readings:

   104°F = ___ °C   50°C = ___ °F
   −40°F = ___ °C   75°C = ___ °F
   75°F = ___ °C   20°C = ___ °F

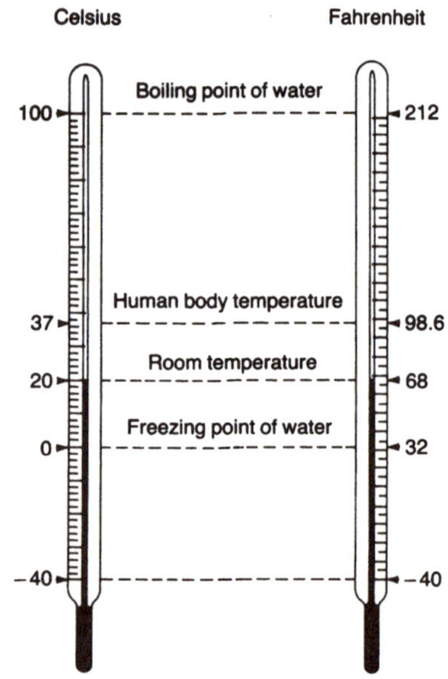

Questions 2–5 refer to the data and diagram below, which represents a beach on a hot summer day at about 2:00 P.M.

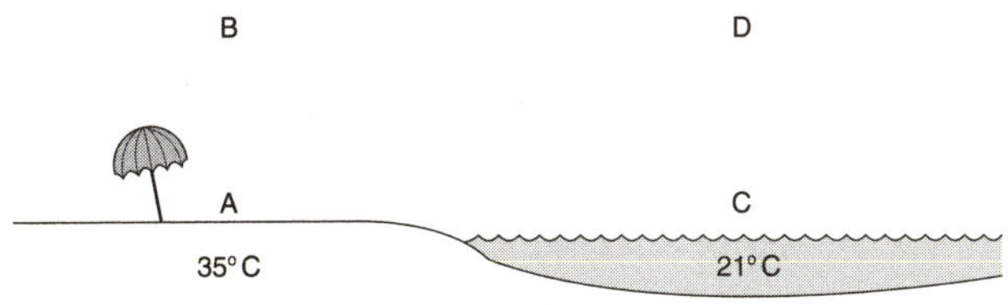

A: the air one meter above the beach sand

B: the air at 1525 meters above the beach sand

C: the air one meter above the ocean water

D: the air at 1525 meters above the ocean water

2. Rising warm air is at location
(1) A (2) B (3) C (4) D

3. There is sinking cool air at location
(1) A (2) B (3) C (4) D

4. There is horizontally moving cool air at location
(1) A (2) B (3) C (4) D

5. If clouds form from rising air, clouds should be forming at location
(1) A (2) B (3) C (4) D

---

6. The source of the energy that sets Earth's atmosphere in motion and causes weather is
(1) volcanism (3) the ocean
(2) gravity (4) the sun

7. Air would be densest and heaviest at a temperature of
(1) 25°C (2) 20°C (3) 15°C (4) 10°C

8. The average weather conditions of an area over many years determine that area's
(1) balance of nature
(2) geologic history
(3) climate
(4) latitude

9. Which temperature is indicated by the thermometer in the diagram?

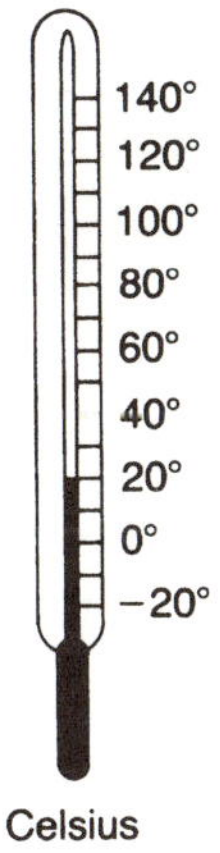

(1) 10°C (2) 20°C (3) 30°C (4) 25°C

10. In the diagram, the most likely reason for City B to be cooler than City A is that City B is

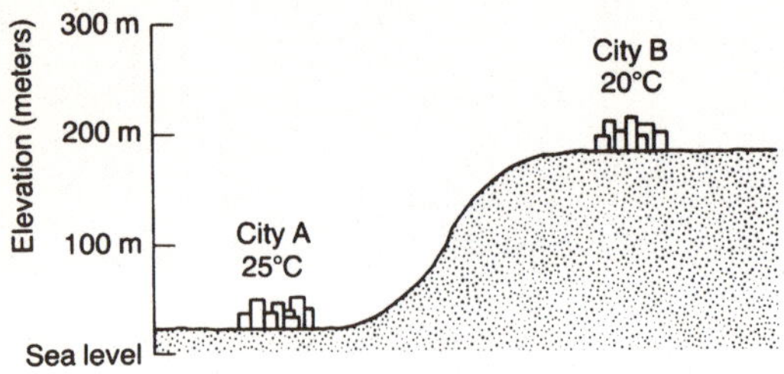

(1) at a different latitude
(2) at a higher altitude
(3) closer to the ocean
(4) closer to the sun

11. Mr. Chang decided to drive his car up Pikes Peak, a tall mountain in Colorado. As he gets higher up the mountain, the air temperature will most likely
(1) decrease
(2) increase
(3) remain the same
(4) first decrease and then increase

12. On a sunny summer day, a thermometer was placed above each of the surfaces shown in the diagram. Which thermometer would most likely have had the highest temperature reading?

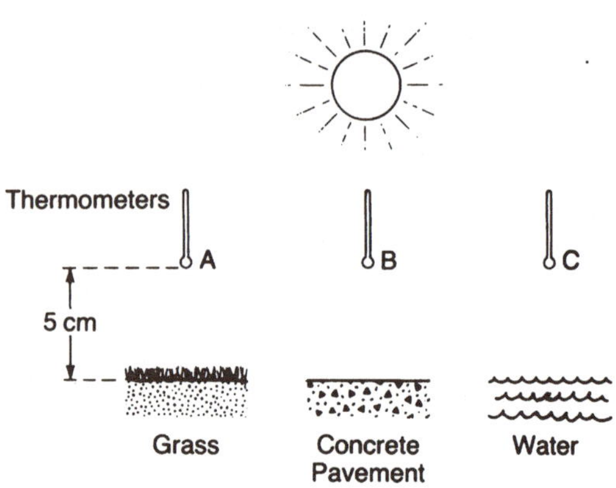

(1) A
(2) B
(3) C
(4) All would have the same reading.

13. The diagram shows two cities, A and B, and their positions on a continent. How will the climates of the cities compare?

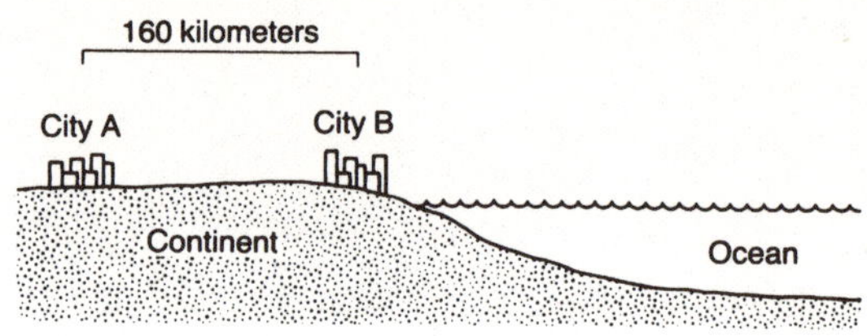

(1) City B will have warmer summers and colder winters than City A.
(2) City B will have cooler summers and warmer winters than City A.
(3) City B will have cooler summers and colder winters than City A.
(4) Both cities will have the same climate.

14. The most likely reason that New York City has a cooler climate than Miami is the difference in

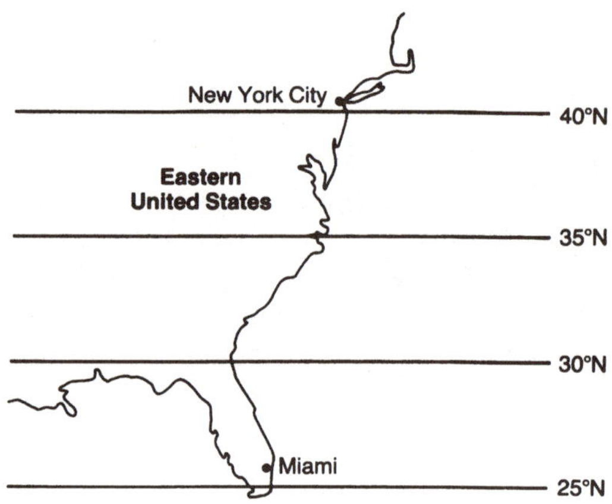

(1) distance from the ocean
(2) altitude
(3) air pressure
(4) latitude

15. In one or two sentences, explain why some people might spend their summer vacation near an ocean or a lake.

16. Which graph indicates the general temperature change as you travel from the North Pole (NP) to the South Pole (SP)?

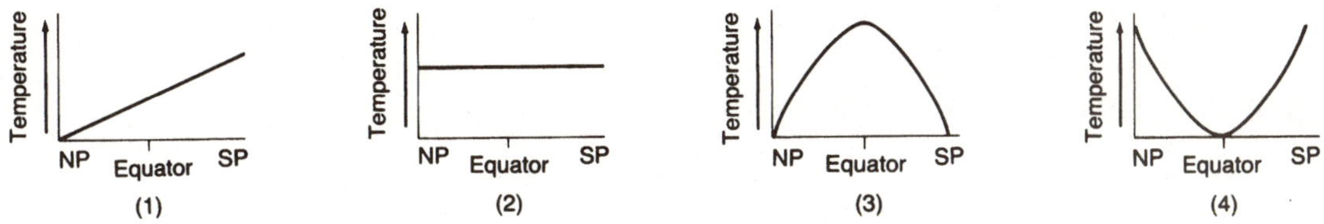

Large-Scale Weather Systems

Air Masses

A large body of air that has uniform temperature and moisture conditions throughout it is called an ***air mass.*** Much of our weather is determined by air masses.

An air mass forms when air stays over a large area of Earth's surface and takes on the temperature and moisture characteristics of that area. An air mass that forms over a warm body of water, such as the Gulf of Mexico, will be warm and moist. Air masses that enter the United States from Canada are usually cold and dry because they formed over a cool land surface. An air mass builds up over an area for a few days and then begins to drift across Earth's surface. There are four different surface conditions that affect the formation of air masses. Air masses that originate over land are dry. Those that form over water are moist. Cold air masses originate near the poles, and warm air masses form near the equator.

The major air masses that affect the continental United States, shown in Figure 7-8, enter the country from the north, west, and south. They are

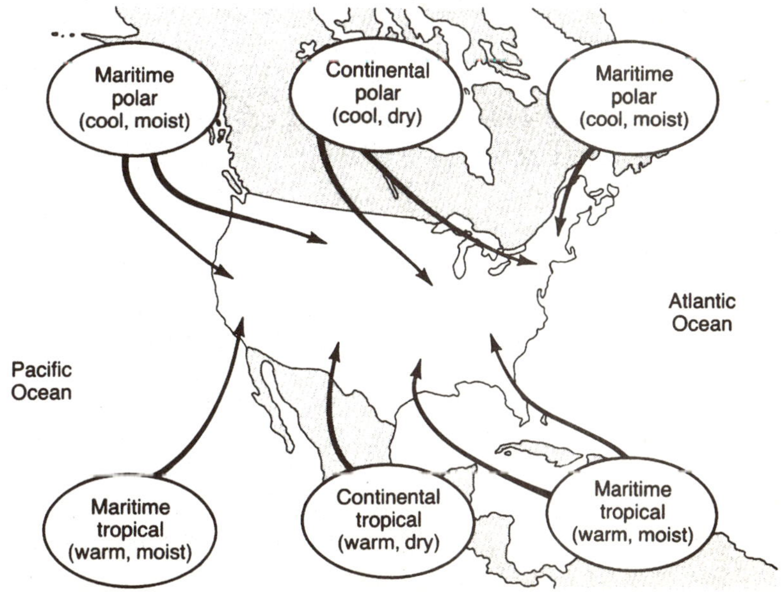

Figure 7-8. Major air masses affecting the United States.

then blown from west to east by the prevailing winds. As an air mass moves, it changes the local weather conditions at the surface below. The weather may become warmer or cooler, wetter or drier, depending on the type of air mass that is passing by.

High-Pressure and Low-Pressure Systems

Surface air pressures are usually highest in the center of air masses, so these areas are called ***high-pressure systems,*** or simply highs. The air in a high tends to sink, and winds blow outward from the center, turning in a clockwise direction, as shown in Figure 7-9(a). High-pressure systems usually bring clear skies, dry weather, and gentle winds.

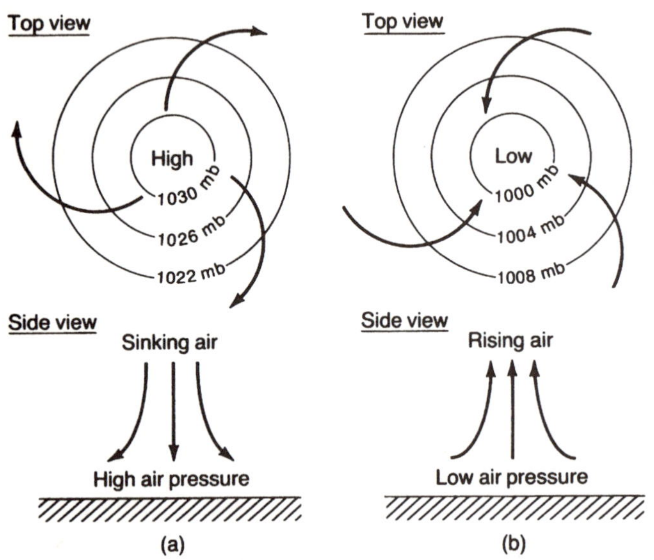

Figure 7-9. Movement of air currents in (a) a high-pressure system and (b) a low-pressure system (the mb stands for millibars).

Surface air pressures are lower toward the edges of air masses. These areas form ***low-pressure systems,*** or lows. The air in a low tends to rise, and winds spiral in toward the center in a counterclockwise direction, as shown in Figure 7-9(b). (Note that in both highs and lows, winds always blow from areas of higher air pressure toward areas of lower air pressure.) Low-pressure systems usually bring cloudy, wet weather, often with strong, gusty winds.

Highs and lows are generally indicated on weather maps by the letters **H** and **L** (see Figure 7-12). The highs on weather maps usually indicate the centers of air masses, while lows are generally found at the edges of air masses, or between air masses. Because highs are large and tend to move slowly, weather changes are usually gradual rather than sudden. Changes in air pressure readings signal the passing of highs and lows.

Fronts

When one air mass comes into contact with another air mass, a boundary called a ***front*** forms between them. Sudden changes in weather conditions can occur across a front. The air masses that meet often differ in temperature, humidity, and density. These differences prevent the air masses from mixing. The cooler, drier air is heavier and remains close to the ground, while the warmer, moister air is lighter and rises upward. This causes areas of low pressure to develop along fronts, often producing clouds, strong winds, and precipitation. These lows are the major storm systems of our latitudes.

Cold fronts or warm fronts are produced, depending on how the air masses come together.

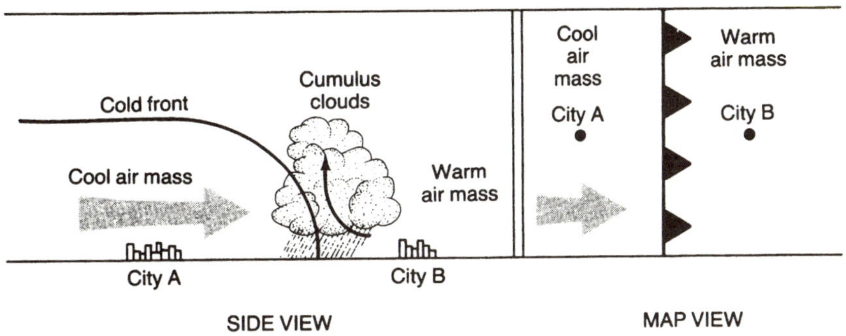

Figure 7-10. A cold front.

1. If a cold air mass pushes into and under a warm air mass, a ***cold front*** is formed (Figure 7-10). Cold fronts usually bring brief, heavy downpours, gusty winds, and cooler temperatures. On hot, humid summer days, the passing of a cold front typically causes thunderstorms, followed by a decrease in temperature and humidity.
2. When a warm air mass pushes into and over a cold air mass, a ***warm front*** is created. The warm air slides up and over the cooler air (Figure 7-11).

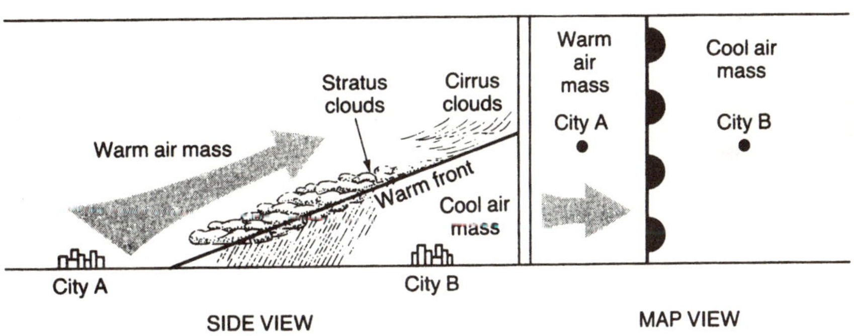

Figure 7-11. A warm front.

Warm fronts bring light precipitation lasting a day or two and warmer temperatures. When the sky is overcast all day with light rain falling, a warm front is most likely present. Different types of clouds are produced, depending on how air rises. If air is pushed straight up, as it is along a cold front, puffy cumulus clouds are produced. If air rises at a low angle, as it does along a warm front, flat layers of stratus clouds are formed. Wispy cirrus clouds, which are made up of ice crystals, form high in the atmosphere; they may look like feathers or tufts of hair. The presence of many high cirrus clouds may indicate that a warm front is approaching (see Figure 7-11).

Process Skill

ORGANIZING INFORMATION INTO A TABLE

Information can often be put into an orderly format by creating a table that summarizes it. For instance, information about air masses can be organized in the form of a table. Air masses are named for the temperature and moisture characteristics of the region over which they form. *Continental* air masses form over land and are dry. *Maritime* air masses form over water and are moist. *Tropical* air masses form in the tropics and are warm, while *polar* air masses form at the poles and are cool. For example, an air mass that forms over Canada would be called continental polar, indicating that it is cool and dry.

Types of Air Masses

Air Mass	*Temperature (Warm or Cool)*	*Humidity (Moist or Dry)*
Continental polar	Cool	Dry
Maritime polar		Moist
	Warm	Dry
Maritime tropical		

The table above represents some of this information. Copy the table and fill in the missing information. Based on your completed table, how would you describe a continental tropical air mass in terms of its temperature and moisture content?

The following incomplete table contains information about air pressure systems. Copy the table and refer to the section in the text about highs and lows to help you fill in the missing information. Then answer the questions below.

Air Pressure Systems

Type of Air Pressure System	*Vertical Air Movement*	*Horizontal Air Movement*	*Type of Weather*	*Location in Air Mass*
High	Sinking		Fair	
Low		In toward center, turning counter-clockwise		Edges

1. The movement of air in a high-pressure system is
   (1) counterclockwise and rising
   (2) counterclockwise and sinking
   (3) clockwise and rising
   (4) clockwise and sinking

2. The weather on the edge of an air mass is usually
   (1) stormy, with low air pressure
   (2) stormy, with high air pressure
   (3) fair, with low air pressure
   (4) fair, with high air pressure

Weather Forecasting

Weather forecasting is an attempt to make accurate predictions of future weather. The accuracy of weather forecasting is improving as technology advances. In addition to weather balloons, thermometers, and barometers, weather forecasters now have a wide array of weather satellites, radar devices, and computer systems at their disposal.

Short-range local forecasts are comparatively easy. They are based mostly on air pressure readings and observations of cloudiness and wind direction. Changes in these weather elements are usually good indications of the weather for the next day or two. For example:

- Decreasing air pressure readings signal the approach of stormy weather, while rising air pressure suggests that fair weather is coming.
- An increase in cloudiness is a sign that a front is approaching, probably bringing precipitation.
- In the Midwest, winds blowing from the west usually bring fair weather, while winds from the south or east often bring wet weather.

Today, weather forecasters also use information from weather satellites and radar to improve their short-range forecasts. This information is used to produce up-to-date weather maps, such as the one shown in Figure 7-12. Such maps can help us predict coming changes. Weather systems generally move from west to east across the United States. Therefore,

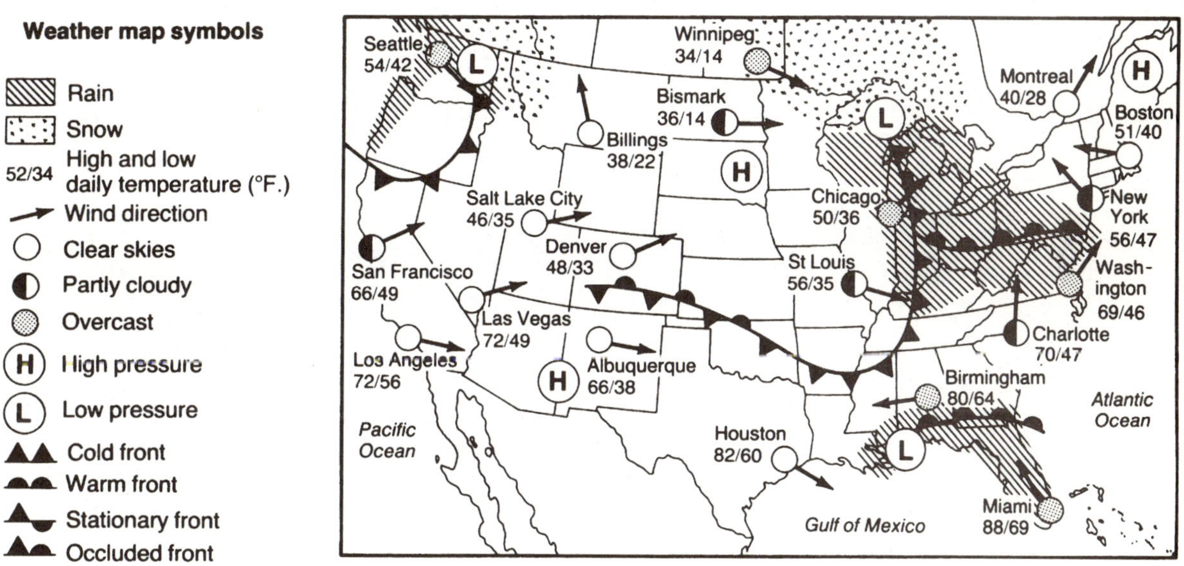

Figure 7-12. A weather map is useful for making predictions.

if a weather map shows a high immediately to our west, we can forecast fair weather for the next day. If the map shows a low to our west, we can expect stormy weather.

Long-range weather forecasting is more difficult. Computers, satellite photographs, and radar images enable forecasters to track the movements of large-scale weather systems like air masses and fronts. With this information, they can make predictions of weather several days in the future.

However, such forecasts are not always accurate. Generally, weather forecasts for the next day or two are reliable, while long-range forecasts usually have to be revised.

EXERCISE 3

1. During winter, air masses that form over northern Canada often affect the weather in Ohio. Such an air mass would be
   (1) dry and warm (3) moist and warm
   (2) dry and cool (4) moist and cool

2. Air pressure within an air mass is usually
   (1) high in the center and low on the edges
   (2) high on the edges and low in the center
   (3) constant throughout
   (4) varying throughout

3. The major storm systems of our latitudes are
   (1) high-pressure systems
   (2) cold, dry air masses
   (3) tropical air masses
   (4) low-pressure systems

4. On a weather map, Mark saw the word "High" over Ohio. This indicates
   (1) high temperature
   (2) high clouds
   (3) high air pressure
   (4) high winds

5. Juanita listened to the weather report on television. The forecaster said a warm front was approaching. Juanita knew that she could expect
   (1) fair weather
   (2) brief thunderstorms
   (3) tornadoes
   (4) light rain for about a day or two

6. At which city in the diagram would the air pressure most likely be the greatest?

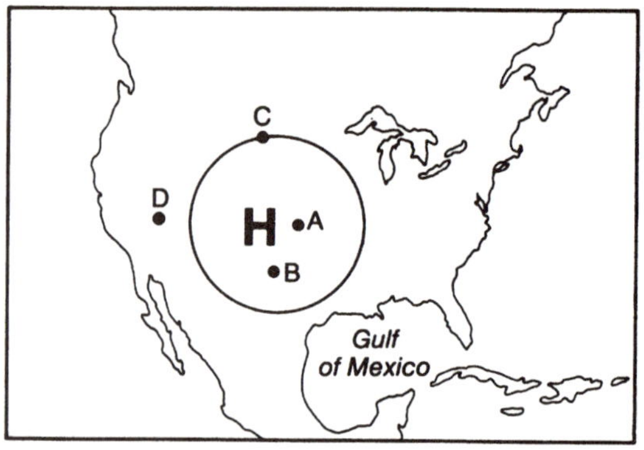

   (1) City A (3) City C
   (2) City B (4) City D

7. Changes in air pressure indicate the
   (1) change of seasons
   (2) passing of highs and lows
   (3) climate is changing
   (4) sun is setting

8. Margaret observed the weather from Friday night to Saturday night. First it was hot and humid, then there were thunderstorms, and finally the air became cooler and drier. These changes were probably due to the passing of a
   (1) warm front (3) hurricane
   (2) cold front (4) wind belt

9. The diagram shows an air mass entering the United States from the northwest. This air mass formed over the North Pacific Ocean, so it would be

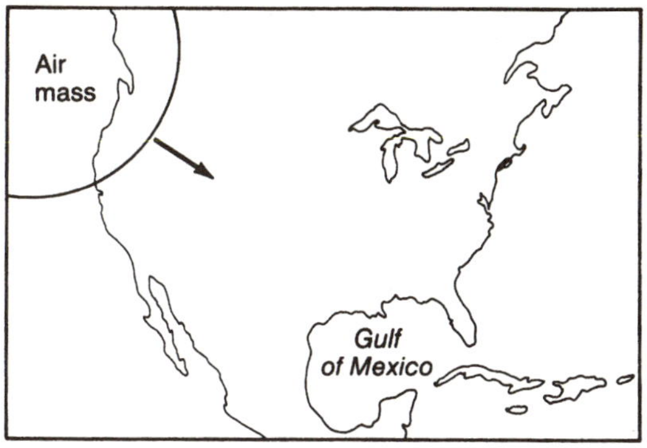

(1) moist and warm
(2) moist and cool
(3) dry and warm
(4) dry and cool

Questions 10 and 11 refer to the following diagram.

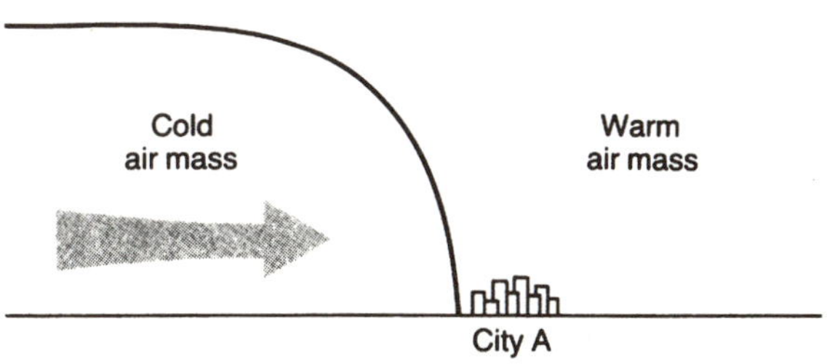

10. The line separating the cold air mass from the warm air mass represents a
    (1) warm front (3) line of latitude
    (2) cold front (4) high-pressure system

11. City A is most likely about to experience
    (1) a light rain, followed by warmer temperatures
    (2) heavy rains lasting several days
    (3) no change in weather conditions
    (4) brief downpours, followed by cooler temperatures

12. In the diagram, the air mass most likely to affect New York State the next day would be

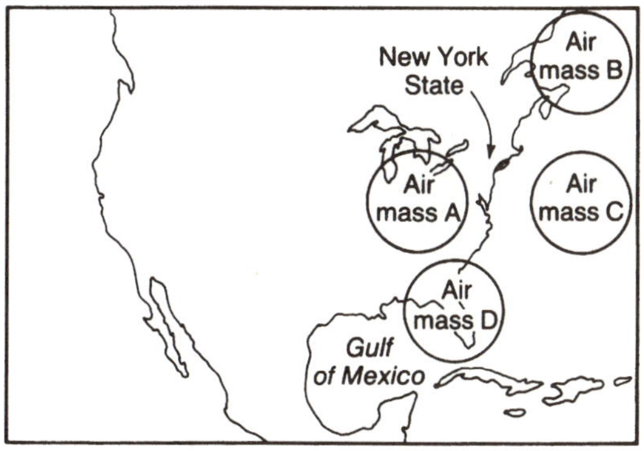

(1) air mass A (3) air mass C
(2) air mass B (4) air mass D

Interaction of Humans and the Atmosphere

Weather Hazards

Weather directly affects our lives. Lack of rain can ruin crops and force emergency measures such as water rationing. Too much rain can cause destructive floods. Storms produce severe weather conditions that can make travel dangerous, force schools and businesses to close, and cause deaths, injuries, and property damage.

Storms

Storms are natural disturbances in the atmosphere that involve low air pressure, clouds, precipitation, and strong winds. The major types of storms are *thunderstorms, hurricanes, tornadoes,* and *winter storms*. Each type has unique characteristics and dangers.

1. ***Thunderstorms*** are brief, intense storms that affect a small area. They are produced when rapidly rising warm air causes cumulus clouds to build upward into a *thunderhead* (see Figure 7-13). Thunderstorms are characterized by thunder and lightning, strong gusty winds, and sometimes hail.

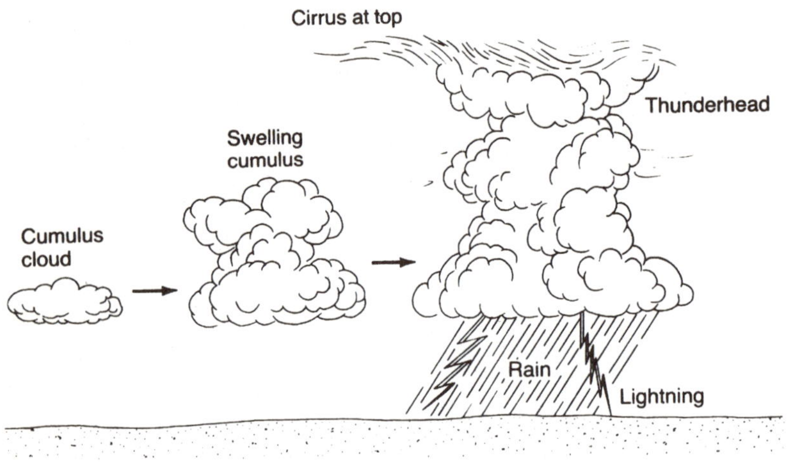

Figure 7-13. Development of a thunderhead; this causes a thunderstorm.

Lightning is a huge electrical discharge, like a giant spark. Lightning strikes are very dangerous and can be fatal. Large hailstones that sometimes fall during thunderstorms also can be dangerous. During a thunderstorm, you should stay indoors and especially avoid hilltops, open fields, beaches, and bodies of water.

2. ***Hurricanes*** are huge rotating storms that form over the ocean near the equator. They produce very strong winds, heavy rains, and large, powerful waves. A calm region in the storm's center is called the eye of the hurricane.

Hurricanes can cause severe flooding and damage from their high winds. A storm surge produced by high winds blowing water toward the shore can cause damage to homes and boats along the shore. People living along the coast and in floodprone regions should leave their homes and move to higher ground when a hurricane strikes.

3. ***Tornadoes*** are violently whirling winds, sometimes visible as a funnel-shaped cloud (see Figure 7-14). They are produced by particularly severe thunderstorms. Tornadoes usually appear suddenly, carve a narrow path of destruction, and disappear as suddenly as they came. Spiraling high-speed winds and extremely low air pressure are the unique features of tornadoes.

A tornado can lift and toss large objects, including cars, into the air. It can destroy houses in its path in a matter of seconds. An underground cellar or basement is the safest place to be during a tornado.

4. ***Winter storms*** include blizzards and ice storms. *Blizzards* are fierce storms with strong winds, blowing snow, and very cold temperatures. *Ice storms* occur when falling rain freezes at Earth's surface, coating everything with ice. Under these conditions, you should remain indoors and not attempt to travel.

Figure 7-14. A tornado funnel cloud.

Pollution

Human activities can affect the atmosphere and the weather. Factories, power plants, cars, and airplanes produce harmful substances called ***pollutants.*** The buildup of pollutants in the atmosphere can cause a number of weather problems.

Smog is a haze formed by the reaction of sunlight with chemicals in automobile exhaust and factory smoke. Smog tends to hang over large cities, giving the air a hazy, dirty look. Inhaling smog is very dangerous for people with breathing problems such as asthma and is harmful to the lungs of even healthy people.

Chemicals in smoke from factories and vehicles can also increase the acidity of the moisture in clouds. When this moisture falls to Earth as ***acid rain,*** it can harm lakes and forests and the creatures that live in them. Because air pollutants are often carried along by the prevailing winds, acid rain may fall far from the source of pollution (see Figure 7-15 on page 93).

Many industries and most forms of transportation produce carbon dioxide. In the atmosphere, carbon dioxide acts like the glass of a greenhouse, trapping heat close to Earth instead of letting it radiate back out to space. Many scientists fear that the buildup of carbon dioxide in the atmosphere (caused by human activities) may lead to global warming, a rise in worldwide average temperatures. This condition, often called the ***greenhouse effect,*** could have disastrous results, making climates hotter and drier and interfering with agriculture. Polar ice caps could melt, raising the sea level and flooding coastal areas and many major cities.

Certain natural events also release pollutants. Forest fires and volcanic eruptions give off huge quantities of dust and ash particles that collect high in the atmosphere, blocking sunlight and causing cooler temperatures on Earth. Plants release irritating pollen into the air, causing health problems for people with hay fever and asthma.

Process Skill

INTERPRETING DATA IN A TABLE

Storms form when the temperature, air pressure, and moisture conditions necessary for their development exist. At certain times of the year, weather conditions are more likely to produce one of the three types of storms—thunderstorms, tornadoes, and hurricanes—than at other times. Therefore, each storm type should have its particular "storm season."

You can determine if there are "storm seasons" by keeping track of the storms of each type that strike during the year and recording which months they occur in, as shown in the table below. With this information, you can determine if each kind of storm takes place mostly within a definite time of year or "storm season."

Number of Storms Per Month

	Type of Storm		
Month	*Thunderstorm*	*Tornado*	*Hurricane*
January	0	1	0
February	1	0	0
March	3	2	0
April	6	4	0
May	9	5	0
June	15	4	0
July	18	2	1
August	16	1	5
September	10	1	8
October	6	0	4
November	2	1	1
December	0	1	0

According to the table, most thunderstorms occur in the months of June, July, and August, so there does appear to be a thunderstorm season. Is there evidence in the table of a season for tornadoes? How about for hurricanes? Use the table to answer the following questions.

1. During which three months do most tornadoes occur?
   (1) August, September, and October
   (2) April, May, and June
   (3) June, July, and August
   (4) May, June, and July

2. Which storm type has the most clearly defined "storm season"?
   (1) thunderstorm
   (2) tornado
   (3) hurricane
   (4) all three types have equally well-defined seasons

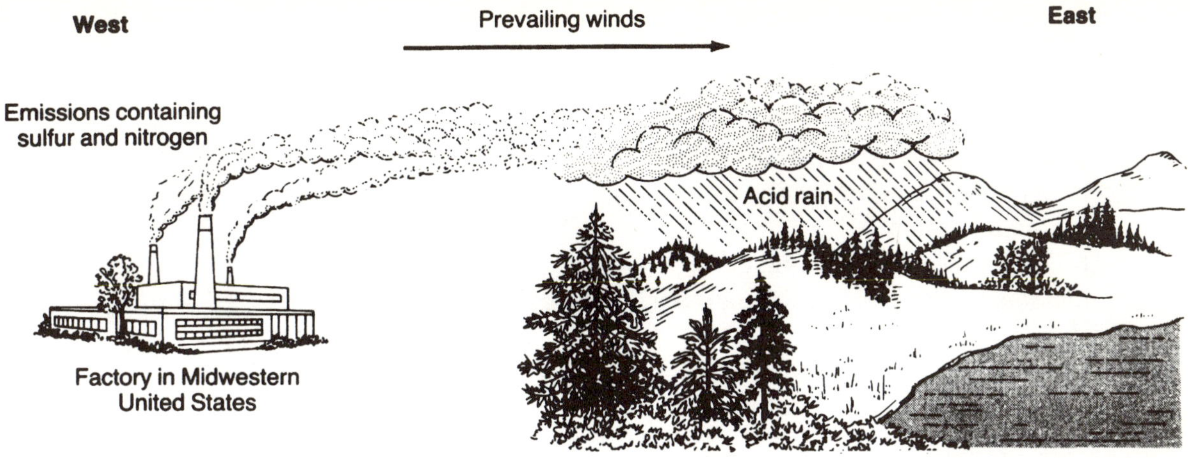

Figure 7-15. Acid rain may affect areas far downwind from the source of pollution.

There is little that we can do to control such natural pollutants. However, we can, and must, control our own activities that cause pollution if we are to avoid harming our planet.

EXERCISE 4

1. The main hazard of a thunderstorm is
   (1) thunder
   (2) funnel-shaped winds
   (3) heavy rain
   (4) lightning

2. The violent windstorm visible as a funnel-shaped cloud in the illustration is called a

   (1) hurricane
   (2) thunderstorm
   (3) tornado
   (4) blizzard

3. Substances released into the atmosphere by the activities of humans
   (1) always have a positive effect on the weather
   (2) can have a harmful effect on the weather
   (3) have no effect on weather
   (4) may cause global cooling

4. John visited the city on a warm, summer day. He noticed that the air was hazy and that his eyes and throat burned. This was probably caused by
   (1) an approaching storm
   (2) smog
   (3) low clouds
   (4) global warming

5. When scientists speak of the greenhouse effect, they are referring to
   (1) the fact that vegetables grown in a greenhouse do not taste as good as vegetables grown outdoors
   (2) the use of green paint on houses to camouflage them
   (3) the idea that pollution caused by human activities may lead to global warming
   (4) the fact that ash from volcanic eruptions can cause cooler worldwide temperatures

6. A hurricane is approaching the east coast of Florida. What dangers should the people there take precautions against?
   (1) cold temperatures, blowing snow, and poor visibility
   (2) funnel-shaped winds that can lift large objects
   (3) lightning and hailstones
   (4) flooding, large waves, and strong winds

7. The map shows a major industrial city and three lakes in the central U.S. Which lake is most likely to be affected by acid rain caused by pollution from the city? Explain why you think your answer is correct.

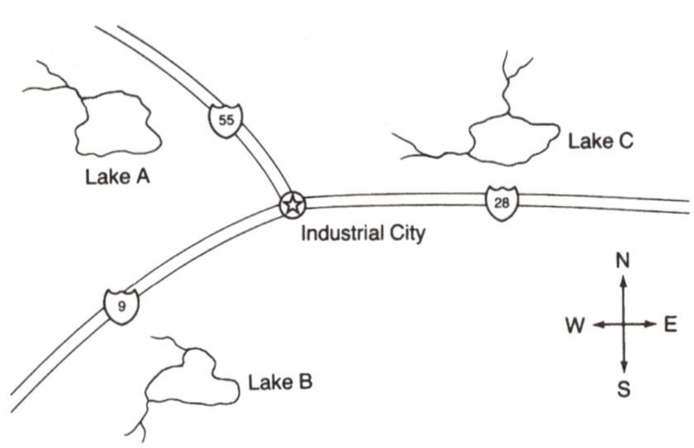

   (1) Lake A
   (2) Lake B
   (3) Lake C
   (4) all three lakes will be affected equally

8. The table below lists sources of air pollutants and the percentage contributed by each source.

Pollutant Source	*Percentage of Total Pollutants*
Transportation	42
Fuel	21
Solid waste disposal	5
Forest fires	8
Miscellaneous	10
Industrial processes	14

Which pie graph correctly represents the data in the table?

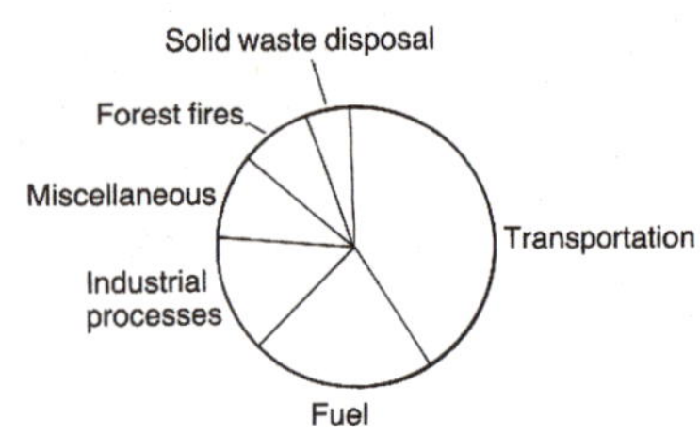

(1)

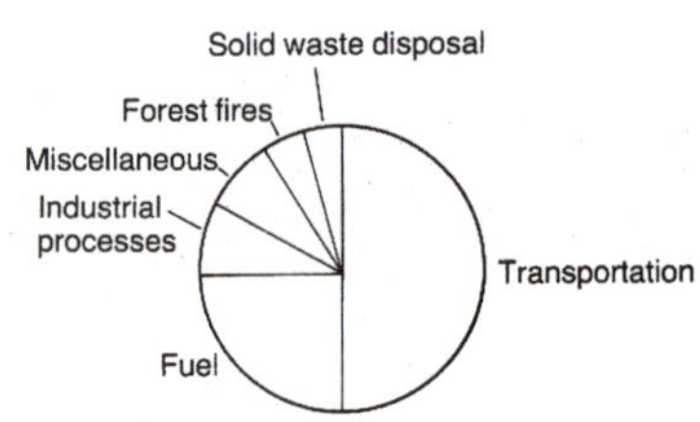

(2)

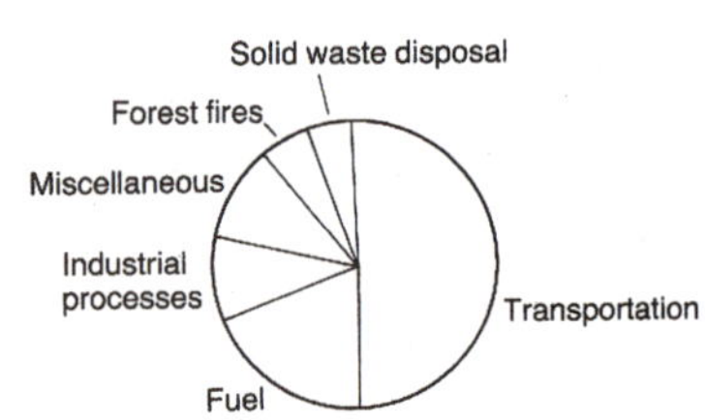

(3)

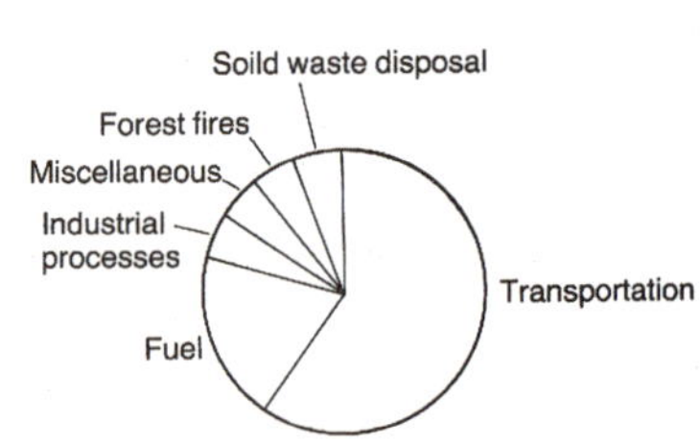

(4)

Hydrosphere

Oceans of Earth

The five oceans cover about 388 million square kilometers of the 518 million square kilometers of Earth's surface. That is, about three-fourths of Earth is covered by the oceans. The largest ocean, the Pacific, covers more surface area than all land combined. The other oceans are the Atlantic, Indian, Arctic, and Antarctic (see Figure 7-16).

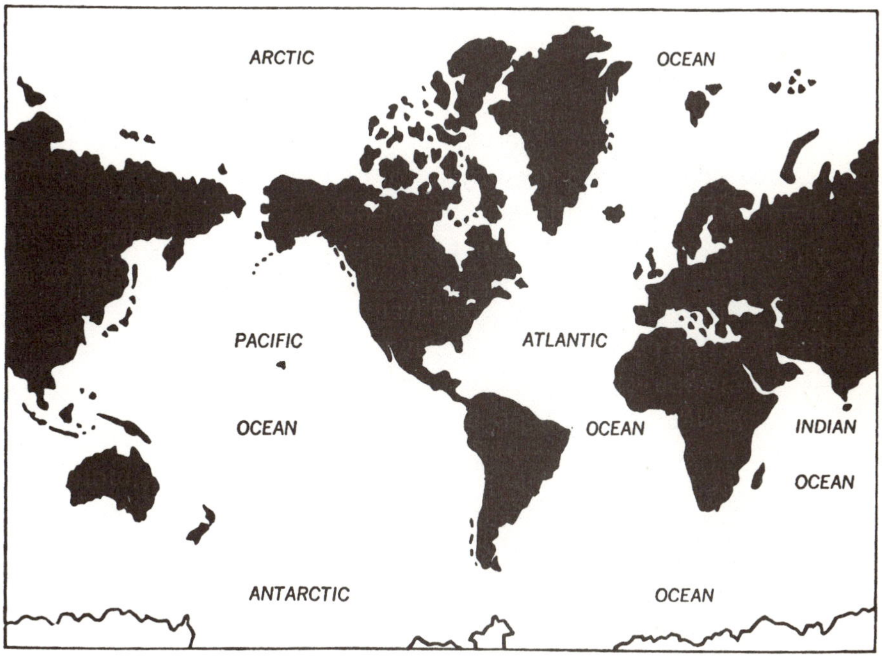

Figure 7-16. The oceans of the world.

Minerals found in seawater were once part of the land. Through the ages, freshwater rivers and streams dissolved the minerals and carried them to the ocean. Plants and animals in the oceans extract some of these minerals to build their shells and skeletons. Minerals that are not removed by marine plants and animals accumulate and increase in concentration.

Motions of Ocean Water

The waters of the oceans are in constant motion. The exchange of energy with the atmosphere plays a major role in the motion of ocean water.

Winds and Waves

Winds produce waves by causing friction at the water's surface. The stronger the wind, the larger the waves produced (see Figure 7-17).

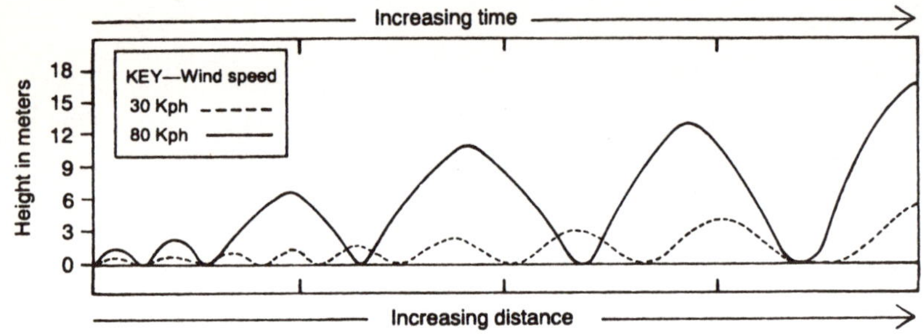

Figure 7-17. The effect of wind speed on wave height: stronger winds cause larger waves.

Tides are the daily rise and fall of ocean waters. The energy necessary to produce tides comes primarily from the gravitational pull of the moon and, to a lesser degree, the sun. Each day, there are two high tides and two low tides, due to the rotation of Earth. Figure 7-18 shows how high tides and low tides are caused by the gravitational pull of the moon and the sun.

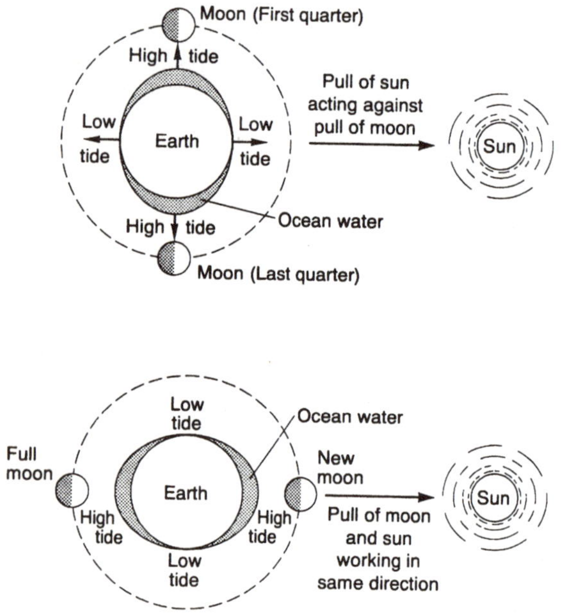

Figure 7-18. Tides are caused by the gravitational pull of the sun and the moon.

Surface currents are produced by global winds blowing great distances across the ocean's surface. For example, the North Atlantic Drift flows northeast from the middle Atlantic coast of North America to Great Britain. This slow, warm current moderates the climate of Great Britain.

Density currents are movements of water caused by differences in its density. Water can have differences in density due to temperature, mineral content, or sediment. Basically, cold water is denser than warm water, salt water is denser than fresh water, and sediment-filled water is heavier than clear water. That is, cold water sinks when mixed with warm water, salt water sinks when mixed with fresh water, and sediment-filled water sinks when mixed with water free of sediment.

Earthquakes, volcanoes, and landslides at sea produce violent ***tsunami waves*** that can travel 800 kilometers per hour across the entire span of an ocean. These huge waves can drown coastal villages.

EXERCISE 5

1. The table below lists Earth's five oceans and the percentage of Earth's total water surface covered by each ocean. Given that Earth's total water surface is about 400,000,000 square kilometers, determine the area of each ocean.

Ocean	*Percent of Water Surface*	*Size in Square Kilometers*
Pacific	37	
Atlantic	25	
Indian	13	
Antarctic	13	
Arctic	3	

Note that the combined percentage of Earth's water surface for the five oceans does not add up to 100 percent. What is the best explanation for this?

(1) Some surface water is in the form of lakes and rivers.
(2) The water cycle places some water in the atmosphere.
(3) Animals drink water.
(4) Some water is in the ground.

2. The following table compares the percentages of dissolved minerals in river water and ocean water. Some minerals are removed from ocean water by animals and plants to make their shells and skeletons. Those minerals that are not removed have accumulated in ocean water over time.

Percentages Dissolved

Mineral	*River Water*	*Ocean Water*
Carbonate	35.2	0.4
Calcium	20.4	1.2
Silcate	11.7	Trace
Chloride	5.7	55.0
Sodium	5.8	31.0
Magnesium	3.4	3.7
Potassium	2.1	1.1
Sulfate	12.1	7.7

Which two minerals have accumulated in seawater the most?

(1) potassium and sulfate
(2) calcium and carbonate
(3) sodium and chloride
(4) sodium and potassium

Questions 3 and 4 relate to the following experiment. Carmen's science teacher slowly poured three different water solutions into a glass tank: (1) red, warm fresh water, (2) clear, cool fresh water, and (3) blue, cold fresh water.

3. Which diagram illustrates the most likely outcome of this experiment?

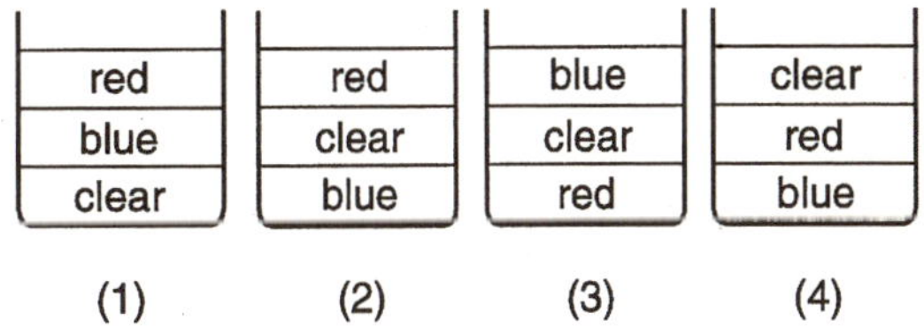

4. The best explanation of this demonstration is that
   (1) warm water is more dense than cold water
   (2) cold water is more dense than warm water
   (3) blue water is more dense than red water
   (4) red water is more dense than blue water

5. Michael lives in Florida along the Atlantic Ocean. Michael's science teacher asked the class to observe the relationship between ocean waves and wind speed. Michael wrote the following notes regarding his observations of ocean waves and wind speed as reported on the radio.

   Day 1: The ocean was calm, no waves were evident. The wind speed was 0 kilometers per hour.
   Day 2: Moderately high waves with foam and blowing spray. The wind speed was 43 kilometers per hour.
   Day 3: Ripples in the water but generally calm. The wind speed was 4.8 kilometers per hour.
   Day 4: Large waves forming with many whitecaps and some spray. The wind speed was 64 kilometers per hour.
   Day 5: Small waves with many whitecaps. The wind speed was 24 kilometers per hour.

Does there appear to be a direct relationship between the waves and the wind speed? Describe how the waves might appear if the wind speed was 96 kilometers per hour.

6. All motions of ocean water are caused mainly by energy interactions between the hydrosphere and lithosphere or atmosphere *except* for
   (1) tsunami waves
   (2) surface currents
   (3) small waves
   (4) high and low tides

Questions 7–9 refer to the table below, which shows the relationship between wind speed and the average height of waves produced by wind blowing at that speed across the water's surface.

Wind Speed (km/hr)	*Average Height (meters)*
19	.3
31	.8
40	1.8
50	3.2
59	5.1
69	7.4
80	10.3
90.1	13.9

7. As the wind speed increases, the average wave height
   (1) increases
   (2) decreases
   (3) increases, then decreases
   (4) remains the same

8. As the wind speed doubles, the average wave height
   (1) remains the same
   (2) doubles
   (3) more than doubles
   (4) less than doubles

9. When the wind speed is 64 kilometers per hour, the average wave height would most likely be
   (1) 5.1 meters
   (2) 7.4 meters
   (3) greater than 5.8 meters but less than 7.4 meters
   (4) greater than 5.1 meters but less than 5.8 meters

10. The oceans on Earth's surface cover
   (1) about the same surface area as land does
   (2) twice as much surface area as land does
   (3) three times as much surface area as land does
   (4) four times as much surface area as land does

11. A teacher places a capped bottle filled with cold salt water containing a red dye into a tank containing warm, clear fresh water. After the bottle is carefully opened, the red water enters the tank. The resulting water would most likely look like

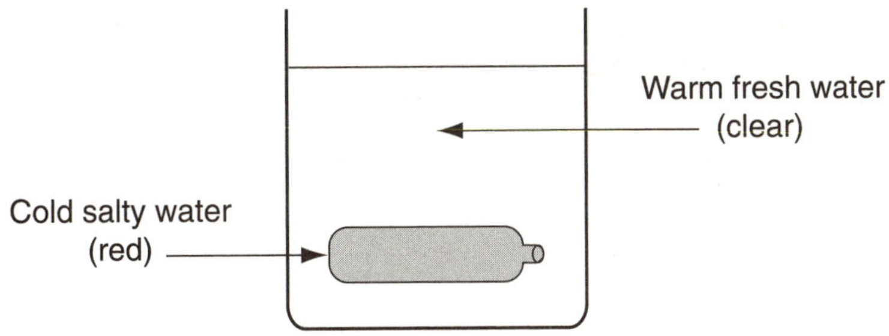

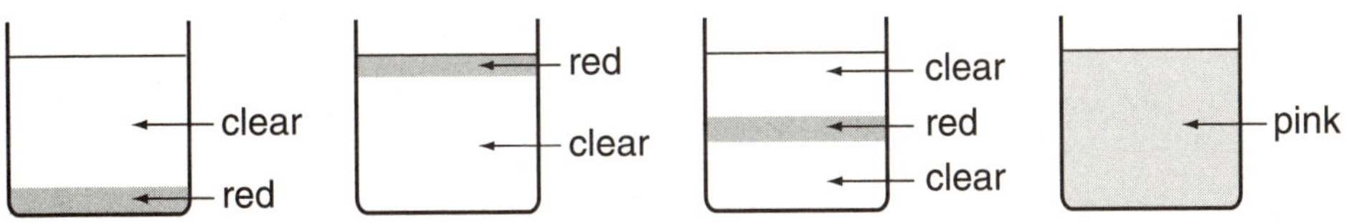

(1) (2) (3) (4)

12. Hurricanes can cause greater damage to a shore community during a high tide than during a low tide. In one or two sentences, explain why this is true.

8 Simple Machines

Points to Remember —

- A simple machine is a device that makes work seem easier. When doing work, a simple machine can change the effort needed and/or the direction or distance over which the effort is applied.
- Types of simple machines are the lever, pulley, inclined plane, wheel and axle, screw, and wedge. Many common tools are examples of simple machines.
- The work put into a machine is always more than the work output of a machine. Loss of work output is due to friction. Friction in the form of heat is produced when parts rub against each other. Lubricants reduce friction.

Work

Any push or pull on an object is called a *force*. Work is done when a force causes an object to move over a distance. The amount of work done depends on the amount of force applied and the distance the object is moved. The relationship between work, force, and distance is given by the formula:

$$\textbf{work} = \textbf{force} \times \textbf{distance}$$

When a force is applied to an object, the force may or may not cause the object to move. If the force does not produce motion, no work is done. As shown in Figure 8-1, a force results in work only if motion is produced.

Figure 8-1. Work is done when a force acts over a distance.

Machines and Work

A ***machine*** is a device that transfers mechanical energy from one object to another object. Machines make work seem easier. They do this by multiplying force and by changing the direction or the distance over which a force is applied. For example, a single pulley changes the direction of a force. A wrench multiplies the force applied to it when removing a tight bolt. A loading ramp attached to the back of a truck changes the distance over which a force is applied. The force a machine has to overcome is called ***resistance,*** and the force applied is called ***effort.*** Using a machine can reduce the amount of effort needed to overcome a given amount of resistance. However, the amount of work done is not made less by using a machine.

An example will make this clearer. Suppose you had to lift a box weighing 200 pounds (lb) up onto a platform 5 feet high. To lift the box straight up by yourself, you would need to apply 200 lb of force over a distance of 5 feet. Using the formula work = force x distance, you get:

$$\text{work} = 200 \text{ lb} \times 5 \text{ ft} = 1000 \text{ ft-lb}$$

However, if you set up a rope and a system of pulleys to change the direction and distance of the force required, you might have to pull in 20 feet of rope using only 50 lb of force:

$$\text{work} = 50 \text{ lb} \times 20 \text{ ft} = 1000 \text{ ft-lb}$$

The pulley system lets you use less effort over a longer distance to do the same amount of work you would have done without the pulleys.

Simple Machines

Many complex modern machines are made up of a number of simple machines working together to perform some task. The *lever* and the *inclined plane* are the most basic simple machines. Most other simple machines are based on either the lever or the inclined plane.

A ***lever*** consists of a rigid bar that can turn around a point called a *fulcrum*, as shown in Figure 8-2. Levers make work easier by multiplying applied forces. Examples of levers are pliers, scissors, and a crowbar.

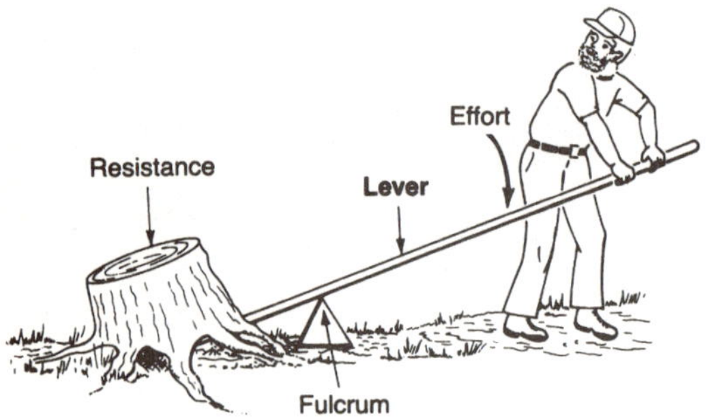

Figure 8-2. A lever multiplies effort, making it easier to uproot a tree trunk.

A *pulley* is a modified form of the lever. Figure 8-3 shows several types of pulleys. A *wheel and axle* is also a modified lever. It consists of a large wheel with a smaller wheel, or axle, in its center. When one wheel is turned,

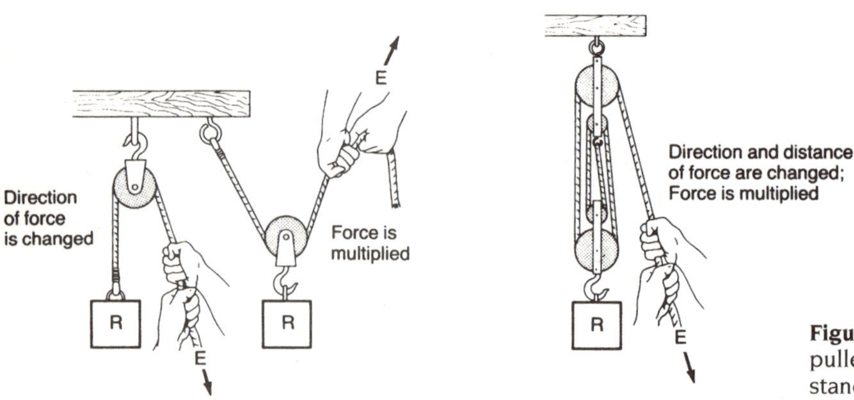

Figure 8-3. The three types of pulleys. (R stands for resistance, E stands for effort.)

so is the other (see Figure 8-4). Common examples of the wheel and axle are a steering wheel, doorknob, screwdriver, and wrench.

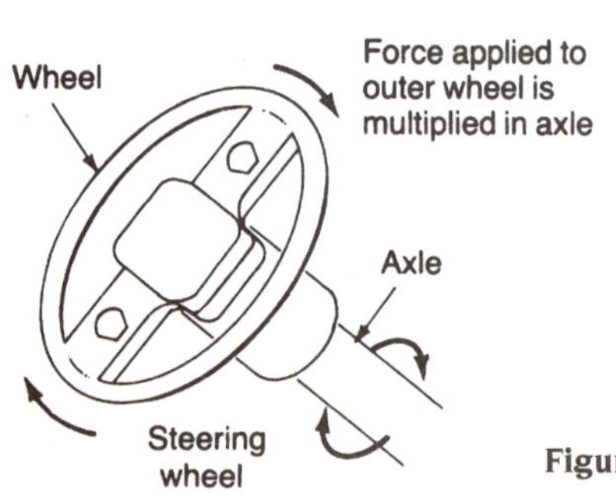

Figure 8-4. A steering wheel is an example of a wheel and axle.

An ***inclined plane*** is a flat surface with one end higher than the other. A wheelchair ramp is an inclined plane; so is a staircase. Figure 8-5 shows how an inclined plane makes work easier by changing the direction and distance of a force needed to do a job.

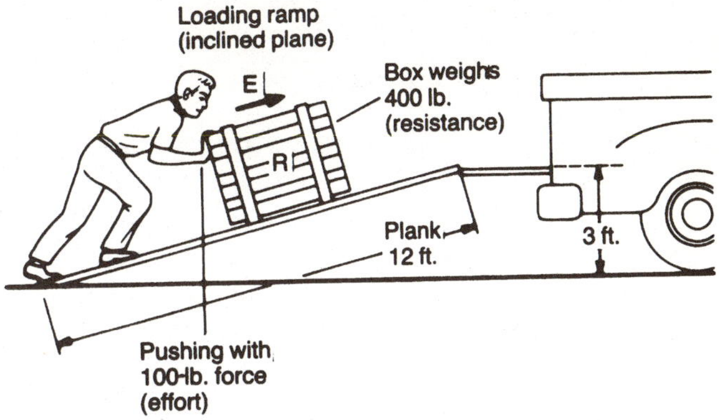

Figure 8-5. A loading ramp is an inclined plane.

The *screw* and the *wedge* are simple machines that are based on the inclined plane (Figure 8-6). A screw is an inclined plane wrapped around a pole. A wedge that is used to split wood consists of two inclined planes placed back to back. Other examples of a wedge are knives, nails, and teeth on saw blades.

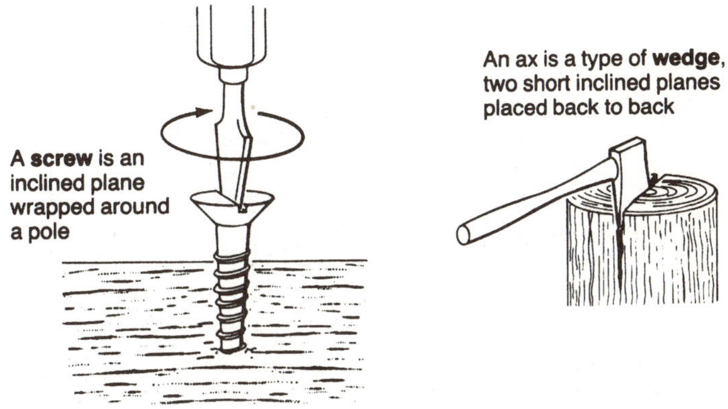

Figure 8-6. The screw and the wedge are based on the inclined plane.

Compound Machines

Most machines used in your daily life are compound machines. That is, they are made up of more than one simple machine. A scissors is an example of a compound machine. It is made of three different simple machines.

Scissors Part	*Simple Machine*
Center screw	Screw
Blade edge	Wedge
Handle and blade	Lever

Another compound machine is a bicycle. It is made of at least four different simple machines.

Bicycle Part	*Simple Machine*
Brake handles	Lever
Wheels	Wheel and axle
Chain	Pulley
Tire bolts	Screw

Efficiency of Machines

Ideally, a machine's work output should equal the amount of work put into the machine. However, in reality, machines are never 100 percent efficient. The amount of work done by any machine is always less than the amount of work put into it. This is because some of the work put into a machine is converted into heat energy and thus wasted. The heat is produced by friction between the machine's moving parts.

Returning to our earlier example, suppose you lift a 200-pound box up 5 feet, using pulleys. Although 1000 foot-pounds of work output were accomplished, you actually had to perform more than 1000 foot-pounds of work. Some of your work input is wasted because friction between the wheel and axle of each pulley creates heat.

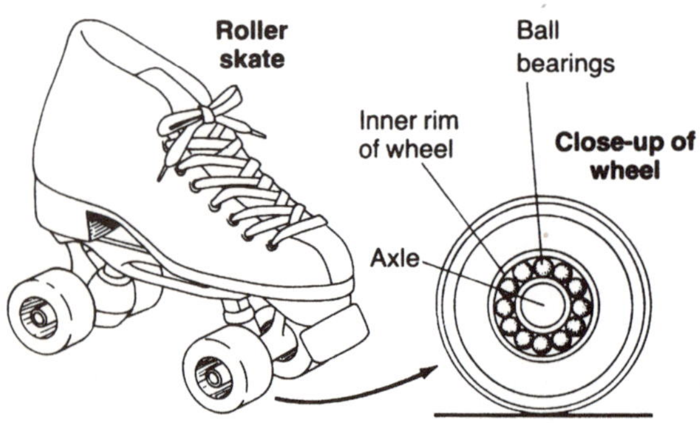

Figure 8-7. Ball bearings in the wheel of a roller skate reduce friction as the wheel turns.

A machine can be made more efficient by reducing friction. A common way to reduce friction is to lubricate the contact surfaces of a machine's moving parts with grease or oil. Other methods are waxing the contact surfaces, sanding the surfaces to make them smoother, or using ball bearings between the surfaces (Figure 8-7).

EXERCISE

1. A screw is a modified form of
   (1) a wheel and axle
   (2) a lever
   (3) a pulley
   (4) an inclined plane

2. A substance that is commonly used to reduce friction between two pieces of metal is
   (1) water (3) oil
   (2) air (4) sand

3. The force that a machine has to overcome is called
   (1) effort (3) friction
   (2) resistance (4) energy

4. The chain of a bicycle is greased in order to
   (1) increase weight
   (2) reduce air drag
   (3) reduce friction
   (4) increase resistance

5. Andrea moved a heavy box using a pulley system. If her work output was 800 foot-pounds, her work input was
   (1) greater than 800 ft-lb
   (2) less than 800 ft-lb
   (3) exactly 800 ft-lb
   (4) no way to tell

6. The pulley on the flagpole in the illustration makes it easier to raise the flag by

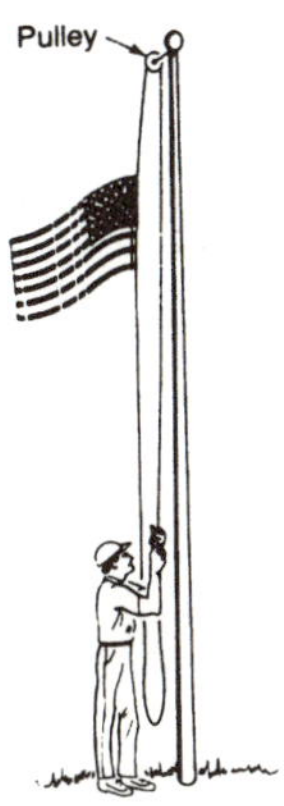

   (1) decreasing the amount of work required
   (2) changing the direction of the force applied
   (3) putting out more work than is put into it
   (4) making the flag lighter

7. Which type of simple machine shown below is being used to split the wood?

   (1) wheel and axle (3) lever
   (2) pulley (4) wedge

8. In science class, Latoya tested the efficiency of four machines and recorded the results in a chart. For which machine must she have made an error?
   (1) Lever I (3) Lever III
   (2) Lever II (4) Pulley

Machine	*Efficiency*
Lever I	75%
Lever II	100%
Lever III	60%
Pulley	30%

9. What three simple machines are being used in the diagram to help move bricks from position A to position B?

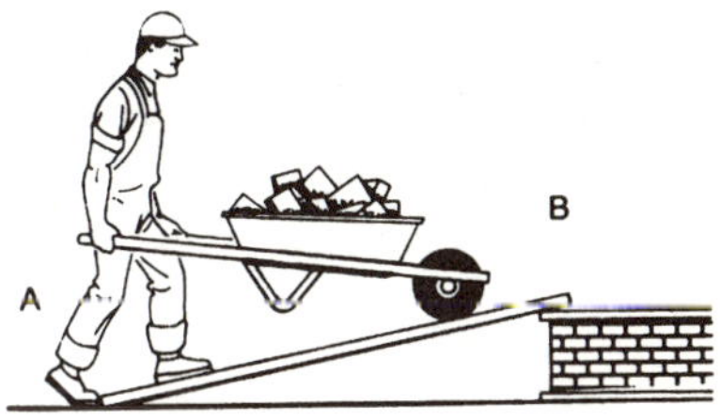

   (1) inclined plane, wheel and axle, wedge
   (2) inclined plane, wheel and axle, lever
   (3) inclined plane, wheel and axle, screw
   (4) inclined plane, pulley, and lever

10. Pablo has a large, heavy box to move up onto a platform 1 meter high. He has two boards available to help him. The lengths of the two boards are 3 meters and 2 meters. Which diagram shows how Pablo should place the boards to slide the box onto the platform using the least amount of effort?

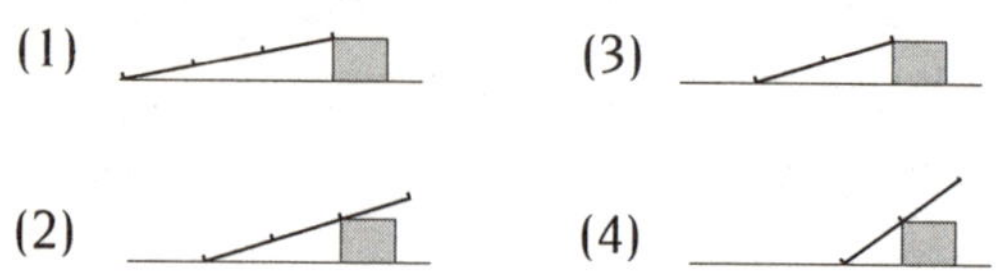

Questions 11–13 refer to the diagram below, which shows a 6-meter board, balanced as shown. When Misako and Tamisha placed boxes of known weight at different locations, the board changed its position.

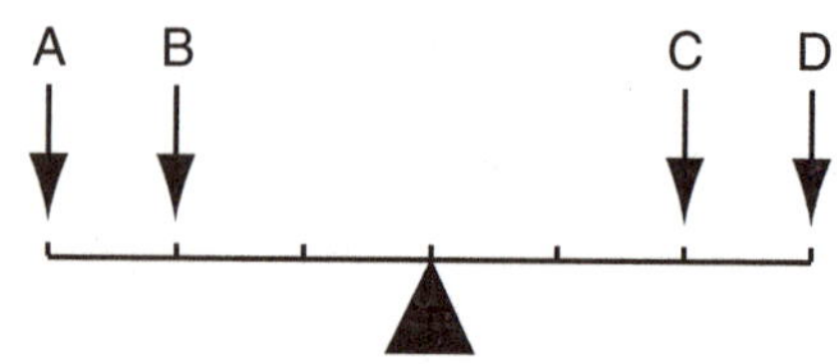

11. When a 10-lb weight was placed at position A and another 10-lb weight at position C, the board probably looked like

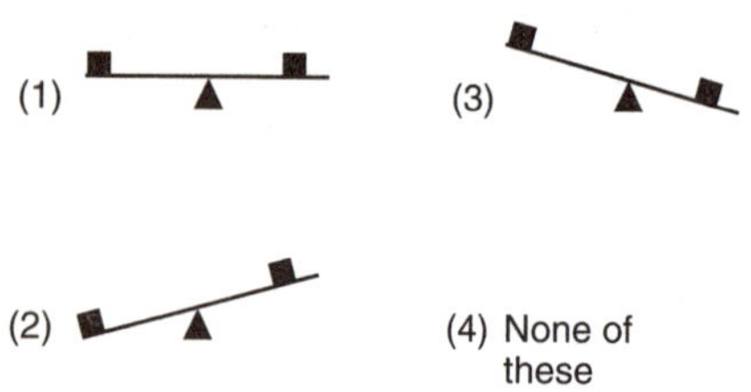

12. When a 10-lb weight was placed at position B and a 10-lb weight at position C, the board probably looked like

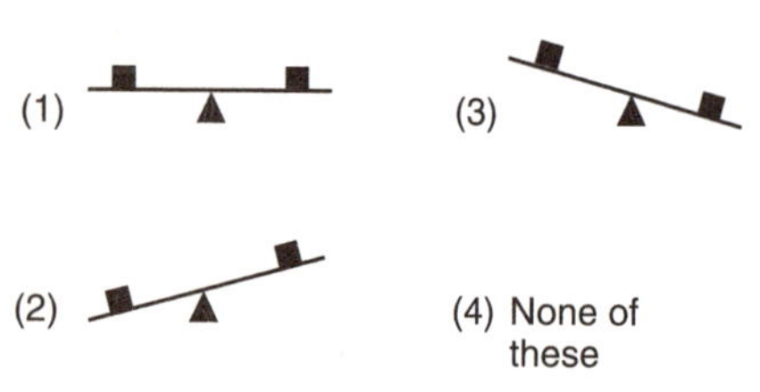

13. When a 40-lb weight was placed at position A and a 10-lb weight at position C, the board probably looked like

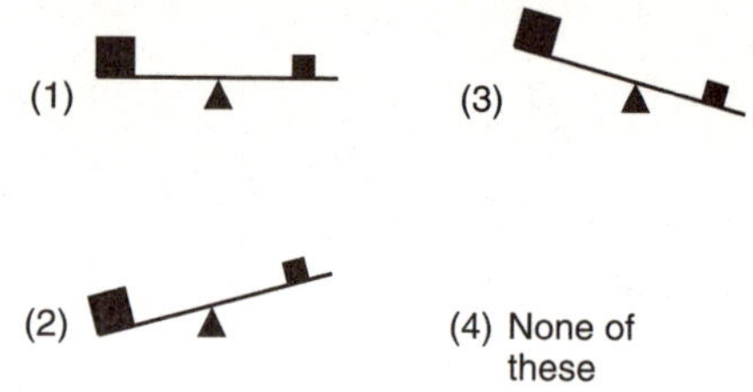

Read the following paragraph and then answer questions 14–16.

Work is the application of a force over a distance. For example, picking up a 10-lb box off the floor and placing it on your desk 3 feet above the floor is doing work. In fact, using the formula work = force × distance, you would have done 10 lb × 3 feet, or 30 ft-lb, of work. If the box weighed 200 lb, it might be difficult for you to lift it. Therefore, you might use a simple machine to help lift the box. Building a ramp would allow you to push the box up an inclined plane to the desktop. (Remember: there will be friction between the box and the inclined plane.)

14. Pushing the box up an inclined plane would make you exert
    (1) more force
    (2) less force
    (3) the same force
    (4) no force

15. The amount of work needed to push the box up the inclined plane would be
    (1) more than 600 ft-lb
    (2) less than 600 ft-lb
    (3) exactly 600 ft-lb

16. The amount of work needed to move the box would be least if
    (1) the surface of the inclined plane was sandy
    (2) wheels were placed under the box
    (3) you were stronger
    (4) two people pushed the box

Questions 17–20 refer to the diagram below, which shows a wrench being used to try to turn a tight bolt. Points A, B, C, and D are locations on the wrench handle where you might apply effort to turn the wrench and loosen the bolt.

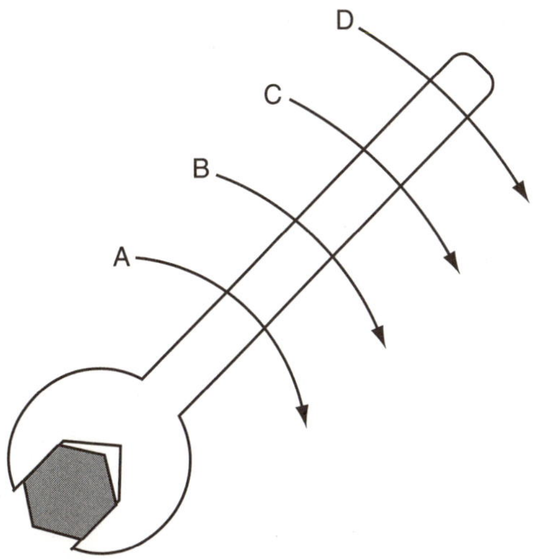

17. The wrench is an example of a simple machine. The simple machine best exemplified in this process is the
    (1) lever (3) inclined plane
    (2) wheel and axle (4) pulley

18. The location on the wrench's handle where the least effort would have to be applied to turn the bolt is point
    (1) A (2) B (3) C (4) D

19. The location on the wrench's handle where the least circular distance would be applied to turn the wrench and bolt once around is point
    (1) A (2) B (3) C (4) D

20. This example of a simple machine demonstrates that
    (1) as greater effort is needed, the distance of motion is increased
    (2) as greater effort is needed, the distance of motion is decreased
    (3) as greater effort is needed, the distance of motion remains the same
    (4) as lesser effort is needed, the distance of motion remains the same

Questions 21–24 refer to the following paragraph and the accompanying diagrams.

Using a smooth wooden board and some books, Lenno and Chin set up a ramp to do an experiment. They attached a 10-lb metal block to a cord and a spring scale, as shown in the first diagram. During the experiment, they made a number of changes as they pulled the block up the ramp, as shown in the two other diagrams.

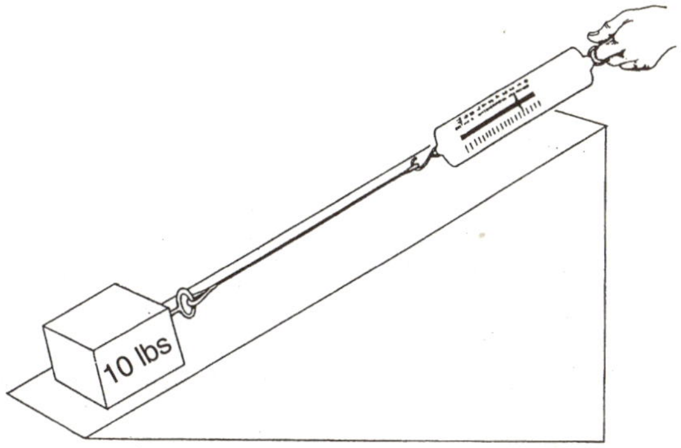

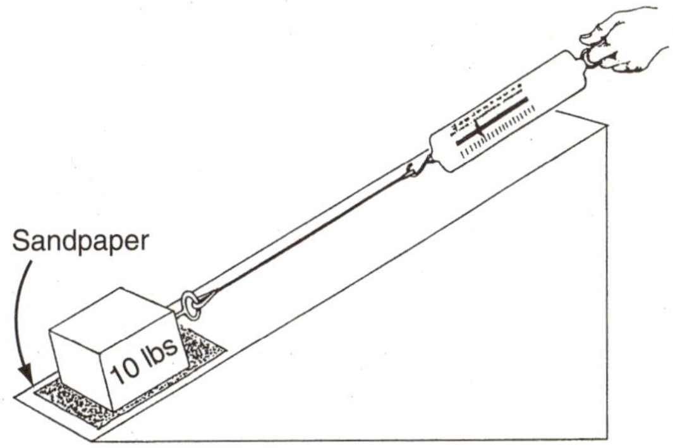

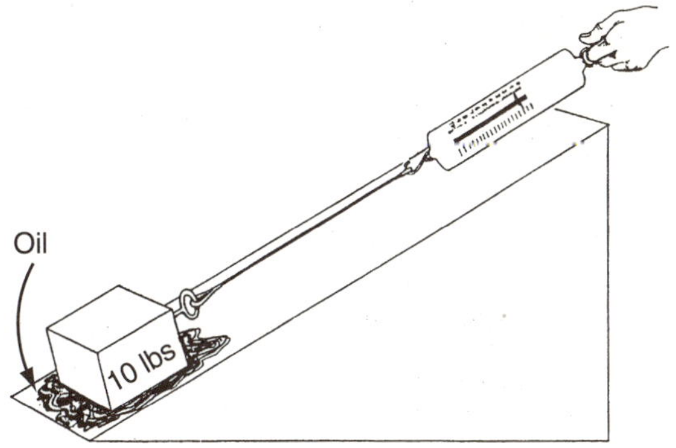

21. Use of the ramp to pull the weight is an example of a simple machine. The simple machine demonstrated in this process is the
    (1) lever (3) inclined plane
    (2) wheel and axle (4) pulley

22. When pulling the weight up the wooden ramp, the spring balance probably indicated an effort of
    (1) 10 lb
    (2) more than 10 lb
    (3) less than 10 lb
    (4) zero lb

23. When Chin attached a sheet of sandpaper to the ramp and pulled the weight up across the sandpaper, the amount of effort required most likely
    (1) increased
    (2) decreased
    (3) remained the same

24. When Lenno placed oil on the surface of the ramp and pulled the weight up across the oil, the amount of effort required probably
    (1) increased
    (2) decreased
    (3) remained the same

25. A person tries to lift a heavy box off the ground, but cannot make it move. Even though the person exerts great effort, no actual work is done. Explain.

Process Skills

OBSERVING AND CLASSIFYING

Many common tools are levers of some kind. For example, scissors, shovels, and salad tongs are levers. All levers can be grouped into three basic classes, depending on the location in the lever of the *resistance*, the *effort* applied, and the *fulcrum* (the point around which the lever turns). Diagram 1 illustrates the three lever classes.

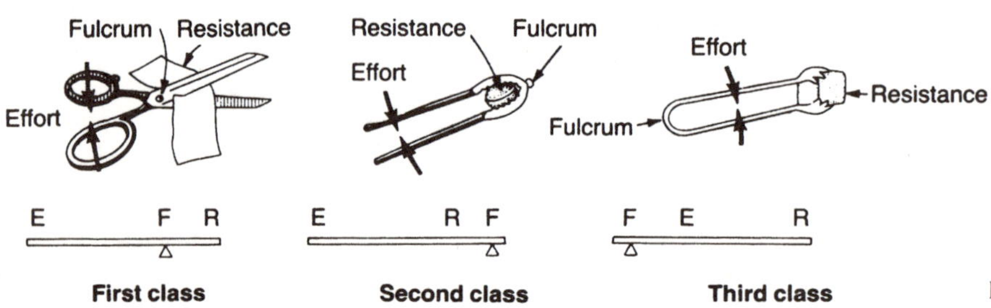

Diagram 1

A *first-class lever*, such as scissors, has the effort (**E**) applied on one end, the resistance (**R**) on the other end, and the fulcrum (**F**) in between. A *second-class lever*, like a nutcracker, has the fulcrum and effort at opposite ends and the resistance in the middle. A *third-class lever*, such as ice tongs, has the resistance and the fulcrum at opposite ends and the effort applied in the middle.

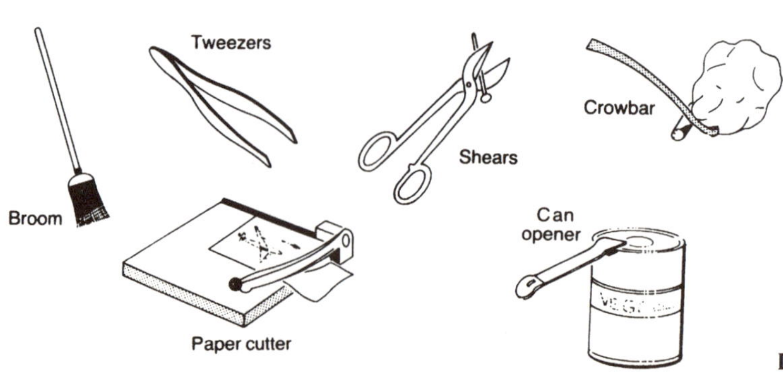

Diagram 2

Diagram 2 shows some examples of levers. Can you determine which lever class each item represents? Drawing a lever diagram for each one will help you do this. First, draw a line to represent the item. Next, think about how you use the item and try to identify the positions of the fulcrum, effort, and resistance. Where does the item meet resistance? Is effort applied to one of the ends of the item, or somewhere in between? Where does the object turn or change direction? Once you have located the fulcrum, resistance, and effort, and labeled them on your lever diagram, you can then classify the item using the definitions of lever classes given above.

1. The broom is an example of a
   (1) first-class lever (2) second-class lever (3) third-class lever

2. The tweezers are an example of a
   (1) first-class lever (2) second-class lever (3) third-class lever

3. The shears are an example of a
   (1) first-class lever (2) second-class lever (3) third-class lever

4. The crowbar is an example of a
   (1) first-class lever (2) second-class lever (3) third-class lever

5. The paper cutter is an example of a
   (1) first-class lever (2) second-class lever (3) third-class lever

6. The can opener is an example of a
   (1) first-class lever (2) second-class lever (3) third-class lever

9 The Laws of Motion

Points to Remember —

- A force is a push or a pull. A force can cause an object to be put in motion, stop motion, change its speed of motion, or change its direction. Objects at rest are affected by balanced forces, and objects in motion are affected by unbalanced forces.
- Mass is the measure of the amount of substance.
- Acceleration is the rate of change of velocity.
- Newton's three laws of motion govern the motion of all objects.
  i. Newton's first law states that an object at rest will remain at rest and an object in motion will remain in motion unless another force affects it.
  ii. Newton's second law states that acceleration of an object depends directly on the force applied and inversely on its mass.
  iii. Newton's third law states that every action has an equal but opposite reaction.

Force, Mass, and Acceleration

A force is a push or pull (see Figure 9-1). Pulling the refrigerator door will open the door. Pushing a computer mouse will move the mouse across the mouse pad. Lifting a heavy log for the fireplace is pulling the log up against the force of gravity. In each example, a force causes motion. Table 9-1 gives some examples of pushing and pulling forces.

Table 9-1. Pushing and Pulling Forces

Pushing Forces	*Pulling Forces*
Hitting a baseball	Pulling a tug-of-war rope
Closing a refrigerator door	Opening a closet door
Moving a shopping cart	Lifting a shovel full of dirt
Wind knocking down a tree	Climbing a rope
Hammering a nail	Gravity pulling an apple to the ground

Figure 9-1. Pulling and pushing forces: (a) opening and (b) closing doors.

A force can also stop motion, change the speed of motion, and change the direction of motion of an object, as the following examples show.

Force stops motion: Ground striking a falling acorn; glove catching a baseball

Force slows motion: Friction slowing a skateboard; car going from a flat road to an uphill road

Force changes direction: Tennis racket striking a tennis ball; wind causing a fly ball to curve

When the forces acting on an object are balanced, the object is at rest. A book on a table is stationary if the force of gravity pulling the book downward is equal to the force of the table pushing upward on the book. If more books are put on the table than the upward force of the table can hold, the forces become unbalanced, the table will break, and the books will fall to the floor. Also, when other forces are introduced, such as you pushing a book across the table, the forces become unbalanced, and the stationary book is put in motion. Figures 9-2 and 9-3 illustrate balanced and unbalanced forces; Table 9-2 gives some examples.

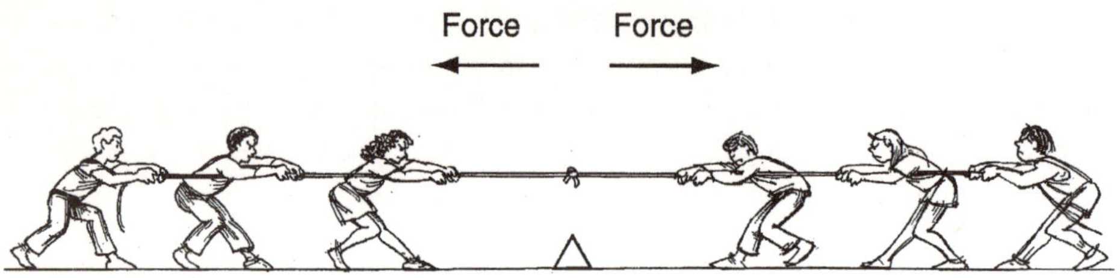

Figure 9-2. This tug-of-war illustrates balanced forces.

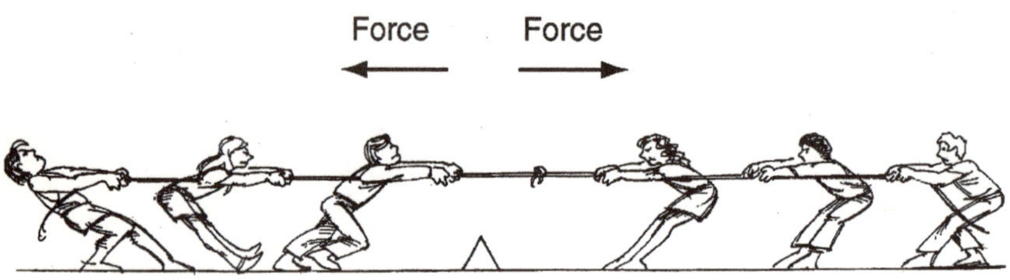

Figure 9-3. This tug-of-war illustrates unbalanced forces.

Table 9-2. Balanced and Unbalanced Forces

Balanced Forces	*Unbalanced Forces*
Car parked in driveway	Car pulling out of driveway
Apple hanging on a tree	Apple falling to the ground
Roller coaster on top of ride	Roller coaster coming down

Mass

Mass is the measure of the amount of substance. The mass of an object, unlike weight, remains constant and does not change anywhere in the universe. A car and a baseball have different masses. The car contains more matter and is more massive than the baseball.

The greater the mass of an object, the greater the force necessary to move it. Thus, it takes a greater force to move a car than it does to move a baseball.

Acceleration

Acceleration is the rate of change of velocity. Velocity and speed are similar. However, technically, velocity has a directional component. You accelerate a car when it increases velocity from 10 kph (kilometers per hour) to 30 kph. If it takes 2 seconds to change (in this case, increase) the velocity 20 kph, the car's acceleration was 10 kph per second.

A car traveling at a steady velocity of 30 kph has no acceleration. If the car increases its velocity to 50 kph, it accelerates. It takes additional force produced by the car's engine to cause it to accelerate.

EXERCISE 1

Questions 1-3 refer to the following paragraph.

Roberto threw a baseball up into the air. It reached a height of 9 meters and then started to come down. William threw the ball to a height of 12 meters.

1. The force that pulled the ball down was
   (1) air friction (3) wind
   (2) gravity (4) mass of the ball

2. William threw the ball higher because
   (1) less air friction affected his ball
   (2) the pull of gravity on his ball was less
   (3) he applied a greater pushing force
   (4) more air friction affected his ball

3. At which point in the arc of the baseball were the forces balanced?
   (1) A (2) B (3) C (4) D

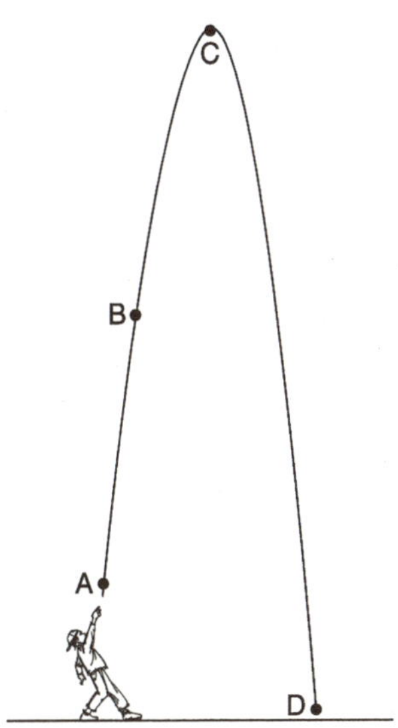

4. If four different balls were on the gymnasium floor, the greatest force would be needed to move the
   (1) golf ball (3) basketball
   (2) baseball (4) bowling ball

5. The mass of the Washington Monument would be
   (1) greatest in Washington, D.C.
   (2) greatest at the North Pole
   (3) greatest on the moon
   (4) the same at all the above locations

Questions 6 and 7 refer to the following paragraph.

A force is a push or a pull. A force can (1) start motion, (2) stop motion, (3) slow motion, or (4) change direction of motion. In volleyball, a person serving the ball tosses it upward to start the ball in motion. The server applies a different, directed force by striking the ball and changing its direction toward the opposing team. When the ball reaches the other team, they first apply a gentle force in an effort to slow and control the ball. Then they apply a stronger force to send the ball back to the serving team. This continues until the point is won and the motion of the ball is stopped.

6. Striking the ball on the serve in volleyball is
   (1) a pushing force
   (2) a pulling force
   (3) not any force
   (4) both a pushing and a pulling force

7. The purpose of striking the ball in a volleyball game is to
   (1) place the ball in motion
   (2) change the direction of the ball's motion
   (3) slow the ball's motion
   (4) all of the above

8. Closing the front door of a house can be done with
   (1) a pushing force
   (2) a pulling force
   (3) either a pushing or a pulling force
   (4) neither a pushing nor a pulling force

9. A rocket traveling in space increases its velocity from 32,000 kph to 40,000 kph. To do this, the rocket must
   (1) increase its mass
   (2) decrease its mass
   (3) increase the force acting on it
   (4) decrease the force acting on it

Laws of Motion

Newton's Laws of Motion

In the mid-1660s, Sir Isaac Newton formulated the three laws of motion. The laws of motion explain how forces affect the state of motion of all objects. Even today, Newton's laws of motion remain as the basis for understanding the motion of all objects.

First Law of Motion

The first law of motion states that **an object at rest will remain at rest and an object in motion will remain in motion until an outside force acts on the object.** There are two parts to this law. First, any object at rest will not move unless some force acts on it. A bicycle will remain leaning against a tree, or a book will remain on a table, until some force is applied that moves the bicycle or the book. Second, any moving object will continue to move unless a force acts on the object to slow it or to change its direction. A thrown ball will move at a constant speed and in a straight line unless a force acts upon it. Air friction slows the ball, and gravity—another force affecting the ball—changes its direction and pulls the ball to Earth (see Figure 9-4).

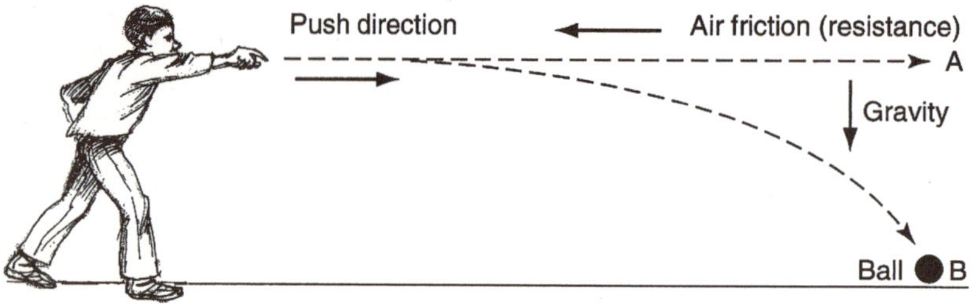

Figure 9-4. Air friction and gravity are forces that pull the ball back to Earth.

Inertia is the tendency of an object at rest to remain at rest, or of an object in motion to remain in motion. In other words, it is the tendency of an object to resist any change in motion. The more massive an object is, the more inertia it has, or the more it resists a change in motion. When riding in a moving car that stops suddenly, you feel your body move forward. Also, when seated in a parked car that suddenly accelerates, you feel your body move backward. In the stopping car, your moving body is resisting the stopping action; in the accelerating car, your body is resisting being put in motion (see Figure 9-5).

Figure 9-5. The first law of motion: Objects resist a change in motion—an effect you can feel in a starting or stopping car.

Second Law of Motion

The second law of motion states the relationship among force, mass, and acceleration. The law is commonly explained using the formula:

$$\textbf{acceleration} = \frac{\textbf{Force}}{\textbf{mass}}, \text{ or } a = \frac{F}{m}$$

An increase in force on a given mass will increase acceleration. If an equal force is applied to two objects of different mass, the object with the lesser mass will have the greater acceleration.

An object with a large mass will not be moved easily by a small force, whereas a large force acting on the same object will move it more easily. A wheelbarrow filled with dirt might be difficult for a child to move, while the same wheelbarrow can be moved more easily by a construction worker. If we increase the mass of the wheelbarrow by adding more dirt, the construction worker will not be able to move it as fast. Figure 9-6 illustrates the second law of motion.

Figure 9-6. The second law of motion: acceleration is equal to force divided by mass; the greater the force (and lesser the mass), the greater the acceleration.

Third Law of Motion

The third law of motion states that **for every action there is an equal and opposite reaction.** A simple demonstration of blowing up a balloon and releasing it shows how this law works. When the air is released from the balloon, the balloon moves in the opposite direction (see Figure 9-7).

Kicking a soccer ball, hot gases moving out the bottom of a rocket, and walking are all motions that produce an equal and opposite reactive force.

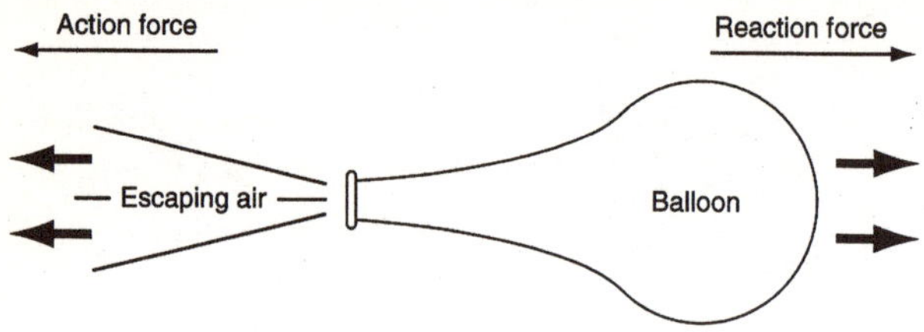

Figure 9-7. The third law of motion: every action has an equal and opposite reaction.

Process Skill

USING MATH TO ANALYZE DATA

Newton's second law of motion states the relationship of force, mass, and acceleration. The relationship is shown in the following formula:

$$\textbf{Force = mass} \times \textbf{acceleration, or acceleration} = \frac{\textbf{Force}}{\textbf{mass}}$$

The following table shows six mathematical examples of how the formula operates.

	Acceleration (meters/second/second)	*Force (newton)*	*Mass (kilogram)*
Example 1	1	1	1
Example 2	2	20	10
Example 3	1	10	10
Example 4	1	20	20
Example 5	0.5	10	20
Example 6	2	X	50

1. If the mass of a body remains the same and the force moving it is doubled, then the body's acceleration
   (1) is doubled (2) is halved (3) remains the same (4) is equal to mass

2. If the force moving an object remains the same and the object's mass is doubled, then the acceleration
   (1) is doubled (2) is halved (3) remains the same (4) is equal to mass

3. In Example 6 in the table, the unknown force X would be
   (1) 50 newtons (2) 100 newtons (3) 150 newtons (4) 52 newtons

4. According to the table, a newton is a
   (1) force that will move 1 kg at 1 meter/second/second
   (2) force that will move 20 kg at 10 meters/second/second
   (3) force that will move 10 kg at 20 meters/second/second
   (4) measure of mass

EXERCISE 2

1. When you walk, your feet push against the ground, and the ground pushes against your feet. This process demonstrates Newton's law that:
   (1) a body at rest remains at rest unless a force affects it
   (2) a body in motion remains in motion unless a force affects it
   (3) a large mass needs a large force to move it
   (4) every action has an equal and opposite reaction

2. When a golf ball is placed on a tee, it will remain there until the golfer strikes the ball. This demonstrates Newton's law that
   (1) a body at rest remains at rest unless a force affects it
   (2) a body in motion remains in motion unless a force affects it
   (3) a large mass needs a large force to move it
   (4) every action has an equal but opposite reaction

3. Morris helped his dad push their car out of the garage to repair it. He pushed as hard as he could, but the car moved only very slowly. This example demonstrates Newton's law that
   (1) a body at rest remains at rest unless a force affects it
   (2) a body in motion remains in motion unless a force affects it
   (3) a large mass needs a large force to move it
   (4) every action has an equal but opposite reaction

4. A pitcher on a baseball team exerts a pushing force when delivering a pitch to a batter. The batter swings and misses the ball. The force that eventually changed the direction of the ball's motion was probably
   (1) air friction (3) the catcher's mitt
   (2) gravity (4) the backstop

---

5. Esther pushed her model sailboat across a pond. According to Newton's first law, the sailboat should have continued across the pond and landed at point A. However, it landed at point B. The most likely reason for this was that

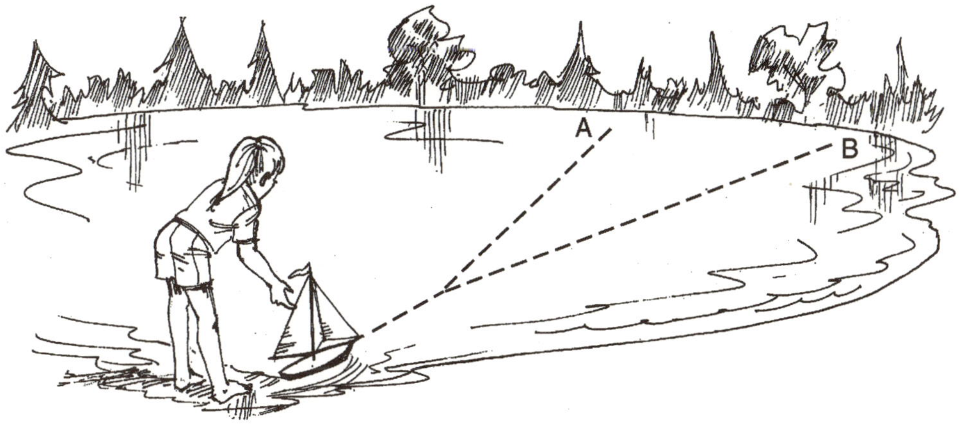

   (1) the force of the wind changed the sailboat's direction of motion
   (2) fish swimming near the sailboat changed the direction of motion
   (3) the force of the water changed the sailboat's direction of motion
   (4) the sailboat accelerated across the pond

6. Jane was on a bus when the bus suddenly started. She immediately lost her balance and almost fell toward the back of the bus. This was because
   (1) Jane was standing in the back of the bus
   (2) Jane was at rest and her body resisted the change in the bus's motion
   (3) Jane was walking through the bus
   (4) the force of gravity was greater than the force of the moving bus

7. According to Newton's first law, when pushing a penny across a table, the penny will slide to the edge of the table and fall off. However, if the force is not strong enough, the penny will stop on the table. The force that stops the penny on the table is
   (1) gravity (3) air
   (2) friction (4) the force of the table

Questions 8–11 refer to the following short paragraph.

A CO_2 cartridge was mounted on the top of a small toy car. When the compressed gas was released from the cartridge, the toy car moved 6 meters across the room. This was repeated using a car that had a mass five times greater than the first car.

8. The reaction of the first car to the release of gas would be to move
   (1) sideways (3) forward
   (2) backward (4) upward

9. The second car most likely moved
   (1) more than 6 meters
   (2) about 6 meters
   (3) less than 6 meters

10. The reaction of the cars to the release of gas from the cartridge
   (1) supports Newton's first law—a body in motion remains in motion unless a force affects it
   (2) supports Newton's third law—every action has an equal but opposite reaction
   (3) supports Newton's first and third laws
   (4) does not support any of Newton's laws

11. The mass of the second car was greater than that of the first car. The force affecting the second car was the same as the force affecting the first car. The acceleration of the second car
   (1) was less than that of the first car
   (2) was greater than that of the first car
   (3) was the same as that of the first car
   (4) cannot be compared to that of the first car

A lawn sprinkler is shown in the illustration below. When the water is turned on, it comes out of the nozzle and sprays onto the lawn. Study the illustration and answer questions 12–14.

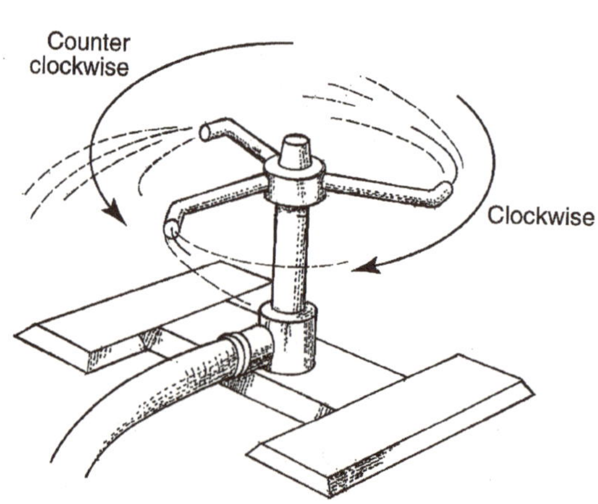

12. When the water comes out of the nozzle, the sprinkler
    (1) rotates clockwise
    (2) rotates counterclockwise
    (3) does not move
    (4) moves in one direction and then reverses to the other direction

13. If the amount of water coming out of the nozzle is increased, the speed of rotation of the sprinkler will
    (1) increase
    (2) decrease
    (3) remain the same
    (4) increase, then decrease

14. The law that governs the direction that the sprinkler moves in is
    (1) an object in motion will remain in motion unless an outside force acts on it
    (2) an object at rest will remain at rest unless an outside force acts on it
    (3) an increase in force will increase motion
    (4) for every action there is an equal but opposite reaction

---

Read the following paragraph and answer questions 15–17.

Gail and Jared set up an experiment that they saw in their science textbook, using a glass, several different coins (penny, nickel, and quarter), and an index card, as shown in the diagram. They flicked the card from under each of the coins to see what would happen.

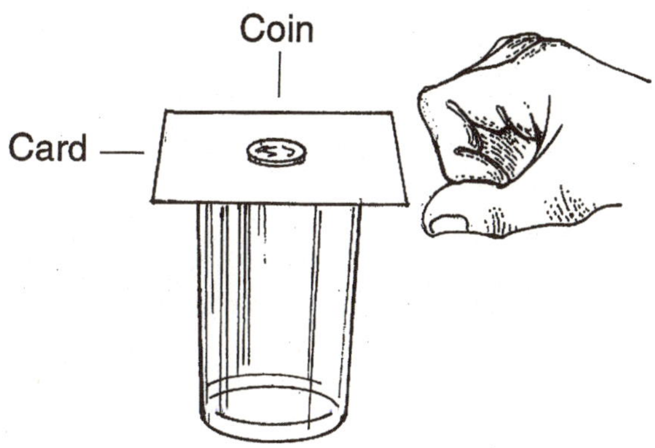

15. When they flicked the card from under each of the coins,
    (1) the coins all stayed with the card and fell on the table
    (2) the coins all dropped into the glass
    (3) the quarter fell into the glass, and the other two coins fell on the table
    (4) the penny fell into the glass, and the other two coins fell on the table

16. This experiment indicated that an object at rest will remain at rest and will resist being put in motion. In other words, this experiment demonstrated
    (1) differences in mass
    (2) acceleration
    (3) inertia
    (4) differences in forces

17. The quarter has the greatest mass of the three coins. Therefore, it would have the greatest
    (1) resistance to a change in motion
    (2) acceleration
    (3) change in direction
    (4) distance to fall

18. According to the second law of motion, an increase in force on a given mass will cause
    (1) a decrease in acceleration
    (2) an increase in acceleration
    (3) a decrease in the mass
    (4) an increase in the mass

19. A bus and a car are both traveling at the same speed. Why does it take longer for the bus to stop than for the car? Which one of Newton's laws of motion best explains your answer?

20. Scientists want to send a lunar explorer to the moon and return it to Earth. Why does it take a large rocket to lift the lunar explorer away from Earth? Why can the lunar explorer return to Earth without using a large rocket to take off?

10 Forms of Energy

Points to Remember —

- Energy is the ability to do work.
- There are two states of energy: potential energy is stored energy based on position or chemical makeup; kinetic energy is the energy of motion.
- Energy occurs in the following forms: mechanical, chemical, nuclear, sound, heat, electrical, and light.
- The Law of Conservation of Energy states that energy cannot be created or destroyed, it can only be transformed into other forms of energy.
- Energy transformation occurs when energy is changed from one form to another form.
- When people transform energy for some purpose, some unwanted form of energy is always produced, which is not useful for that purpose.

Energy Basics

Energy

Energy is the ability to do work. A flowing river has the ability to move a boat. A moving car has the ability to carry people from one place to another. Therefore, the river and the car possess some form of energy.

States of Energy

There are two basic states of energy: *potential* and *kinetic*.

1. ***Potential energy*** is stored energy that an object has because of its position or its chemical composition. A rock on a cliff has potential energy because of its position. A lump of coal contains potential energy in its chemical makeup.

2. ***Kinetic energy*** is energy that an object has when it is moving. A rock falling off a cliff has kinetic energy. The heat given off by a burning lump of coal is also a form of kinetic energy. The faster an object moves, the more kinetic energy it has. Figure 10-1 shows examples of potential and kinetic energy.

Figure 10-1. Some examples of potential energy and kinetic energy.

Potential energy may be changed into kinetic energy when motion is produced. Water held back by a dam has potential energy but no kinetic energy. Releasing the water and letting it flow changes its potential energy into kinetic energy.

Kinetic energy may also be changed into potential energy. When a ball is thrown straight up into the air, its kinetic energy of motion is changed into potential energy as the ball rises higher above the ground. At the ball's highest point, it is motionless and has only potential energy. As the ball falls back to the ground, this potential energy is changed back into kinetic energy.

Forms of Energy

Both potential and kinetic energy exist in many forms. For example, *mechanical energy* is the energy with which moving objects perform work. A hammer striking a nail, a jack lifting a car, and pedals turning the wheel of a bicycle are examples of things using mechanical energy. **Sound** is a particular type of mechanical energy. Table 10-1 lists some examples of different forms of energy.

Table 10-1. Different Forms of Energy

Form of Energy	*Example*
Mechanical	Fan
Sound	Bell
Chemical	Candle
Nuclear	Nuclear reactor
Heat	Toaster
Electrical	Generator
Light	Lamp (bulb)

Chemical energy is energy stored in certain substances because of their chemical makeup. When these substances are burned, the energy is released. Coal, oil, propane gas, and foods are examples of substances that contain chemical energy.

Nuclear energy is the energy stored within the nucleus (center) of an atom. This energy can be released by joining atoms together or by splitting atoms apart.

Heat energy is produced by the molecular motion of matter. All matter contains heat energy. Heat energy is produced by rubbing your hands together, burning a match, or burning fuel oil in a home heating system.

Electrical energy is produced by the flow of electrons through a conductor, such as a wire. Computers, lightbulbs, and washing machines are all operated with electrical energy. A generator produces electrical energy.

Light is a form of radiant energy that moves in waves. Light as a form of energy can be demonstrated by using a magnifying glass to burn a hole in a leaf, or using a laser beam to burn a hole in a steel plate.

EXERCISE 1

1. A flowing stream of water is an example of
   (1) potential energy
   (2) kinetic energy
   (3) nuclear energy
   (4) chemical energy

2. The best example of an object that possesses potential energy is a
   (1) piece of coal (3) rolling ball
   (2) falling rock (4) burning log

3. What type of energy is contained in gasoline?
(1) potential energy (2) kinetic energy (3) sound energy (4) mechanical energy

4. In the diagram below, a skier is about to start a slide (position A), ski down the hill (B), and stop at the bottom (C). At which position would the skier have the most kinetic energy?

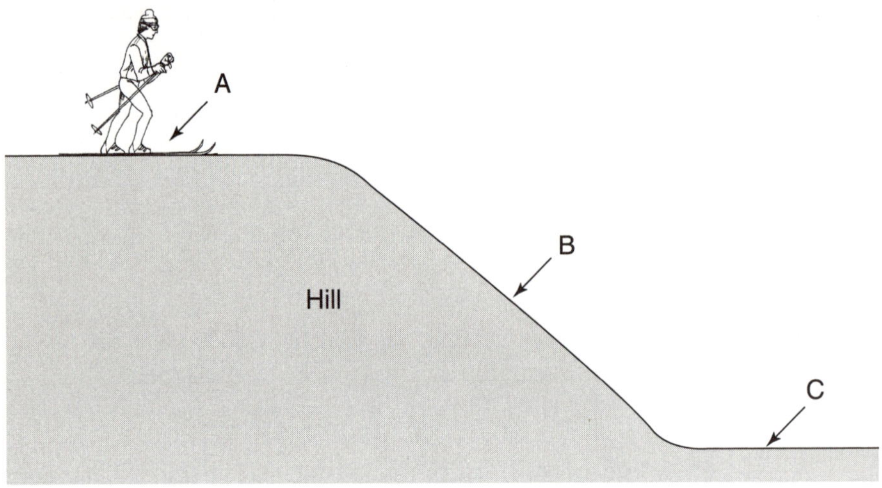

(1) A (2) B (3) C

Questions 5 and 6 refer to the following sentences.

Angie stretches a large rubber band and then releases one of its ends. The rubber band snaps back to its original shape.

5. When the rubber band was stretched, it contained
(1) potential energy
(2) kinetic energy
(3) no energy
(4) motion energy

6. As the rubber band was released and snapped back, it contained
(1) potential energy
(2) kinetic energy
(3) no energy
(4) position energy

7. Three items that represent potential chemical energy are
(1) lightbulb, oil, wood
(2) lightbulb, motor, wood
(3) battery, coiled spring, motor
(4) coal, oil, wood

Questions 8 and 9 refer to the following sentences and the list of items below.

Energy is the ability to do work. Each of the items listed below produces energy to do work around the house.

toaster	can opener
house fan	hair dryer
vacuum cleaner	lawn mower
leaf blower	chain saw
telephone	stove

8. Items that produce mechanical energy to do work are the
(1) toaster, house fan, telephone
(2) telephone, lawn mower
(3) house fan, telephone, can opener
(4) can opener, vacuum cleaner, chain saw

9. Items that use heat energy to do work are the
(1) toaster, hair dryer, stove
(2) toaster, leaf blower, stove
(3) toaster, vacuum cleaner, lawn mower
(4) can opener, vacuum cleaner, stove

Nature of Energy Transformation

Law of Conservation of Energy

The ***Law of Conservation of Energy*** states that **energy can neither be created nor destroyed** but can only be transformed from one type of energy into one or more other types of energy. When electricity passes through a lightbulb, not all of the electricity is converted into light energy; some of the electricity is converted into heat energy. Nevertheless, the total amount of energy entering the lightbulb in the form of electricity is equal to the total amount of energy being released from the lightbulb in the form of light and heat.

Energy Transformations

Energy can be changed, or ***transformed,*** from one form into another. For instance, when you take a bus to school, chemical energy in gasoline is changed into mechanical energy that turns the wheels of the bus. At school, when the bell rings between classes, electrical energy is transformed into sound energy. And at night, when you turn on a reading light, electrical energy is changed into light energy. Figure 10-2 shows two common energy transformations.

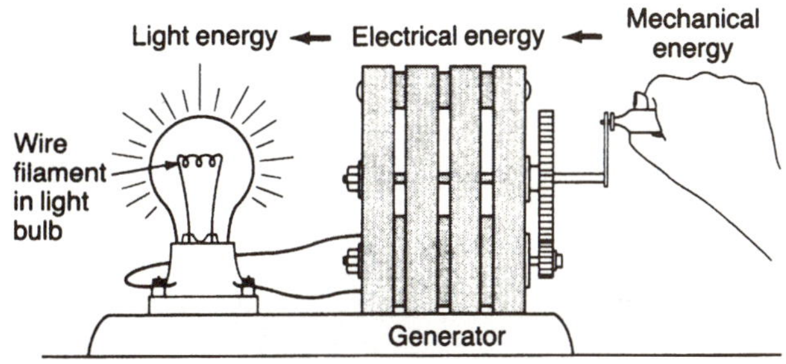

Figure 10-2. This hand-operated generator transforms mechanical energy into electrical energy, which is then transformed into light energy.

Unwanted Energy

Sometimes, when people transform energy for some purpose, an unwanted form of energy is produced. For example, a car's motor is designed to change chemical energy into mechanical energy. However, a running motor eventually becomes hot, due to the friction of the motor's parts rubbing against each other. In other words, some of the original chemical energy is transformed into unwanted heat energy instead of mechanical energy.

When we burn wood in a fireplace, we change the chemical energy in the wood into heat energy for the purpose of heating the room. Although most of the energy produced is in the form of heat, some energy is also released in the form of light energy.

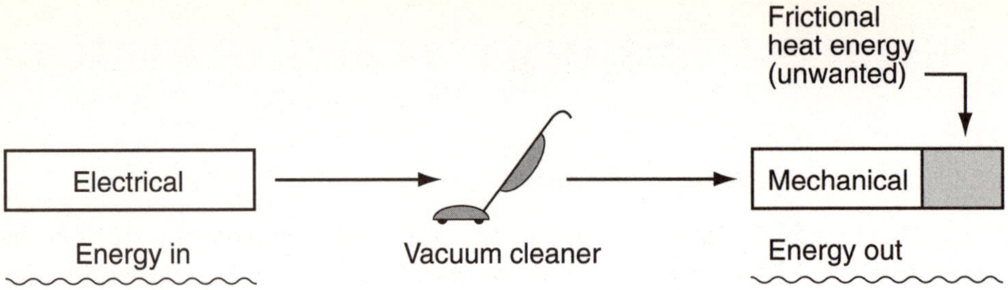

Figure 10-3. Law of Conservation of Energy: "unwanted" energy may be released during a transformation, but no energy is lost.

Sometimes this unwanted heat energy from friction in a car's engine or light energy from a burning log is referred to as "lost energy." However, it is not lost in terms of the Law of Conservation of Energy. It is just a form of energy produced that may not be useful for our purposes. This is illustrated in Figure 10-3.

EXERCISE 2

1. Which type of energy change is represented in the diagram below?

   (1) chemical energy to sound energy
   (2) sound energy to chemical energy
   (3) electrical energy to sound energy
   (4) sound energy to electrical energy

Questions 2 and 3 refer to the following sentences. In an effort to break a strip of aluminum metal, Charles bent it back and forth many times. He was surprised to find the aluminum getting warm at the point of the bend.

2. This is an example of a transformation of
   (1) heat energy to mechanical energy
   (2) chemical energy to mechanical energy
   (3) chemical energy to heat energy
   (4) mechanical energy to heat energy

3. The heating of the metal strip in this case is
   (1) a gain of energy
   (2) the creation of energy
   (3) the loss of energy
   (4) an unwanted energy transformation

4. The diagram below represents three items that transform electrical energy into mechanical energy. The three items that best fit into boxes A, B, and C are the

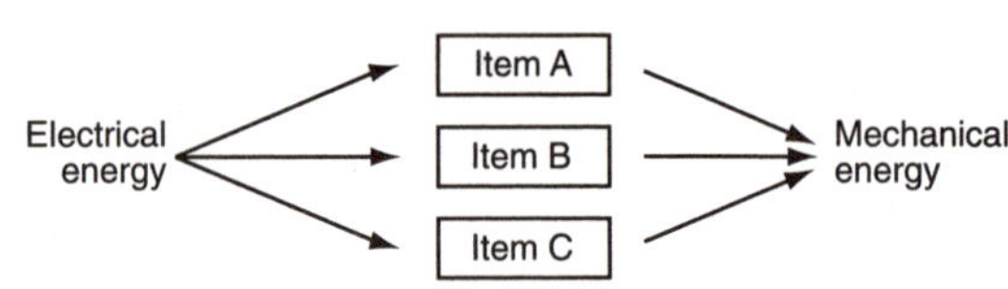

   (1) vacuum cleaner, table saw, fan
   (2) television, fan, refrigerator
   (3) radio, clock, fan
   (4) lawn mower, hair dryer, telephone

5. Many appliances in the home receive energy from a distant source. The appliances use the energy to do work. The table below shows the path of the energy.

Dam and Hydroelectric Generator ⟶	*Wires* ⟶	*Vacuum Cleaner*
Transform mechanical energy to electrical energy.	Transmit electrical energy.	Motor transforms electrical energy to mechanical energy.

If 10 joules of electrical energy are produced at the dam, by the time the energy reaches the vacuum cleaner at home and is transformed into mechanical energy,

(1) slightly more than 10 joules of energy are produced by the vacuum cleaner
(2) less than 10 joules of energy are produced by the vacuum cleaner
(3) exactly 10 joules of energy are produced by the vacuum cleaner
(4) much more than 10 joules of energy are produced by the vacuum cleaner

Questions 6–8 refer to the following paragraph and the list of items below.

Energy transformations are occurring all around you. For example, each of the following items involves an energy transformation. That is, one form of energy goes into making the item work, and another form of energy is produced by the item.

battery in a flashlight	battery in a toy car
house fan	hair dryer
lamp	lawn mower
guitar	television
telephone	radio

6. Chemical to mechanical energy transformations occur in the
(1) battery in a flashlight and battery in a toy car
(2) battery in a toy car and lamp
(3) battery in a toy car and power lawn mower
(4) battery in a flashlight and power lawn mower

7. Sound to electrical energy transformation occurs in the
(1) telephone (3) television
(2) radio (4) guitar

8. Which item transforms 100 percent of the energy it receives into useful energy?
(1) house fan
(2) battery in a flashlight
(3) hair dryer
(4) none of the items

9. Dwayne opened the hood of his car after a long drive. He felt the warmth of the engine. The heat energy released by the engine in the process of moving the car is
(1) the primary energy output expected from a car engine
(2) the source of energy to make the car move
(3) produced by friction between the wheels and the road
(4) produced as unwanted energy in the energy transformation process

10. Hitting a nail with a hammer involves using mechanical energy (swinging the hammer) to produce mechanical energy (moving the nail into the wood). As you hit the nail, you hear a banging sound, and if you touch the nail, it feels warm. The sound and heat energy generated when hitting the nail are
(1) helping to move the nail
(2) unwanted forms of energy
(3) not part of the energy transformation process
(4) newly created energy

11. Five-year-old Tawana got a toy wind-up dog for her birthday. Turning the key on the side of the toy made a metal spring tightly coil inside the dog. When the toy dog was released, it walked across the floor. The motion of the dog was produced by
    (1) heat energy transformed into mechanical energy
    (2) chemical energy transformed into mechanical energy
    (3) mechanical energy transformed into mechanical energy
    (4) mechanical energy transformed into sound energy

Questions 12–14 refer to the following sentence and diagram.

Lighting matches on a matchbook illustrates several different energy transformations.

12. When you strike a match on the rough strip of a matchbook cover, you transform
    (1) mechanical energy to heat energy
    (2) mechanical energy to chemical energy
    (3) mechanical energy to light energy
    (4) heat energy to chemical energy

13. The burning match involves the transformation of
    (1) chemical energy to light and sound energy
    (2) chemical energy to light and heat energy
    (3) chemical energy to heat energy only
    (4) heat energy to light energy only

14. The matches in the matchbook contain
    (1) potential chemical energy
    (2) potential light energy
    (3) potential heat energy
    (4) kinetic heat energy

Questions 15 and 16 refer to the following paragraph and diagrams.

Four car engines were tested for their efficiency in transforming the chemical energy in gasoline into mechanical energy to move a car. The bar graphs below show the relative amounts of mechanical energy and wasted frictional heat energy produced by each engine.

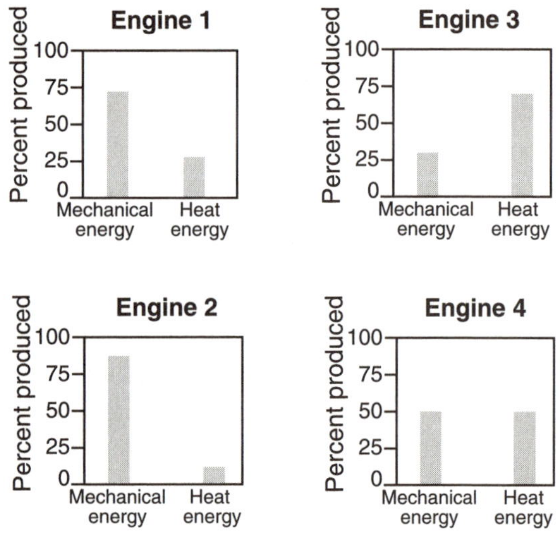

15. Which engine was the least wasteful in its use of energy?
    (1) Engine 1 (3) Engine 3
    (2) Engine 2 (4) Engine 4

16. Which engine generated the most wasted heat energy?
    (1) Engine 1 (3) Engine 3
    (2) Engine 2 (4) Engine 4

17. In two or more sentences, explain how the energy of moving water can be converted into light energy.

11 Sound and Light

Points to Remember —

- Sound is a form of energy produced by a vibrating object. Sound travels in concentric waves of compressed particles.
- Sound waves travel through different media at different speeds. The denser the medium, the faster the sound wave speed. Sound waves cannot travel through a vacuum.
- The ear is an organ that transforms sound vibrations into nerve impulses, which are sent to the brain for interpretation.
- Light is a visible form of energy that travels in waves. Light waves are the visible portion of a larger group of energy waves that make up the electromagnetic spectrum.
- Light travels at about 300,000 kilometers per second. When light strikes an object, it may be reflected, absorbed, or transmitted through the object.
- The eye is an organ that captures light waves and transforms them into nerve impulses, which are sent to the brain for interpretation.

Sound

Sound Energy

Sound is a form of energy produced by a vibrating object. When an object vibrates, it moves rapidly back and forth. This motion pushes and pulls the surrounding air, producing alternating compressed and expanded layers of air particles called ***sound waves*** (Figure 11-1). These sound waves spread outward in all directions from their source, somewhat like the circular ripples that are produced when you toss a pebble into a calm pool of water.

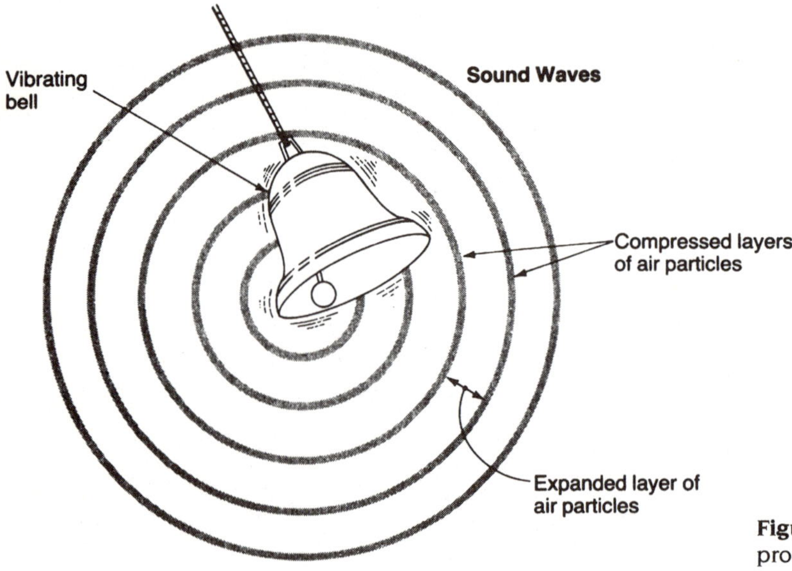

Figure 11-1. A vibrating object produces sound waves.

Objects that can produce sound include bells, radio speakers, guitar strings, or any other thing that can vibrate. For instance, the sound of your voice is caused by vibrating vocal cords in your throat. If you place your hand on your throat while you speak, you can feel the vibrations that produce the sound.

Sound waves can travel only through *matter*, whether in the form of a solid, a liquid, or a gas. Sound cannot travel through a *vacuum*, an area that contains no matter. The substance that sound travels through is called its *medium*.

Sound Waves

Sound waves can be represented by a wavy line, as shown in Figure 11-2. The crest, or top, of each wave represents the compressed particle portion of the wave, and the trough, or bottom, of each wave represents the expanded (rarefaction) particle portion. The distance from one point in a wave to a corresponding point in the next wave is a wavelength. The distance between a midline drawn through a wave and the crest or trough of the wave is the amplitude of the wave. The larger the amplitude, the louder the sound.

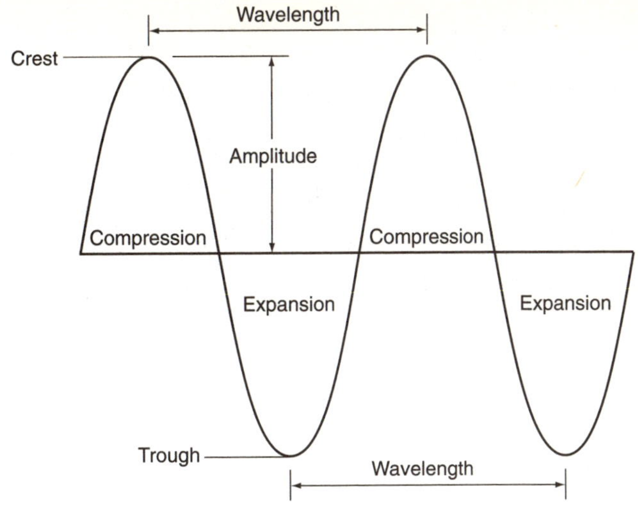

Figure 11-2. A representation of sound waves.

The number of waves that pass a point in a given amount of time is called the wave frequency. The pitch describes how high or low the sound is. A high-frequency sound has a high pitch, and a low-frequency sound has a low pitch. A violin produces a high-pitch sound, and a bass drum produces a low-pitch sound.

To a motionless observer, the frequency of sound waves produced by a moving train or ambulance increases or decreases, depending on whether the sound source is moving toward or away from the listener. This change in sound frequency is recognized as a change in pitch and is called the Doppler effect. When an ambulance siren is moving toward you, the sound waves are crowded together, producing a higher frequency and a higher pitch. When the siren is moving away from you, the sound waves spread out, producing a lower frequency and a lower pitch (see Figure 11-3).

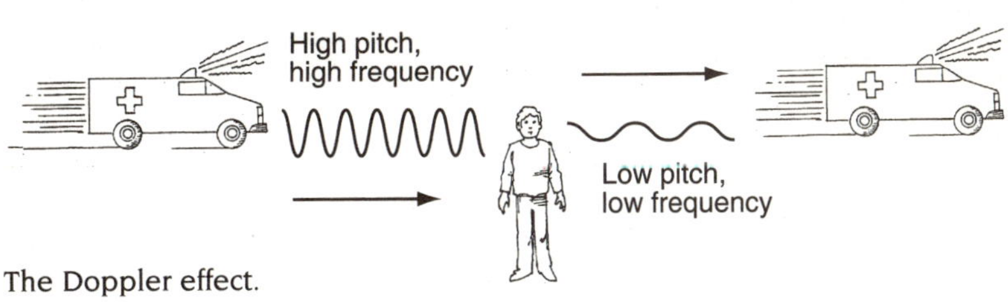

Figure 11-3. The Doppler effect.

The Speed of Sound

The speed of sound depends mostly on the density of the substance, or medium, through which it is passing. The more dense the medium is, the faster the sound waves can travel through it. Generally, sound travels fastest through solids, which have the greatest density, and slowest through gases, which have the least. Table 11-1 gives the speed of sound through several substances.

Although the speed of sound can vary, it is always much slower than the speed of light. During a thunderstorm, for instance, a lightning bolt produces a flash of light and a clap of thunder at the same time. The speed of light is so fast that the light reaches us almost instantly. The sound of the thunder travels much more slowly, so we usually hear the thunder after a pause of several seconds.

Table 11-1. Speed of Sound Through Different Substances (at 25°C)

Medium	*State (phase)*	*Speed (m/sec)*
Iron	Solid	5200
Glass	Solid	4540
Water	Liquid	1497

Sensation of Sound

The human ear has the ability to capture sound waves in the air, process the sound waves, and send them to the human brain for interpretation. As shown in Figure 11-4, the ear consists of three parts: (1) outer ear, (2) middle ear, and (3) inner ear.

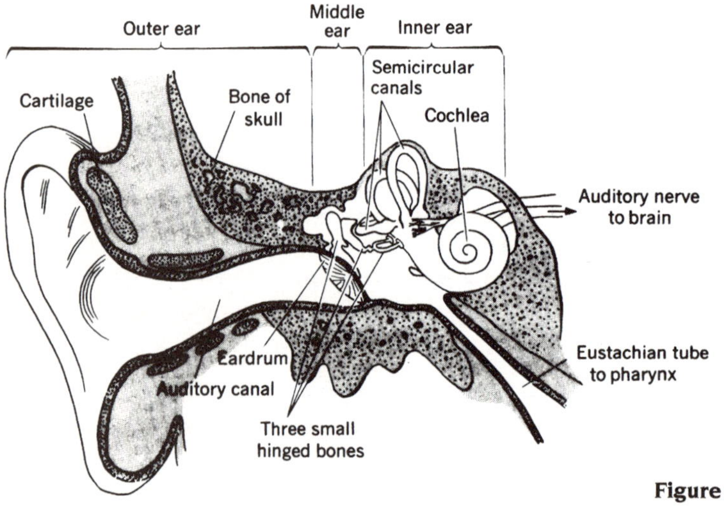

Figure 11-4. Structure of the human ear.

1. The ***outer ear*** contains a funnel-shaped organ that captures sound waves and directs them to the auditory canal. This canal leads to the eardrum, also called the tympanic membrane. When the sound waves strike the membrane, they cause it to vibrate.

2. Within the ***middle ear*** are three tiny bones that are connected like links in a chain. These bones are called the hammer, anvil, and stirrup, because their shapes resemble these objects. The bones pass vibrations from the eardrum to another membrane at the entrance to the inner ear.

3. The ***inner ear*** contains a canal called the cochlea, which is shaped like a spiral snail shell. The cochlea is filled with a liquid and contains tiny nerve endings that are connected to the auditory nerve, which leads to the brain. The vibrations enter the cochlea and are transmitted through the fluid. The nerve endings receive the impulses and send them along the auditory nerve to the brain. The brain then translates the nerve impulses into the sensation of sound and interprets the sound to give it meaning.

Process Skill

INTERPRETING A GRAPH

Sound travels at different speeds through different substances. The speed of sound is faster in solids, such as stone or metal, and slower in liquids and gases, such as water and air. The speed of sound through air is also affected by air temperature. The graph below shows the relationship between air temperature and the speed of sound. Study the graph and answer the following questions.

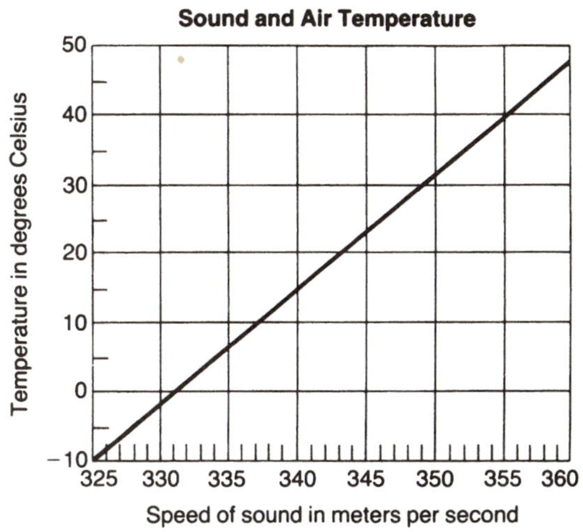

1. Sound travels at a speed of 340 meters per second at about which temperature?
   (1) 22°C (2) 10°C (3) 6°C (4) 15°C

2. At a temperature of 31°C, sound travels at about
   (1) 345 meters per second
   (2) 350 meters per second
   (3) 355 meters per second
   (4) 340 meters per second

3. What does the graph suggest about the relationship between air temperature and the speed of sound?
   (1) As air temperature decreases, the speed of sound increases.
   (2) As air temperature increases, the speed of sound remains the same.
   (3) As air temperature increases, the speed of sound increases.
   (4) As air temperature increases, the speed of sound decreases.

Music and Noise

Two people strum guitars. The first person produces ordered sounds that are related in terms of pitch, or tone, and timing, or rhythm. The sounds have an overall unity or continuity; we call such sounds music. The second person produces random sounds without tones and rhythmic relationships, which lack an overall unity or continuity. Such sounds are termed noise.

Sound Can Be Harmful

Loud noises, especially if prolonged, can damage the hearing abilities of humans and other living things. The damage may be temporary or permanent. Many people work in places where they are exposed to frequent loud noises, such as factories with noisy machinery, construction sites with periodic explosions and loud jackhammers, and even concert halls or recording studios where loud music is played. Workers in such places should protect their hearing by putting earplugs or cotton in their ears.

EXERCISE 1

1. The bar graph below shows the average speed of sound through solids, liquids, gases, and a vacuum. Based on the graph, through which medium would sound travel fastest?

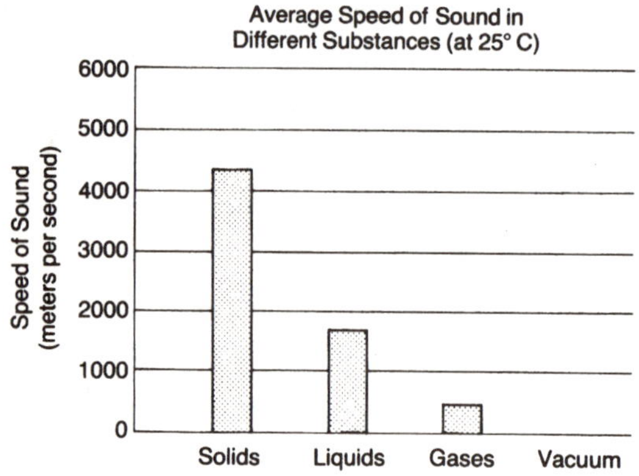

(1) air (3) water
(2) rock (4) outer space

2. Sound waves can travel through
(1) water only (3) all matter
(2) air only (4) a vacuum

3. Sound travels in waves that can be represented in the diagram below. The top of a wave is called the crest, and the distance from crest to crest is called the wavelength. The diagram shows

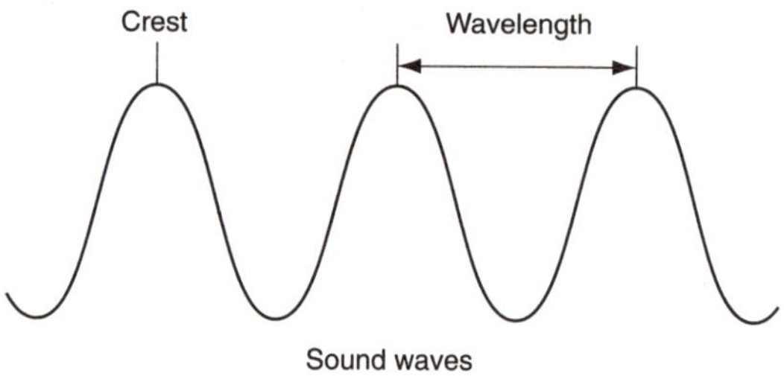

(1) one wave (3) three waves
(2) two waves (4) an incomplete wave

Read the following short paragraph and study the figure above. Then answer questions 4–6.

Sound travels about 340 meters/second in air. Josh yelled "hello" across a canyon. The sound traveled across the canyon and returned as an echo. Josh heard the echo two seconds after he yelled "hello."

4. The distance the sound traveled was about
   (1) 340 meters (3) 1360 meters
   (2) 680 meters (4) 170 meters

5. The distance across the canyon was about
   (1) 340 meters (3) 1360 meters
   (2) 680 meters (4) 170 meters

6. If the canyon had been 1020 meters across, Josh would have heard the echo in about
   (1) 5 seconds (3) 3 seconds
   (2) 6 seconds (4) 12 seconds

7. The piano, guitar, saxophone, and drums are all musical instruments that produce sound. All these instruments have
   (1) strings
   (2) a column of air
   (3) a surface that is struck
   (4) a vibrating object

8. When using the apparatus in the diagram below, the student could not hear the ringing bell after the air was pumped out of the bell jar. This demonstrates that sound
   (1) can travel through the glass bell jar
   (2) cannot travel through the glass bell jar
   (3) cannot travel through a vacuum
   (4) can travel through a vacuum

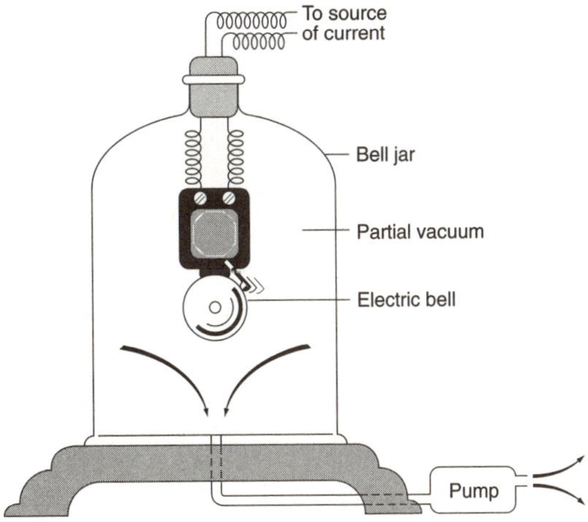

9. The amplitude of a sound wave indicates the loudness of a sound. The greater the amplitude, the louder the sound. The wave that represents the loudest sound is

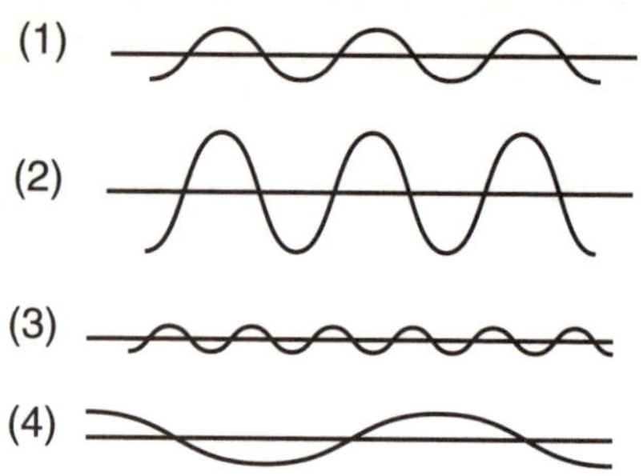

10. A passing train blows a whistle at a constant pitch as it approaches, crosses, and leaves a road crossing. A person standing by the road crossing hears the pitch of the whistle rise as the train approaches and then get lower as the train passes by. This is caused by the
(1) wavelength
(2) Doppler effect
(3) amplitude of the sound
(4) different whistles on the train

Read the following sentences and study the diagram, then answer questions 11 and 12.

A police car's siren is at full blast as the car races down the street to an emergency. The sound wave of the siren when the car is standing still looks like this:

11. The sound wave of the siren when the car is approaching you would look like

Frequency of four waves approaching

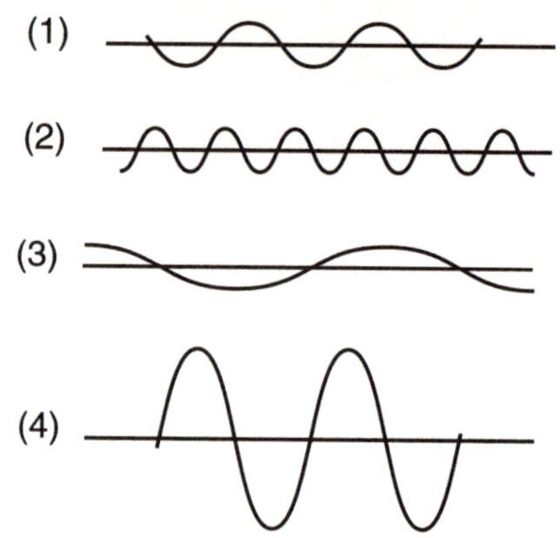

12. The sound wave of the siren after the car has passed you would look like

Frequency of four waves retreating

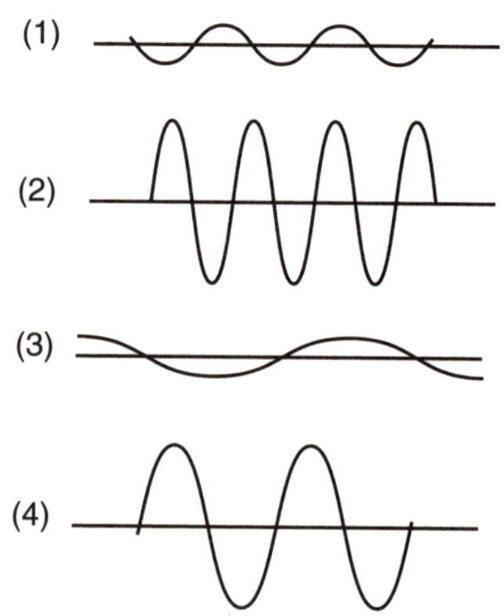

Light

Light Energy

Light is a visible form of energy. Like sound, light travels in waves that move outward in all directions from its source. But unlike sound, light waves move in straight paths called rays, as shown in Figure 11-5. Unlike sound waves, which can curve around corners and objects, light rays cannot curve around objects. This is why objects block out light rays and cast shadows. However, light can travel through a vacuum, something that sound cannot do. In fact, the sun's light travels through the vacuum of space to reach Earth.

The speed of light is extremely fast, about 300,000 kilometers per second. That is almost a million times faster than the speed of sound! Light that travels over distances we commonly encounter on Earth arrives in just a small fraction of a second.

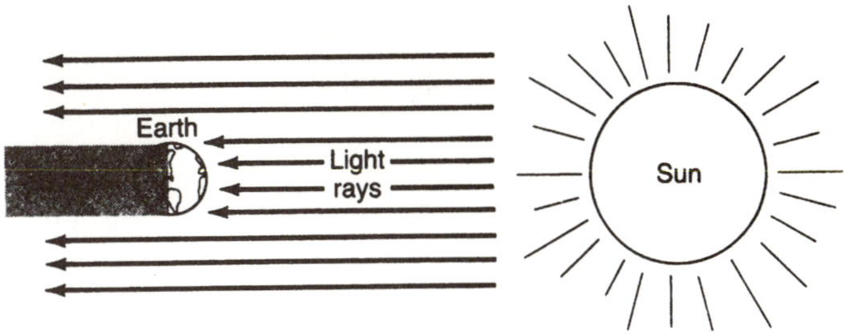

Figure 11-5. Light waves travel in straight paths.

The sun is our main source of light energy. Fire and lightning are other sources of natural light. Light can also be produced artificially, as it is in a lightbulb.

Light Can Be Reflected, Absorbed, or Transmitted

When light strikes the surface of an object, three things can happen, as shown in Figure 11-6. Some light may be bounced back, or ***reflected,*** off the surface. Some light may be ***absorbed*** as heat energy. And some light may be ***transmitted,*** passing through the object. A shiny, metal surface reflects a lot of the light that strikes it. Much of the light that strikes a blacktop road is absorbed as heat. Clear glass allows most light to be transmitted through it.

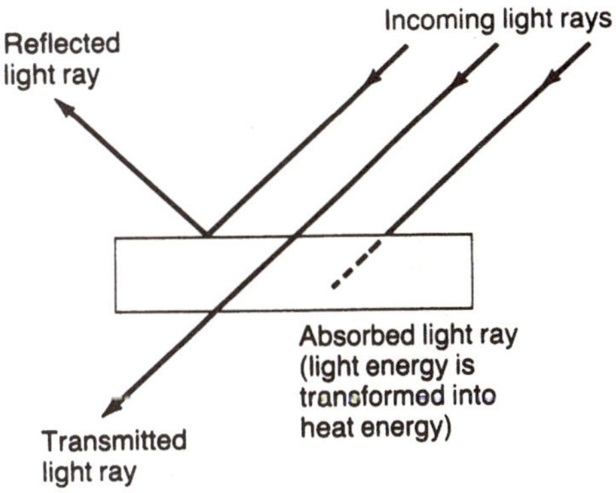

Figure 11-6. When light strikes a surface, some may be reflected, some absorbed, and some transmitted.

We see objects because their surfaces reflect light. The smoother the surface, the more accurate the reflection. A mirror gives an accurate reflection because it has a smooth, shiny surface. A wall produces a much different kind of reflection, because its rougher surface scatters the light.

Objects of different colors absorb light to varying degrees. Dark-colored objects absorb more light as heat energy than do light-colored objects, which reflect more light. For this reason, people usually wear light-colored clothing to keep cool during hot, sunny weather. Materials also differ in their ability to transmit light. *Transparent* materials, such as window glass, permit almost all of the incoming light to pass directly through them. *Translucent* materials, such as wax paper, let some light pass through, but they scatter the light rays so that images are not transmitted clearly. *Opaque* materials, like wood and iron, do not allow any light to pass through them.

Lenses

Sometimes, when light passes from one transparent substance into another, such as from air into water, the light rays are bent, or refracted. This is why a pencil in a glass of water looks broken or bent where it enters the water (see Figure 11-7). The light rays being reflected from the pencil are refracted as they pass from the water into the air. This fact has been put to use in the making of lenses.

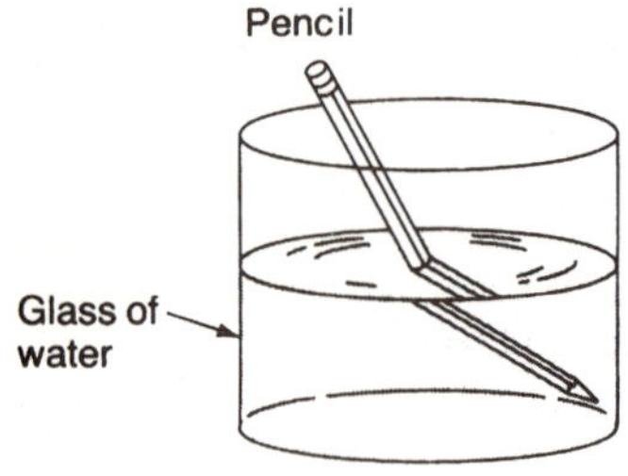

Figure 11-7. Bending of light rays makes the pencil look bent, or broken.

A lens is a piece of transparent glass or plastic that has curved surfaces. The curved surfaces refract light rays that pass through the lens. The shape of a lens determines how it bends light, as shown in Figure 11-8. A lens with surfaces that curve outward bends light rays so that they are focused in toward a common point. A lens with surfaces that curve inward bends light rays so that they spread out.

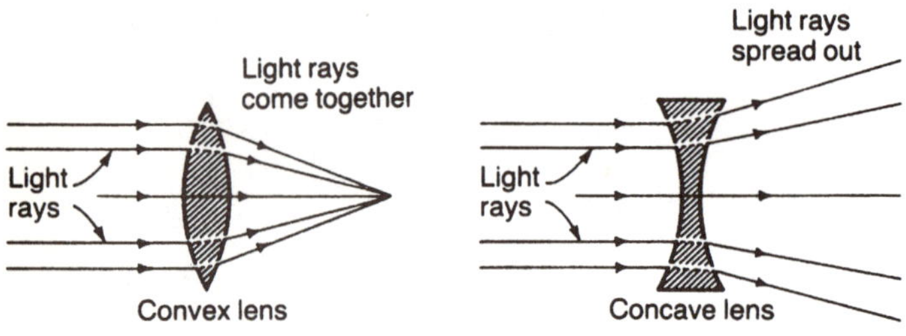

Figure 11-8. The shape of a lens determines how it bends light.

Images of objects seen through lenses may be larger, the same size, or smaller than the object itself. For instance, the lens of a camera forms smaller images of objects. A photocopy machine has a lens that forms images the same size as the original object. Binoculars contain lenses that magnify objects, making them appear larger.

The Electromagnetic Spectrum

Light waves are part of a larger family of energy waves that can travel through a vacuum at the speed of light. These are called ***electromagnetic waves.*** They include radio waves, microwaves, infrared waves, visible light,

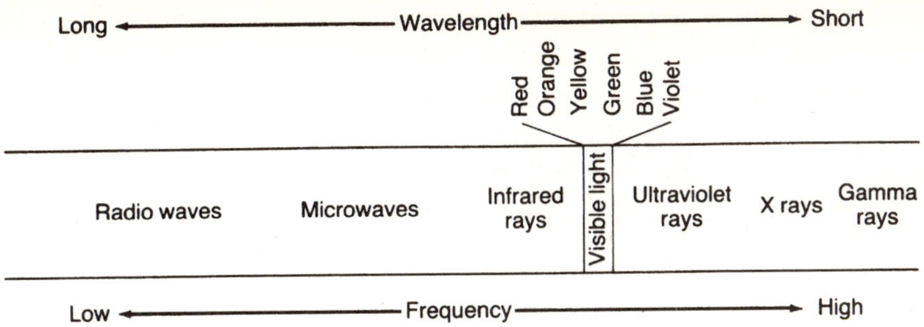

Figure 11-9. The electromagnetic spectrum.

ultraviolet rays, X rays, and gamma rays. Together, these energy waves make up the electromagnetic spectrum (see Figure 11-9).

We are all familiar with some uses of electromagnetic waves. Radio waves are used for radio and television broadcasting. Microwaves are used in communications and in microwave ovens. X rays are used to diagnose illnesses and injuries.

Ordinary white light is actually composed of a range of visible electromagnetic waves with different frequencies and wavelengths. The different wavelengths appear as the colors of the visible spectrum, from red (with the longest wavelength) to violet (with the shortest wavelength). The color that you see when you look at an object, for example, a green plant, is actually the wavelength of light that is reflected by that object; the other visible wavelengths are absorbed by it.

A laser light is an intense beam of light produced by a single wavelength of visible electromagnetic energy. Because the waves travel aligned, the laser light can travel great distances as a concentrated light beam. Lasers can be used for delicate eye operations, or for cutting, drilling, and welding materials in industry.

Overexposure to electromagnetic waves can be harmful to living things. We should be especially careful to limit our exposure to certain types of electromagnetic radiation, such as X rays and ultraviolet rays, which have been linked to genetic mutations and cancer.

Sensation of Sight

Light waves make up the visible portion of the electromagnetic spectrum. The human eye has the ability to capture light waves, process the light waves, and send them to the brain for interpretation. Light rays pass

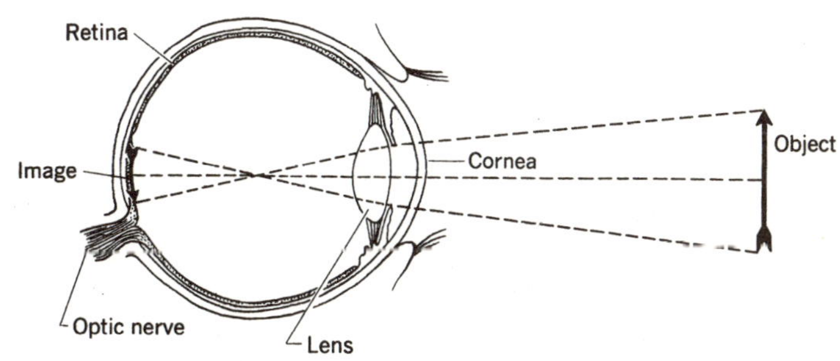

Figure 11-10. Light rays pass through the cornea and are focused by the lens. The image strikes the light-sensitive retina at the back of the eye.

into and through the eye much as light enters a camera and strikes the light-sensitive film in the camera. (See Figure 11-10.)

The path of light into the eye takes it first through the cornea, a protective membrane around the eye. The light then enters the black opening into the eye, the pupil. Surrounding the pupil is the iris, which gives your eye its color. Light passes through the lens, which focuses the image by

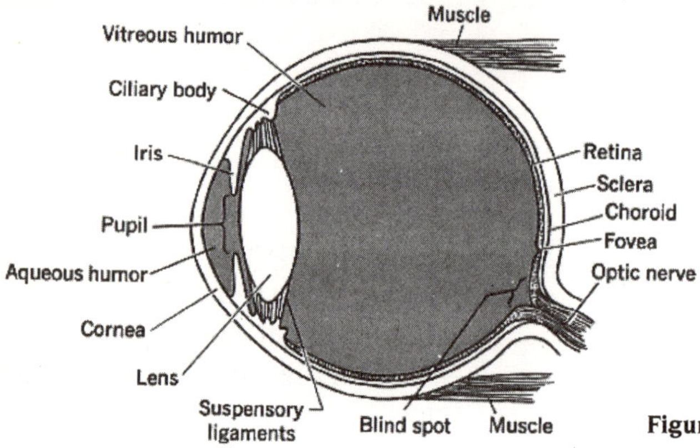

Figure 11-11. Structure of the human eye.

using tiny muscles surrounding the lens to change its thickness. After passing through a fluid that fills the eye, the light rays strike the retina, a sensitive membrane in the back of the eye. Tiny nerve endings in the retina send the sensation of light to the optic nerve, which sends the image to the brain for interpretation. (See Figure 11-11.)

EXERCISE 2

1. A blast of dynamite set off by a work crew produced a bright flash of light and a loud explosion. A person standing 2 kilometers away, with a clear view of the work site, would
   (1) hear the sound first, then see the flash
   (2) see the flash first, then hear the sound
   (3) see the flash and hear the sound at the same time
   (4) hear the sound but not see the flash

2. Compared to the speed of light, the speed of sound is
   (1) much faster (3) the same
   (2) much slower (4) a little slower

3. Lenses can produce images that are
   (1) larger than the original object
   (2) the same size as the original object
   (3) smaller than the original object
   (4) all of the above

4. We can see objects mainly because they
   (1) bend light (3) absorb light
   (2) reflect light (4) transmit light

5. When using the apparatus shown in the diagram below, the student could see the flame only if all three holes were lined up. What property of light does this demonstrate?

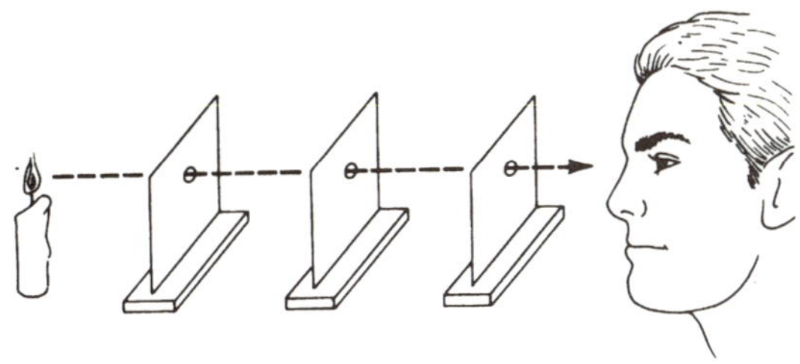

(1) Light rays are reflected from smooth, shiny surfaces.
(2) Light rays are absorbed as heat by dark-colored surfaces.
(3) Light rays travel in straight paths.
(4) Transparent objects transmit most of the light that strikes them.

6. The accompanying diagram shows three ways that light can behave when striking a sheet of colored glass. The light rays at location C have been

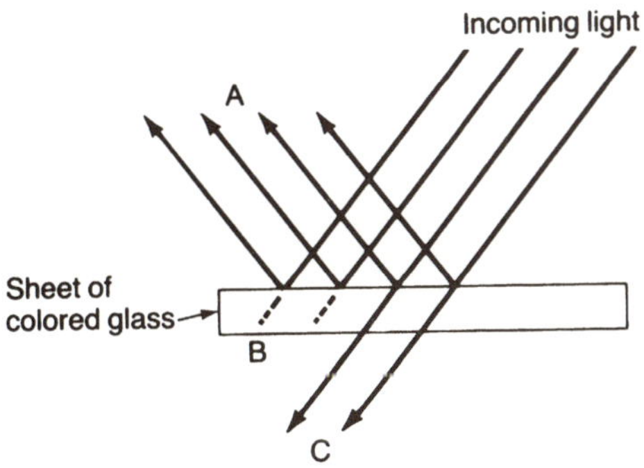

(1) reflected by the glass
(2) transmitted by the glass
(3) absorbed by the glass
(4) blocked out by the glass

7. X rays are often used for
(1) radio and television broadcasting
(2) communications and cooking food
(3) diagnosing illnesses and injuries
(4) all of the above

8. Overexposure to electromagnetic waves
(1) has no effect on living things
(2) always has a good effect on living things
(3) can have harmful effects on living things
(4) cannot occur in a vacuum

9. The eye functions most like which item?
(1) television
(2) camera
(3) videocassette recorder
(4) computer

Questions 10–12 refer to the following information.

Luz made a series of observations during a thunderstorm. She recorded the time difference between each flash of lightning and the thunder that followed it. The table below shows her findings.

Lightning Bolts	*Time Difference (seconds)*
A	12
B	9
C	7
D	3
E	5
F	8
G	10
H	15

Luz knows that the speed of light over relatively short distances is almost instantaneous (taking a small fraction of a second) and that sound travels about 340 meters/second in air. She estimated that for every 5-second difference in time between the flash of lightning and the thunder, the lightning bolt was about 1.6 kilometers away. From the information in the table, Luz could draw some conclusions.

10. The lightning bolt that was closest to Luz was
(1) B (3) D
(2) C (4) H

11. The lightning bolt that was about 1.6 kilometers from Luz was

(1) C	(3) G
(2) E	(4) H

12. From the trend of the data, Luz concluded that the thunderstorm
    (1) moved away from her and then toward her
    (2) moved toward her and then away from her
    (3) was continuously moving away from her
    (4) was continuously moving toward her

13. When white light strikes a green wall, all of the light is
    (1) absorbed
    (2) absorbed except green light, which is reflected
    (3) reflected
    (4) reflected except green light, which is absorbed

14. In one or two sentences, tell why plant leaves appear green.

15. Use one or two sentences to tell why sound travels faster through metal than through water.

16. In one or two sentences, explain how two children can talk to each other over a distance—without shouting—by using two metal cans and a taut string.

17. Suppose a star exploded in space. In one or two sentences, explain why you could see the explosion, but not hear it.

12 Matter

Points to Remember —

- Matter is anything that has mass and occupies space.
- Matter can exist as a solid, liquid, or gas.
- Changes in matter may be physical or chemical.
- During a chemical change, new substances are formed.
- Elements are substances that cannot be broken down into simpler substances.
- Two or more elements can combine to form a compound.
- Atoms are composed of protons, neutrons, and electrons.
- Matter can neither be created nor destroyed in a chemical reaction.
- Energy can either be absorbed or released during a chemical change.
- Factors such as temperature and particle size influence the rate of a chemical reaction.

Matter

Defining Matter

Look around you. The objects you see, such as this book, your desk and chair, and the walls and ceiling, are all composed of *matter.* The air that surrounds you, which is a mixture of gases, is also made of matter. In fact, every solid, liquid, and gas is a form of matter.

Matter is defined as anything that has *mass* and takes up space. ***Mass*** is the total amount of material in an object. We measure mass with an equal-arm balance, as shown in Figure 12-1. Notice that a balloon filled with air has a greater mass than an empty balloon, since air has mass. The amount of space an object occupies is called its ***volume.*** The air in the filled balloon in Figure 12-1 takes up space, giving the balloon a greater volume than the empty balloon.

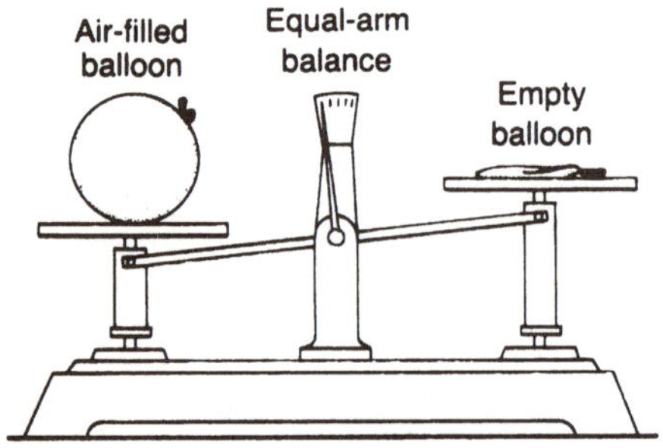

Figure 12-1. The air-filled balloon is heavier and takes up more space than the empty balloon because air is matter.

Is there anything that is not made of matter, that has no mass and takes up no space? Figure 12-2 shows that shining a light on a balance has no effect on the balance. This is because light is a form of energy. Energy is not matter, since it has no mass and no volume. Some other forms of energy are heat and sound.

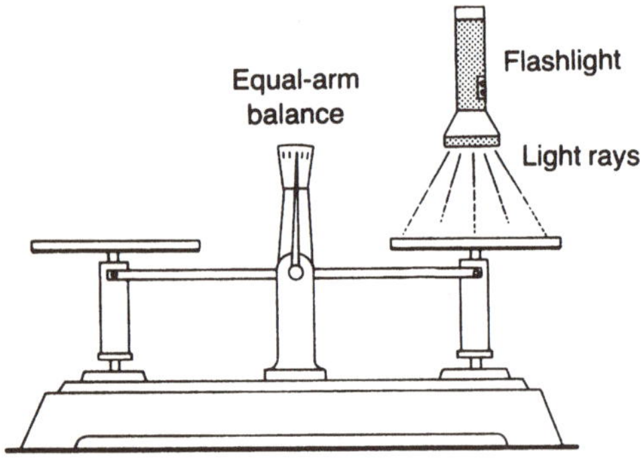

Figure 12-2. The balance is unaffected by the light shining on it, because light is not matter.

Elements

The basic building blocks of matter are called ***elements.*** All substances are made up of one or more elements. Oxygen, hydrogen, gold, and iron are examples of elements. Each element is represented by a symbol made up of one or two letters. For example, the symbol for hydrogen is H, oxygen is O, and gold is Au. There are 110 known elements. (Note: The number of elements changes as new elements are created in research laboratories.) However, less than half of them occur commonly in nature. Table 12-1 lists the most common elements found in Earth's crust.

Table 12-1. Most Common Elements in Earth's Crust

Element	*Chemical Symbol*
Oxygen	O
Silicon	Si
Aluminum	Al
Iron	Fe
Calcium	Ca
Sodium	Na
Potassium	K
Magnesium	Mg

The smallest particle of an element that has the properties of that element is called an ***atom.*** All atoms of a particular element are alike, but they are different from the atoms of any other element. For instance, all hydrogen atoms are alike, but they differ from oxygen atoms. Since there are 110 elements, there are 110 different kinds of atoms.

Atomic Structure

All atoms are composed of smaller *subatomic* particles. These particles are called *protons, neutrons,* and *electrons.* Protons, neutrons, and electrons differ in their mass, electrical charge, and location in the atom. Protons and neutrons have roughly the same mass, while electrons are much lighter. Protons have a positive (+) charge, and electrons have a negative (–) charge. Neutrons have no electrical charge; they are electrically neutral.

Protons and neutrons are found in the center, or *nucleus,* of the atom. Electrons orbit the nucleus, moving very rapidly. Table 12-2 summarizes the properties of the subatomic particles.

Table 12-2. Properties of the Subatomic Particles

Particle	*Mass (AMU)**	*Charge*	*Location*
Proton	1	+	Nucleus
Neutron	1	0	Nucleus
Electron	0.00054	–	Outside the nucleus

*The Atomic Mass Unit (AMU) is a special unit created for measuring the mass of very small particles.

The atoms of different elements differ in the number of protons in their nucleus. Oxygen has 8 protons, while carbon has 6, and uranium has 92. The number of protons in the nucleus is called the *atomic number.* The periodic table of the elements, discussed in Chapter 1, arranges the elements according to their atomic number.

Compounds

Millions of different substances are known to scientists. How is this possible if there are only 110 elements? Elements can combine to form new substances. A substance that is formed when two or more different elements combine is called a *compound*. Since many different combinations of elements are possible, many different compounds can exist. The common substance water is a compound formed when the elements hydrogen and oxygen combine.

A compound is represented by a chemical formula that indicates which elements have combined, and in what proportions. The chemical formula for water, H_2O, indicates that water contains two atoms of hydrogen to every atom of oxygen. Table 12-3 lists some common compounds and their chemical formulas.

Table 12-3. Some Common Compounds and Their Chemical Formulas

Compound	*Formula*	*Elements*
Table salt	$NaCl$	Sodium, Chlorine
Water	H_2O	Hydrogen, Oxygen
Sugar (sucrose)	$C_{12}H_{22}O_{11}$	Carbon, Hydrogen, Oxygen
Quartz	SiO_2	Silicon, Oxygen
Ammonia	NH_3	Nitrogen, Hydrogen

The smallest particle of a compound is called a ***molecule.*** A water molecule is composed of two hydrogen atoms and one oxygen atom, as shown in Figure 12-3. Atoms of the same element can also combine to form molecules. For example, two oxygen atoms combine to form a molecule of oxygen gas, O_2.

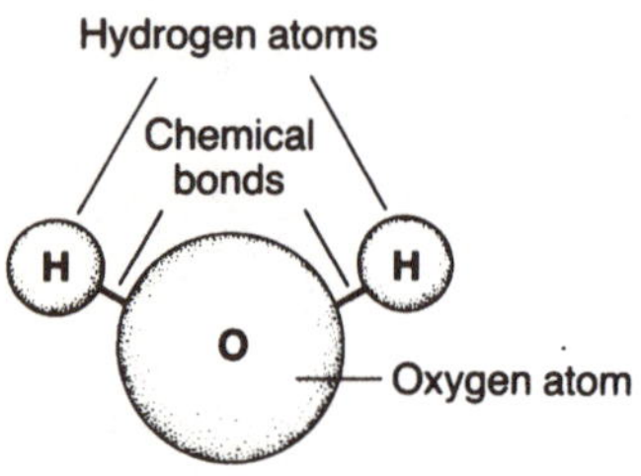

Figure 12-3. The arrangement of atoms in a molecule of water.

Atoms and molecules are extremely small. To get an idea of just how small, consider that 1 teaspoonful of water contains about 175 *sextillion* water molecules. (That would be written as 175 followed by 21 zeros!)

Chemical Bonds

Atoms in a molecule are joined together by a special link called a ***chemical bond*** (see Figure 12-3 above). These bonds contain chemical energy. Sometimes this energy can be released by a chemical reaction. Burning is one type of ***chemical reaction*** that releases energy. When wood is burned, energy stored in the chemical bonds within the wood is released as heat and light. Respiration is another chemical reaction that releases energy from bonds.

Mixtures

When two or more materials are put together and do not form a new substance, a mixture has been formed. Salt water, for example, is a mixture of salt and water. Sand is a mixture of minerals. Blood is a mixture of different cells, water, and other nutrients. Air is a mixture of several gases.

Unlike compounds, mixtures cannot be represented with a chemical formula. Table salt—a compound—is always NaCl. However, salt water—a mixture—can be more or less salty and still be salt water. For example, salt in Utah is exactly the same as salt in New York, but the salt water in the Great Salt Lake in Utah is quite different from the salt water in the Atlantic Ocean off New York.

Physical Properties

Have you ever mistaken salt for sugar? To the eye, they look very much alike. How might you tell them apart? Scientists faced with similar problems identify substances by examining their *properties*.

A difference in taste helps you distinguish salt from sugar. A difference in color (as well as taste) helps you distinguish salt from pepper. Taste and color are ***physical properties***—properties that can be determined without changing the identity of a substance. All substances have unique physical properties by which they can be identified. Table 12-4 lists some physical properties often used for this purpose.

Table 12-4. Examples of Physical Properties of Substances

Property	*Example*
Phase	Mercury is a liquid at room temperature.
Color	Sulfur is yellow.
Odor	Hydrogen sulfide smells like rotten eggs.
Density	Lead is much denser than aluminum.
Solubility	Salt dissolves in water.
Melting point	Ice melts at 0°C.
Boiling point	Water boils at 100°C.

Phases

One obvious physical property of a substance is whether it is a solid, a liquid, or a gas. These three forms of matter are called ***phases***. The phase of a substance is determined by the arrangement and motion of the molecules within it.

1. In *solids*, the molecules are close together, move relatively slowly, and remain in fixed (unchanging) positions. A solid has a definite shape and a definite volume; that is, its shape and size do not depend on the container it is in.

2. In *liquids*, the molecules are usually farther apart and faster moving than the molecules in solids. The molecules in a liquid can change position and flow past each other. A liquid has no definite shape; it takes on the shape of its container. However, liquids do have a definite volume. A given quantity of a liquid takes up the same amount of space regardless of the shape of its container.

3. In *gases*, the molecules are much farther apart and move even faster than in liquids, and they can move anywhere within their container.

A gas has no definite shape or volume but expands or contracts to fill whatever container it is in. Figure 12-4 shows how molecules are typically arranged in solids, liquids, and gases.

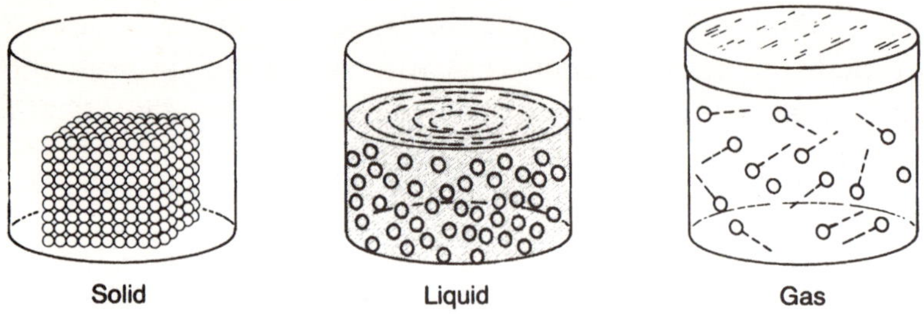

Figure 12-4. The three phases of matter: solid, liquid, and gas.

Changes in Phase

Since the phase of a substance depends on the arrangement of its molecules, a change in this arrangement can bring about a change in phase.

1. To change a solid into a liquid, the molecules must generally be moved farther apart, out of their fixed positions. This is called ***melting.*** Energy must be added to a substance to separate its molecules, so energy is absorbed during melting.
2. ***Freezing*** is the opposite of melting. When a liquid freezes into a solid, the molecules come together and bond more tightly into fixed positions. This process releases energy.
3. Changing a liquid into a gas, by ***boiling*** or ***evaporation,*** requires that the molecules of the liquid be separated even farther. Energy is therefore absorbed when a liquid changes into a gas.
4. The change from a gas to a liquid is called ***condensation.*** During condensation, molecules of a gas move closer together to form a liquid, and energy is released. Figure 12-5 illustrates the energy changes associated with changes in phase.

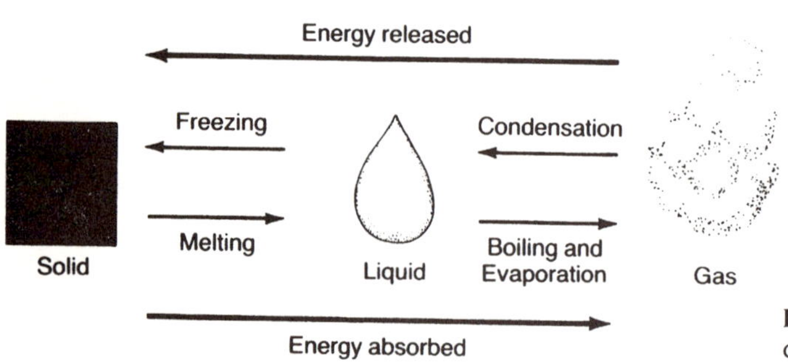

Figure 12-5. Energy changes occur during changes in phase.

For each substance, the change in phase from a solid to a liquid occurs at a particular temperature called its ***melting point.*** The melting point of ice, the solid form of water, is 0°C, or 32°F. The temperature at which a liquid freezes into a solid is called its ***freezing point.*** The freezing point of water is 0°C, or 32°F. The freezing point and melting point of a substance are always the same.

The temperature at which a liquid boils and changes rapidly into a gas is called its ***boiling point.*** The boiling point of water is 100°C, or 212°F. This is also the temperature at which cooling water vapor begins to condense into liquid water.

While a substance is changing phase, its temperature remains constant. For example, while you are boiling water, the temperature remains at 100°C even though you are constantly supplying heat. The heat is used to cause the phase change rather than a change in temperature.

Process Skill

INTERPRETING DATA IN A TABLE

By using melting point and boiling point information, we can determine what phase a substance will be in at a given temperature. If the temperature of a substance is below its melting point, the substance is a solid. If the temperature is above its boiling point, the substance is a gas. If the temperature is between its melting and boiling points, the substance is a liquid. For example, at room temperature (20°C), water is a liquid, because 20°C is between the melting point and boiling point of water.

The table below lists the melting points and boiling points of some common substances. What phase would table salt be in at a temperature of 1000°C? Since 1000°C is above the melting point of salt but below its boiling point, table salt would be a liquid at that temperature. Use the same kind of reasoning to answer the following questions.

Melting Points and Boiling Points of Some Common Substances

Substance	*Melting Point (°C)*	*Boiling Point (°C)*	*Phase at 20°C (room temperature)*
Water	0	100	Liquid
Alcohol	–117	78	Liquid
Table salt	801	1413	Solid
Oxygen	–218	–183	Gas

1. At a temperature of –190°C, oxygen is in the form of a
   (1) gas (2) liquid (3) solid

2. If alcohol is in the form of a liquid, which could *not* be its temperature?
   (1) –100°C (2) 32°C (3) 100°C (4) 77°C

3. The only substance listed that could be a liquid at a temperature of 80°C is
   (1) table salt (2) water (3) alcohol (4) oxygen

EXERCISE 1

1. What does the diagram below show about matter?

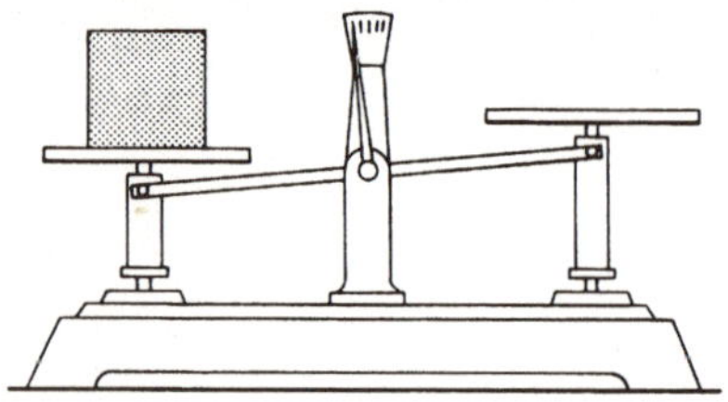

(1) Matter is made up of elements.
(2) Matter takes up space.
(3) Matter is a solid.
(4) Matter has mass.

2. Which is *not* an example of matter?
(1) water (2) air (3) gold (4) sound

3. The amount of space an object occupies is called its
(1) volume (3) weight
(2) mass (4) length

4. Atoms in a molecule are joined together by
(1) chemical bonds (3) electricity
(2) magnetism (4) gravity

5. The circles in the closed jars shown below represent particles of matter. Which jar most likely contains a solid?

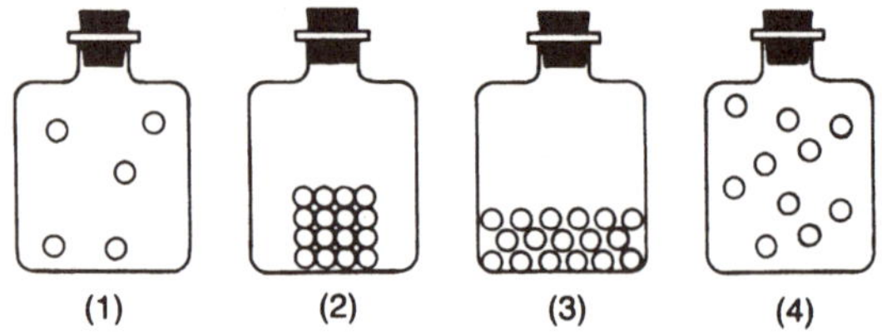

6. In which phase of matter are the particles farthest apart and moving the fastest?
(1) solid (2) liquid (3) gas

7. When you pour water from a beaker into a flask, there is a change in its
(1) mass (3) shape
(2) volume (4) density

8. Why are atoms electrically neutral?
(1) Subatomic particles have no charge.
(2) The number of negative electrons equals the number of positive protons.
(3) Atoms contain only neutrons, which have no charge.
(4) Protons are found in the nucleus, while electrons are found outside the nucleus.

9. Condensation refers to a phase change from
(1) solid to liquid (3) liquid to gas
(2) liquid to solid (4) gas to liquid

10. The temperature at which a substance melts is the same temperature at which it
(1) boils (3) condenses
(2) freezes (4) evaporates

11. In two or more sentences, explain the differences between a mixture and a compound. Give two examples of each.

12. A student opens a bottle of perfume in the back of the classroom. Although you do not see the bottle being opened, you become aware of the scent. In one or two sentences, tell how this effect happens.

Changes in Matter

Physical Changes

As you know, the chemical formula for water is H_2O because each water molecule is made up of two atoms of hydrogen and one atom of oxygen. What is the formula for ice? When water freezes, the arrangement of its molecules changes, but the molecules themselves do not change. They are still H_2O. A change of phase, such as freezing or melting, does not produce any new substances. A change that does not result in the formation of any new substances is a ***physical change.*** All changes of phase are physical changes. Crushing ice cubes into small pieces is also a physical change, since both crushed ice and ice cubes are made of the same substance.

Similarly, when you dissolve sugar in water, the sugar still tastes sweet and the water is still wet. No new substances have been formed, so dissolving is a physical change. Figure 12-6 shows why boiling, melting, and dissolving are physical changes.

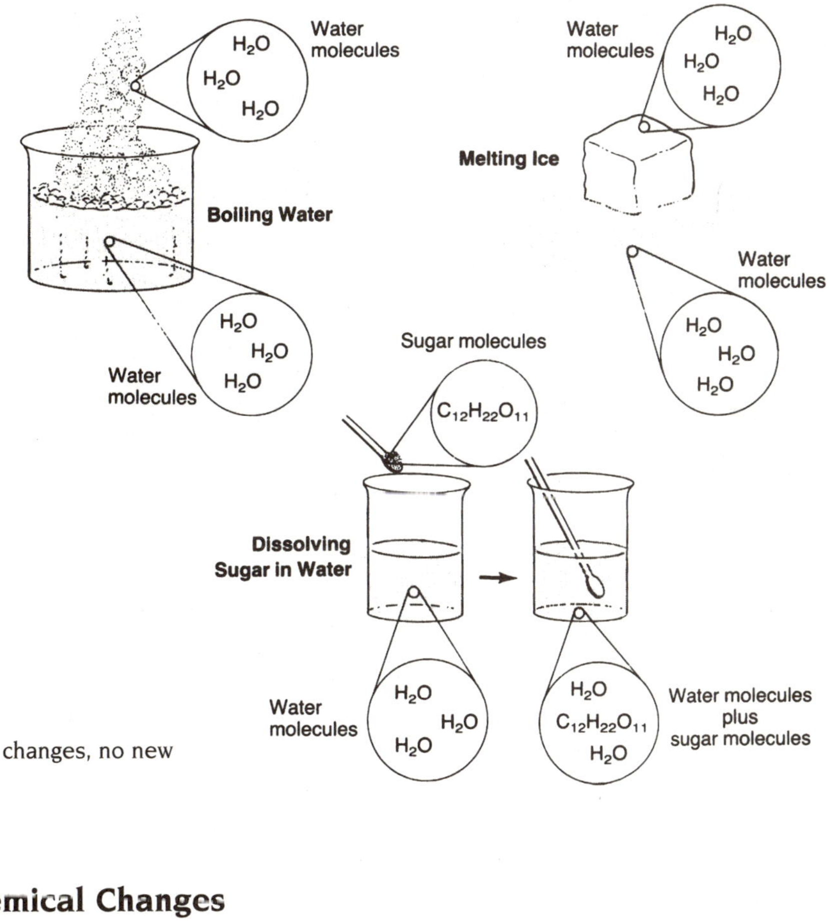

Figure 12-6. During physical changes, no new substances are formed.

Chemical Changes

What happens if you forget to put a carton of milk back into the refrigerator? First, the milk gets warm. This is a physical change. However, if you leave the milk out too long, it turns sour. The sour taste is caused by the production of a new substance called *lactic acid*. A change that produces one or more new substances is called a ***chemical change.*** When a chemical change occurs, we say there was a chemical reaction. Burning paper produces smoke and ash, both of which are new products. Burning is always a chemical change.

Forming a compound always involves a chemical change, whereas forming a mixture involves only physical changes. Similarly, it requires a chemical change to break apart a compound. Mixtures, however, can be separated through physical changes. For example, salt water can be boiled, thus leaving the salt behind; and blood can be spun in circles (centrifuged), thus separating it into its various components.

Chemical changes can be represented by chemical equations. A chemical equation uses formulas and numbers to keep track of a chemical change. The starting materials, called the *reactants*, are listed on the left side of the equation. The final materials, called the *products*, are listed on the right side. An arrow separates the two sides. The equation for the burning of coal would be written as:

$$C + O_2 \rightarrow CO_2$$

A chemist reads this equation as, "carbon plus oxygen yields carbon dioxide." In this reaction, carbon and oxygen are the reactants, and carbon dioxide is the product. Table 12-5 gives some examples of chemical changes.

Table 12-5. Examples of Chemical Changes

Chemical Change	*Reactants*	*Products*	*Equation*
Burning coal	Carbon (C) + oxygen gas (O_2)	Carbon dioxide gas (CO_2)	$C + O_2 \rightarrow CO_2$
Rusting of iron	Iron (Fe) + oxygen gas (O_2)	Rust (Fe_2O_3)	$4Fe + 3O_2 \rightarrow 2Fe_2O_3$
Tarnishing of silver	Silver (Ag) + sulfur (S)	Tarnish (Ag_3S)	$3Ag + S \rightarrow Ag_3S$
Photosynthesis	Carbon dioxide gas (CO_2) + water (H_2O)	Glucose ($C_6H_{12}O_6$); Oxygen gas (O_2)	$6CO_2 + 6H_2O \rightarrow C_6H_{12}O_6 + 6O_2$

Properties and Chemical Changes

The new substances produced by a chemical change have their own properties. These properties differ from those of the original substances that reacted, since those substances are no longer present. For example, the element sodium is a metal that explodes on contact with water. The element chlorine is a poisonous green gas. When sodium and chlorine combine in a chemical reaction, they produce sodium chloride, commonly known as table salt. The new substance formed has completely different properties from those of the original materials, which no longer exist. During a chemical reaction, the atoms are rearranged to form new substances. This involves the breaking of existing chemical bonds and the formation of new bonds.

Both physical and chemical changes occur in nature. The wearing away of a mountain by streams is an example of a physical change called *erosion*.

Erosion is the physical wearing away of rock material at Earth's surface. The Grand Canyon in Arizona was formed by this physical change over millions of years.

The Statue of Liberty in New York City is made of copper but does not look copper colored. This is due to a chemical reaction between the copper and the air, which produces a new, green-colored substance. The chemical wearing away of a metal is called *corrosion*. Corrosion, which forms a new substance, is a chemical change. Erosion, which only moves substances around, is a physical change.

Conservation of Matter

In a chemical change, no atoms are created and no atoms are destroyed. Every atom that is present before a reaction takes place is still there after the reaction takes place. What has changed is the way the atoms are arranged. Chemical reactions change only the way atoms are bonded to one another.

Figure 12-7 shows what happens when hydrogen and oxygen combine to form water in a chemical reaction. How many atoms of hydrogen are there before the reaction takes place? How does this compare with the number of hydrogen atoms after the reaction takes place? There are four hydrogen atoms before and after the reaction. How are the starting substances (the reactants) different from the substances formed (the products)?

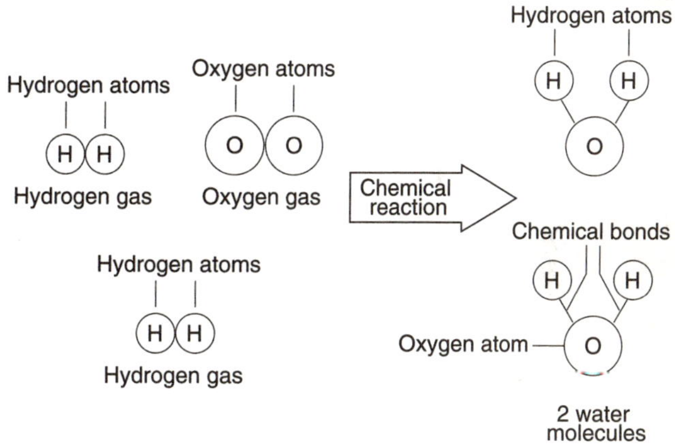

Figure 12-7. The Law of Conservation of Matter: In a chemical reaction, such as the formation of water, there are the same number of atoms before and after the reaction.

Before the reaction, each hydrogen atom was bonded to one other hydrogen atom; after the reaction, each hydrogen was bonded instead to an oxygen atom. How would the mass of the starting materials compare to the mass of the materials formed? The mass remains the same, since no atoms were created or destroyed. This is an example of the *Law of Conservation of Matter*, which states that matter can neither be created nor destroyed in a chemical reaction. It can, however, be changed from one form to another. It is important to remember to account for all the substances before and after a reaction. In particular, it may be easy to forget about gases in the air, since it is difficult to capture and weigh them.

Process Skill

PREDICTING RESULTS

After studying Table 12-5, Maria performed an experiment on the rusting of an iron nail. She carefully measured the mass of the nail before rusting occurred and recorded her observation.

Initial mass: 3.0 grams Final mass: ? grams

1. What prediction should Maria make about the final mass of the iron nail after rusting occurs?
   (1) The mass will decrease because the iron is destroyed.
   (2) The mass will stay the same because of the Law of the Conservation of Matter.
   (3) The mass will increase because rust combines iron atoms with oxygen atoms from the air.
   (4) The mass will increase because iron atoms get heavier as they rust.
2. In another experiment, Maria measured the mass of a piece of wood before and after it was burned. The mass of the wood decreased after it was burned. What would be a possible explanation for this observation?
   (1) Wood does not obey the Law of the Conservation of Matter.
   (2) Maria's measurements were incorrect.
   (3) Atoms become lighter when heated.
   (4) Burning the wood produced gases that were not included in Maria's final weighing.
3. How could Maria improve the experiment described in question 2?
   (1) She should use a material that does not burn.
   (2) She should perform the experiment in an open container.
   (3) She should perform the experiment in a closed container.
   (4) She should measure the mass of the heat released during the reaction.

Energy and Chemical Changes

As you have learned, new substances are formed during a chemical change. An example is making table salt from sodium and chlorine. However, simply mixing sodium and chlorine together does not produce table salt. Energy is needed to start the chemical reaction. Likewise, a match does not start to burn until you strike it. The friction caused by striking the match provides the heat energy needed to start the chemical reaction of burning.

Many chemical changes must be started by the addition of energy, in the form of heat, light, or electricity. However, some chemical changes do not require the addition of energy to get them started. The rusting of iron and the tarnishing of silver are examples of such reactions.

As a chemical reaction proceeds, energy is either absorbed or released. For example, the burning of a match releases energy in the form of heat and light. The chemical reaction that occurs in a battery releases electrical

Chloroplast =

organelle

Chlorophyll –

fills the chloroplast

ATP is released

4-307

BONANZA HIGH SCHOOL
REQUEST FOR STUDENT

DATE 12/10

PLEASE SEND Katy Robles

TO HEALTH OFFICE (circled) ____ TEACHER CONVENIENCE

X IMMEDIATELY

10/14 RC

SIGNATURE P Sousa

EXCUSED TO RETURN TO CLASS AT 10:45
Time

SIGNATURE P Sousa

Revised 03-05

Photosynthesis	Cellular Respiration
• How plant cells produce Engergy	• How animal (cells) produce Energy
*Plants produce energy by converting ~~[illegible]~~ sunlight	* Production of energy is through the conversion of OXYGEN
- LIGHT Dependent (Depends on sunlight to survive)	- Light independent (Does not NEED sunlight

energy. On the other hand, when food is cooked, heat energy is absorbed by the chemical changes taking place. Table 12-6 gives some examples of chemical changes that absorb energy and chemical changes that release energy.

We can use chemical reactions to supply us with heat when we need it. For example, campers often use chemical hand warmers in cold weather. When they open the packet, the chemicals in the hand warmer react with oxygen in the air to release heat.

Reactions that absorb heat are also useful. A cold pack contains two chemicals that absorb heat when they react with each other. To start the reaction, you simply break the seal that separates the two chemicals.

Table 12-6. Energy and Chemical Changes

Chemical Changes That Release Energy	*Type of Energy Released*	*Chemical Changes That Absorb Energy*	*Type of Energy Absorbed*
Burning of wood	Light, heat	Cooking an egg	Heat
Battery powering a flashlight	Electricity	Recharging a battery	Electricity
Decomposing of organic matter	Heat	Photosynthesis	Light

Reversible Reactions

Many physical changes, such as melting and dissolving, can be reversed. Water can be turned into ice by chilling it, and ice can be turned back into water by warming it. Sugar can be dissolved in water, and the water can be evaporated to separate it from the sugar. These are examples of changes in which substances can be returned to their original form.

Most chemical changes, however, are very difficult to reverse. It is impossible to "unburn" a match. However, some chemical reactions are reversible. An important example is a rechargeable battery. When you recharge a battery, you are reversing the chemical reaction used by the battery to produce electrical energy. Reversing a chemical change usually requires much more energy than does reversing a physical change.

Rate of Reactions

As discussed earlier, milk turns sour when left out of the refrigerator. However, milk that is refrigerated eventually turns sour, too, though it takes longer to occur. The chemical reaction of souring, which happens rapidly at room temperature, occurs much more slowly in a cold refrigerator. In fact, most chemical reactions take place faster at higher temperatures. Frying an egg takes less time on a high flame than on a low flame. An increase in temperature increases the ***rate*** (speed) of a reaction.

Another factor that affects the rate of a chemical reaction is the *size* of the reacting particles. In general, the smaller the particles, the faster the reaction. For instance, a log burns more slowly than does an equal quantity of sawdust.

The same factors that influence the rate of a chemical change also affect the rate of many physical changes. Powdered sugar dissolves more rapidly than a cube of sugar does, and sugar dissolves more quickly in hot tea than it does in iced tea.

Process Skills

MAKING PREDICTIONS; DETERMINING A QUANTITATIVE RELATIONSHIP; GRAPHING DATA

A student learned in chemistry class that temperature is a major factor in determining the rate of a chemical reaction. To investigate the effects of temperature on reaction rate, the student decided to time a chemical reaction at several temperatures. The student's results are presented in the table below. Study the table and answer the following questions.

Temperature and Reaction Rate

Trial Number	*Temperature*	*Time for Completion of Reaction*
1	20°C	80 seconds
2	30°C	40 seconds
3	40°C	20 seconds

1. The data seem to indicate a trend: for each 10°C increase in temperature, the time needed to complete the reaction was
   (1) doubled (3) decreased by 20 seconds
   (2) cut in half (4) increased by 20 seconds

2. Assuming that the observed trend remains constant for all temperatures, how long would the reaction take at 50°C?
   (1) 15 seconds (2) 5 seconds (3) 10 seconds (4) 20 seconds

3. Construct a graph by copying the numbered axes provided below onto a separate sheet of graph paper. Then enter the data from the table, as well as your answer to question 2, into the graph. To do this, mark a point at the intersection of a temperature line and a time line for the result of each trial. (The result of Trial 1 has been marked as an example to guide you.) Finally, connect the points with a smooth curve.

Effect of temperature on time needed to complete a chemical reaction

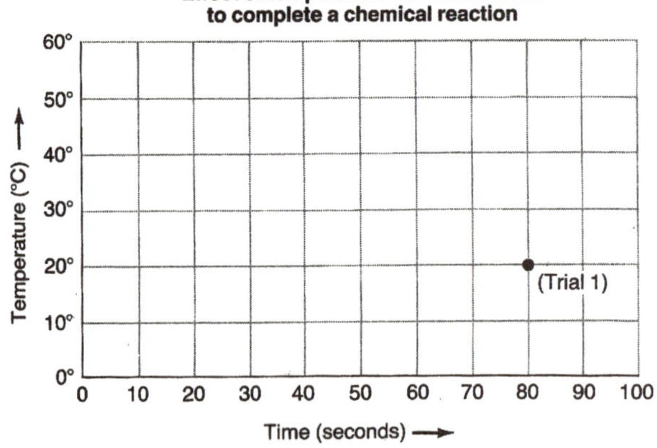

EXERCISE 2

1. Which is only a physical change?
   (1) the souring of milk
   (2) the burning of oil
   (3) the melting of ice
   (4) the rusting of iron

2. Which process involves a chemical change?
   (1) photosynthesis
   (2) boiling water
   (3) freezing water
   (4) melting ice

3. A chemical change always
   (1) forms one or more new substances
   (2) absorbs heat
   (3) releases heat
   (4) absorbs electricity

4. In making an omelet, which process involves a chemical change?
   (1) melting butter
   (2) chopping onions
   (3) frying eggs
   (4) adding salt

5. A chemist mixed sodium and chlorine, but no reaction took place. A probable explanation for this outcome is that
   (1) the reaction only releases energy
   (2) the reaction only absorbs energy
   (3) these substances cannot react
   (4) energy must be added to start the reaction

6. During a chemical change,
   (1) energy is always released
   (2) energy is always absorbed
   (3) energy is either absorbed or released
   (4) energy is neither absorbed nor released

7. When making iced tea, Sayed noticed that there was less ice after he mixed it with the hot tea. On observing this, Sayed remembered that melting is a
   (1) chemical change in which energy is absorbed
   (2) chemical change in which energy is released
   (3) physical change in which energy is absorbed
   (4) physical change in which energy is released

8. Hydrogen gas is produced in a chemical reaction between zinc and an acid. Which setup would most likely have the fastest reaction rate?

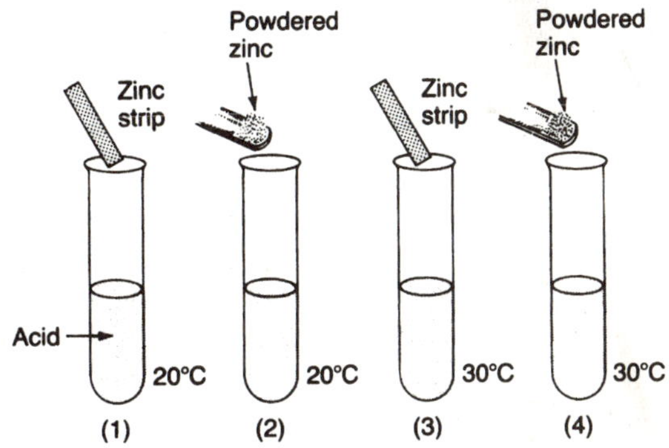

   (1) zinc strip at 20°C
   (2) powdered zinc at 20°C
   (3) zinc strip at 30°C
   (4) powdered zinc at 30°C

9. Food cooks faster at a higher temperature. This is because, as the temperature increases, the rate of a chemical reaction usually
   (1) increases
   (2) decreases
   (3) remains the same

10. The diagram shows four samples of wood, each weighing one kilogram. Which sample would most likely burn the fastest?
    (1) log (3) toothpicks
    (2) planks of wood (4) sawdust

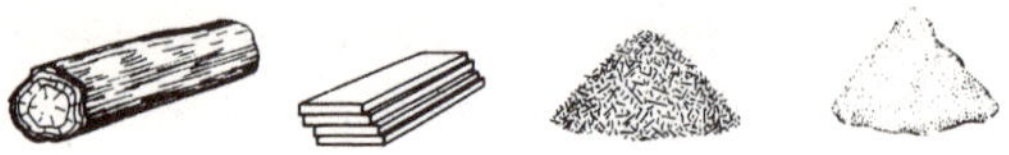

11. Which change would be the most difficult to reverse?
    (1) melting an ice cube, because it is a physical change
    (2) dissolving sugar in water, because it is a chemical change
    (3) burning a match, because it is a chemical change
    (4) rusting an iron nail, because it is a physical change

12. George cracks open an egg and finds that it has a very bad smell. The egg has turned rotten. This change is best described as
    (1) physical, because no new substances are formed
    (2) physical, because a new substance was formed
    (3) chemical, because no new substances are formed
    (4) chemical, because a new substance was formed

13. A molecule of carbon dioxide (CO_2) is made of one atom of carbon and two atoms of oxygen. Carbon dioxide can best be classified as
    (1) an element because it is made of one type of atom
    (2) an element because it is made of two types of atoms
    (3) a compound because it is made of one type of atom
    (4) a compound because it is made of two types of atoms

14. Which subatomic particles are found in the nucleus of an atom?
    (1) positively charged protons
    (2) positively charged electrons
    (3) negatively charged neutrons
    (4) negatively charged electrons

15. The diagram below represents an atom of helium. Based on this diagram, which statement is true?

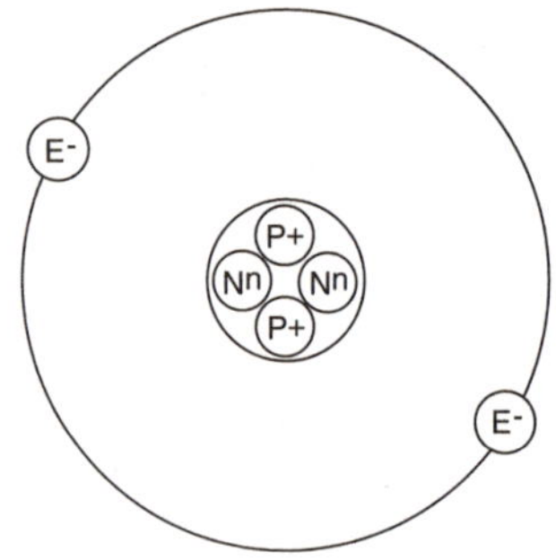

Helium atom

    (1) The positively charged protons are found outside the nucleus.
    (2) The positively charged protons are found inside the nucleus.
    (3) The negatively charged protons are found outside the nucleus.
    (4) The negatively charged protons are found inside the nucleus.

16. What occurs during a physical change? List two physical changes. What occurs during a chemical change? List two chemical changes.

13 Organisms and the Environment

Points to Remember

- Living things interact with their surroundings, called the environment.
- Living things and the nonliving factors of their surroundings make up an ecosystem.
- There is a continual exchange of materials between an organism and its environment; this includes food, water, oxygen, and wastes.
- Green plants use sunlight to make their own food. They are called producers.
- Animals depend on other organisms for food. They are called consumers.
- Decomposers are organisms that break down the remains of dead plants and animals. Decomposers return nutrients to the environment.
- All organisms get their energy directly or indirectly from the sun.
- Producers, consumers, and decomposers may be linked in a sequence called a food chain.
- Disturbing any part of a food chain affects other organisms in the food chain.

Organisms and Their Environment

Look at Figure 13-1. How can we classify all the things shown in the diagram into two categories? One way is to separate the living things from the nonliving things. For example, birds, fish, plants, and insects are living things; water, rocks, air, and soil are nonliving things. Living things are called ***organisms.*** They share certain characteristics that set them apart from nonliving things. In particular, organisms carry out life processes, which will be reviewed in detail in Chapter 16.

Figure 13-1. Living things have characteristics that set them apart from nonliving things.

Living things interact constantly with their surroundings, called the ***environment***. An organism's environment includes all living and nonliving things around it. Organisms obtain food, water, and oxygen from the environment. In turn, they release wastes, such as carbon dioxide (from animals). Thus, there is a continual exchange of materials between an organism and its environment.

Communities and Ecosystems

The particular environment in which an organism normally lives is called its ***habitat.*** There are many different types of habitats in the world. Here is a list of a few of them:

Desert	Salt marsh	Lake
Jungle	Coral reef	Stream
Prairie	Rain forest	Tundra

Habitats vary in their amount of water, light, temperature, wind, and so on. Different types of organisms live in these differing habitats, but they all depend on other organisms for survival. All the different organisms within a habitat make up a ***community.*** When you set up an aquarium that contains plants, catfish, and guppies, you create a small community.

To set up an aquarium, you must provide more than just the fish and the plants. You need water, a source of oxygen, and light. You must also maintain a proper temperature. These nonliving factors, together with the living members of the community, make up an ***ecosystem.*** The members of the community get the materials they need to survive from the ecosystem. In return, they give materials back, such as wastes and dead, decaying bodies.

Process Skills

INTERPRETING A TABLE; ANALYZING GRAPHS

A student performed an experiment in which five plants were placed in sand and five plants were placed in soil. All ten plants were the same type, given equal amounts of water, and exposed to equal amounts of sunlight. The experiment lasted for two weeks. The table below shows the growth of each of the plants.

Plants in Soil	Increase in Height (centimeters)	Plants in Sand	Increase in Height (centimeters)
1	2.0	1	0.5
2	1.9	2	0.6
3	2.2	3	0.4
4	2.1	4	0.7
5	1.9	5	0.6

1. What conclusion may be drawn from this experiment?
   (1) Plants grow just as well in soil as in sand.
   (2) Plants grow taller in sand than in soil.
   (3) Plants grow taller in soil than in sand.

2. What might explain the results of this experiment?
   (1) Plants make their own food.
   (2) Plants produce oxygen.
   (3) Plants get more nutrients from soil than from sand.
   (4) Plants get more nutrients from sand than from soil.

3. Which bar graph correctly represents the *averaged* results of this experiment?
   (1) 1 (2) 2 (3) 3 (4) 4

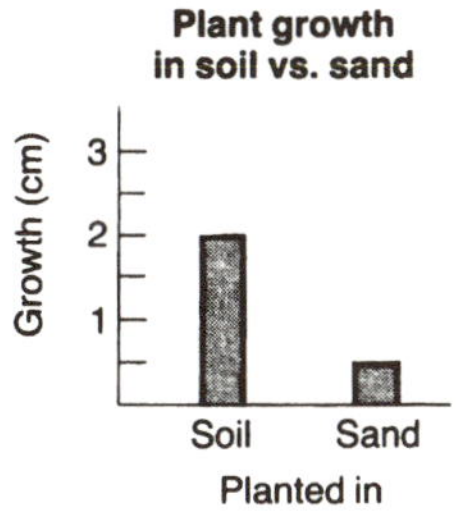

(1)

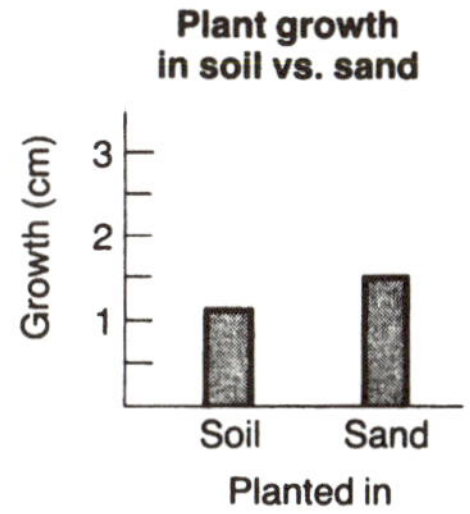

(2)

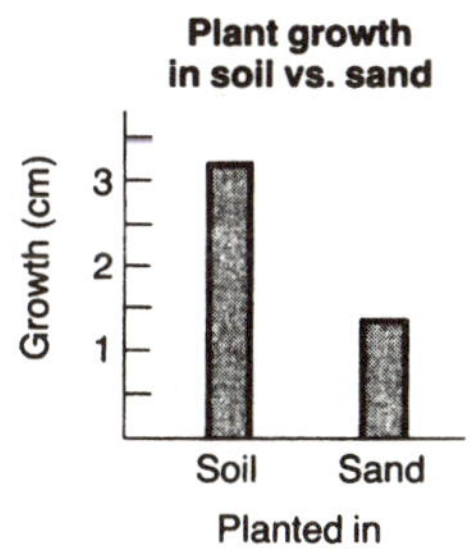

(3)

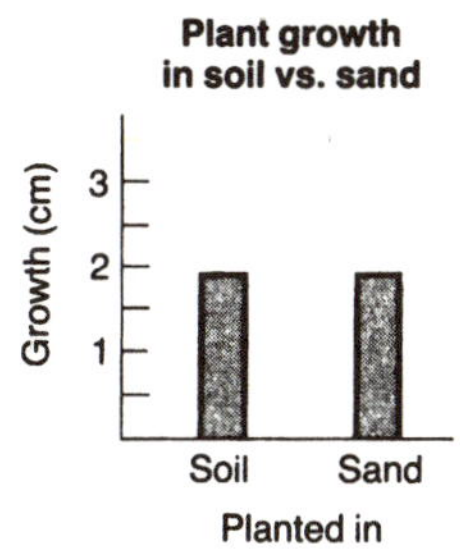

(4)

Materials are constantly being recycled within an ecosystem. Figure 13-2 shows how oxygen and carbon dioxide are recycled. In our environment, oxygen is provided mainly by green plants, through photosynthesis. Energy, however, is not recycled and must be provided by an outside source, such as the sun.

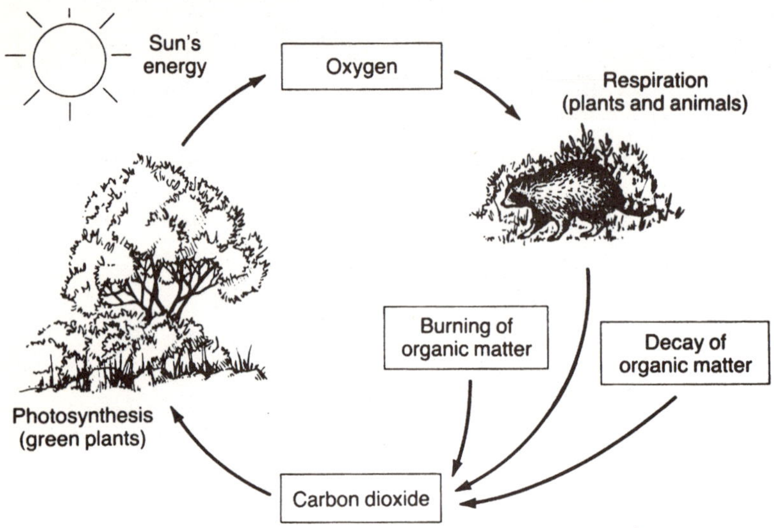

Figure 13-2. Oxygen and carbon dioxide are constantly recycled in an ecosystem.

Nutrition

Every organism needs food to stay alive. Food provides an organism with *nutrients*, which are used for growth and repair and for producing energy. Energy can be stored in chemical bonds. The energy stored in the chemical bonds of the food is released to provide an organism with energy.

Green plants make their own food. Through a process called *photosynthesis* (Figure 13-3), plants use energy from sunlight to change carbon dioxide and water drawn from their environment into sugar. The sun's energy is thereby stored in the sugar. Photosynthesis also produces oxygen, which is released into the environment. The green pigment *chlorophyll*, present in plant leaves, is necessary for this process to take place.

Animals obtain nutrients by eating plants or by eating other animals that feed on plants. The sugar in plants is used by animals to produce energy. The original source of energy in all food is the sun. Figure 13-4 shows one way to get food energy from the sun.

Food Chains

Plants make their own food. Plant-eating animals, called herbivores, get their energy from the plants they eat. Meat-eating animals, called carnivores, also get energy from plants, but indirectly. For instance, when a lion eats a zebra, it obtains nutrients from the meat of the zebra. The lion gets its energy from these nutrients. The zebra had obtained these nutrients from the plants that it had eaten. Animals such as the lion, which must hunt and kill for food, are called carnivores. The animals that they hunt are called their prey.

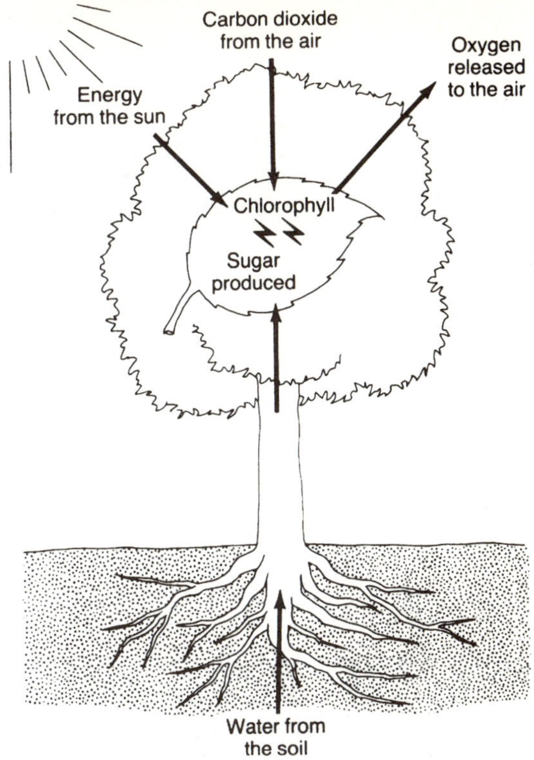

Figure 13-3. Green plants use the energy of sunlight to make their food.

Both the lion and the zebra depend on other organisms for their food. Green plants produce their own food. Therefore, green plants are called ***producers,*** while animals are called ***consumers.*** Nearly every animal depends, either directly or indirectly, on green plants for food and oxygen.

The nutrients in green plants get passed along from one organism to another in a sequence called a ***food chain.*** For example, grass produces food during the process of photosynthesis; a zebra eats the grass to get its nutrients; and then a lion, in turn, eats the zebra. When the lion dies, its body decays. Special organisms called ***decomposers*** break down the lion's remains and return its nutrients to the soil. These nutrients can then be used again by plants, such as grass.

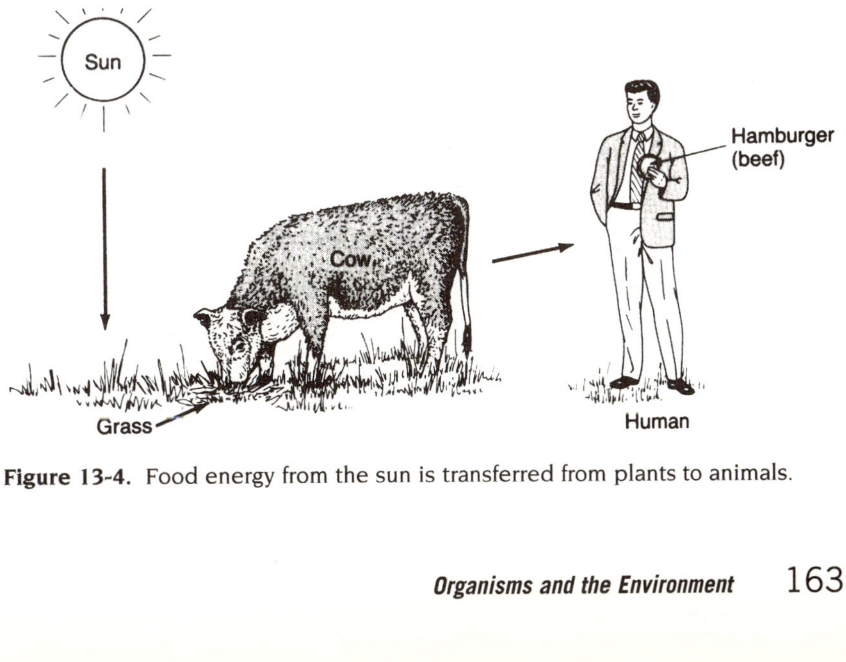

Figure 13-4. Food energy from the sun is transferred from plants to animals.

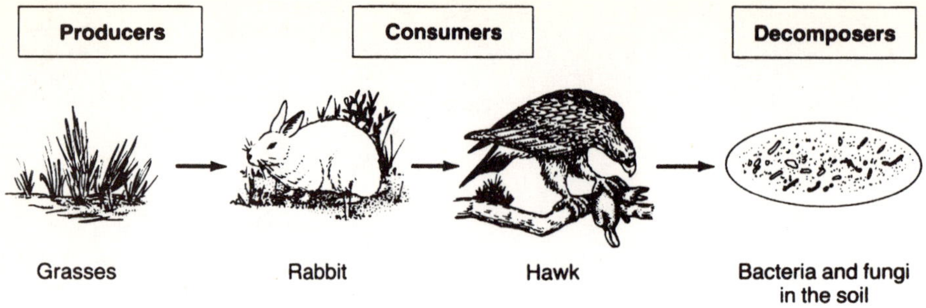

Figure 13-5. An example of a food chain.

Decomposers include fungi, such as mushrooms and molds, and some bacteria. Fungi and decay bacteria cannot make their own food. They therefore depend on other living things for food. Decomposers are the last link in any food chain. Figure 13-5 shows an example of a food chain.

Suppose that, suddenly, there were no more zebras. How would the lion get its nutrients? A lion can also eat other animals. If something wiped out the zebra population, the lion would eat more of the other animals. Thus, the removal of one species would affect many other species.

Food Webs

Most ecosystems contain a number of food chains that are interconnected to form a ***food web,*** as shown in Figure 13-6. There is a delicate

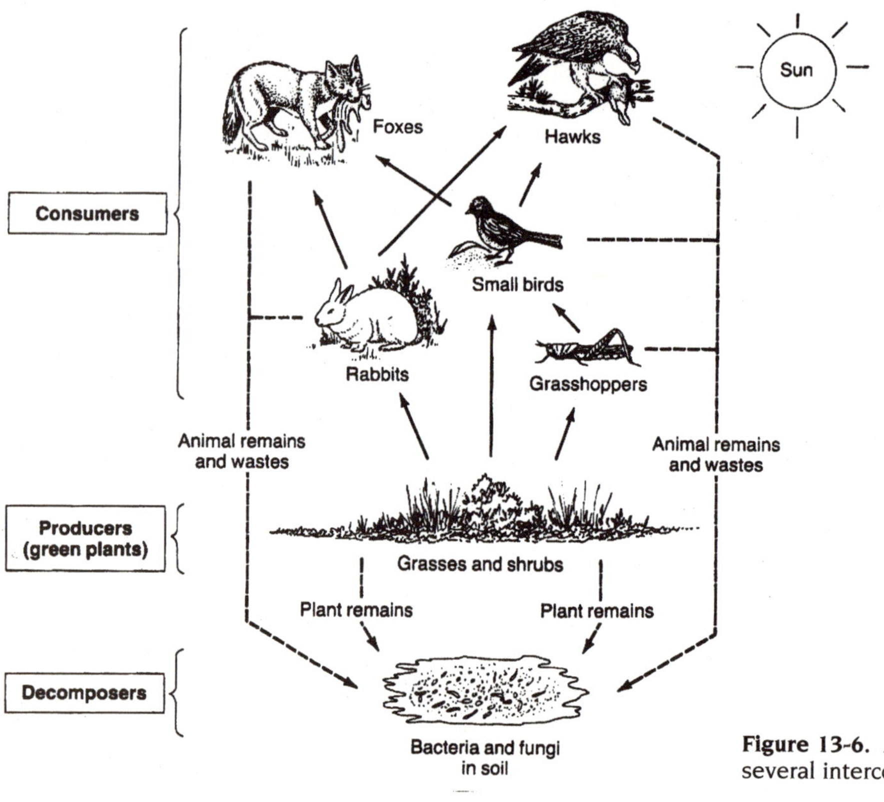

Figure 13-6. A food web consists of several interconnected food chains.

balance in an ecosystem among its producers, consumers, decomposers, and their environment. If this balance is disturbed, it may change the entire ecosystem.

Food Pyramids

What happens to you when you exercise? Exercising causes you to "burn" a lot of food for energy. As you do this, your body temperature increases. Heat is being created and lost to the environment. Every organism uses some of the energy it consumes and stores the rest. The energy that is used is lost to the environment in the form of heat. Only the stored energy is available to the next consumer in the food chain. This results in a decreasing amount of energy available at each step of the food chain. As you move up the food chain, the amount of available energy in the same amount of food becomes less and less. This can be represented by a pyramid, which gets smaller and smaller toward the top.

Figure 13-7 illustrates an energy, or food, pyramid. What type of organism is always at the base? A food pyramid always has green plants (producers) at its base.

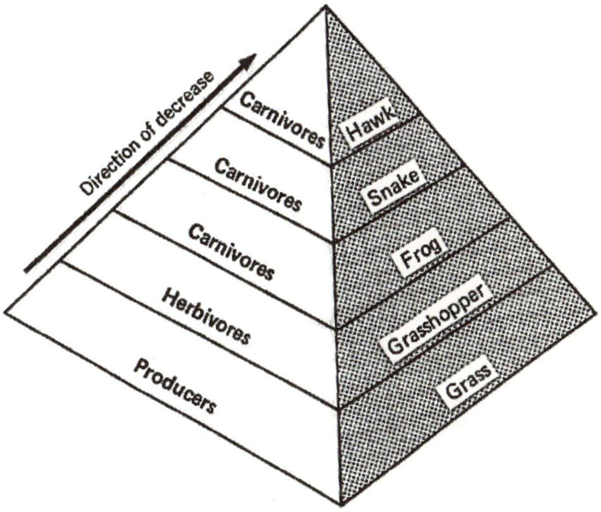

Figure 13-7. A food pyramid shows the relationship of producers to consumers; the producers (plants) are always at its base.

Symbiosis

When different organisms live together, they may interact in a variety of ways. At least one organism always benefits from the relationship. The other organism, however, may or may not benefit. Symbiosis, or a ***symbiotic*** relationship, occurs when two or more different organisms live in close association with one another, that is, when one organism lives on or inside another one. For example, a flea living on the skin of a dog has a symbiotic relationship with the dog; this type of symbosis is called parasitism. Table 13-1 describes three types of symbiotic relationships between organisms in a community.

Table 13-1. Types of Symbiotic Relationships Between Organisms

Relationship	*Description*
Mutualism	Both organisms benefit from the relationship.
Commensalism	One organism benefits while the other is not affected.
Parasitism	One organism benefits while the other is harmed.

Upsetting the Balance of Nature

Humans sometimes interfere with the balance of nature. For example, early settlers in the northeastern United States killed off all the wolves in the region, because the wolves preyed on farm animals. However, the wolves were the only natural enemies of the deer living in the area. Without the wolves to hold their numbers in check, the deer population increased to the point where many deer starved to death each winter.

The actions of people are not the only things that can disturb the balance of nature. Sometimes, the delicate balance is upset by natural events, such as floods and forest fires. In 1980, for instance, a volcano called Mount St. Helens, in the state of Washington, erupted violently. The explosion destroyed almost 100,000 acres of forest. Nevertheless, the trees and forest organisms have begun to return to Mount St. Helens through a series of natural changes in the ecosystem.

EXERCISE

Base your answers to questions 1–3 on the following diagram, which represents a food chain.

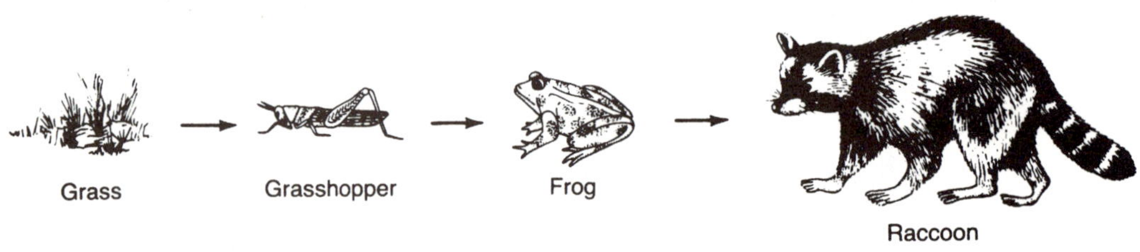

1. Which organism is the producer in this food chain?
   (1) grass (2) grasshopper (3) frog (4) raccoon

2. In this food chain, the frog is a
   (1) producer (2) consumer (3) decomposer

3. Which type of organism is *not* shown in this diagram?
   (1) producer (2) consumer (3) decomposer

Base your answers to questions 4–9 on Figure 13-6, page 164, which shows a food web.

4. A decrease in the number of small birds would most likely result in an increase in the number of
   (1) rabbits (3) foxes
   (2) hawks (4) grasshoppers

5. The organisms that return nutrients to the soil are the
   (1) producers (3) decomposers
   (2) consumers (4) green plants

6. The producers in this ecosystem get their energy from the
   (1) rabbit (3) decomposers
   (2) consumers (4) sun

7. Which two organisms are in competition for the same food resource?
   (1) foxes and rabbits
   (2) small birds and grasses
   (3) foxes and hawks
   (4) grasses and rabbits

8. Which sequence illustrates how energy may flow in this ecosystem?
   (1) sun → grasses → rabbits → hawks
   (2) hawks → rabbits → sun → grasses
   (3) grasses → hawks → foxes → rabbits
   (4) grasses → grasshoppers → rabbits → hawks

9. Animals that eat only plants are called herbivores. Animals that eat only meat are called carnivores. Animals that can eat both meat and plants are called omnivores. According to the food web, which organism may be considered an omnivore?
   (1) rabbit (3) small bird
   (2) hawk (4) fox

Base your answer to question 10 on the diagram below, which represents a food pyramid.

10. Which organism is considered a producer?
    (1) the human, because it is at the top of the pyramid
    (2) the pig, because it changes the energy in plants into food for the human
    (3) the corn, because it changes energy from the sun into food
    (4) the corn, because it cannot be eaten by the human

11. Which organism makes its own food?
    (1) frog (3) tree
    (2) bird (4) snake

12. Photosynthesis is the process by which plants make their own food. The energy for photosynthesis comes from
    (1) the sun (3) water
    (2) oxygen (4) wind

13. All of the organisms that live in a pond make up the
    (1) habitat (3) environment
    (2) community (4) ecosystem

Base your answers to questions 14 and 15 on the diagram below, which represents a small ecosystem.

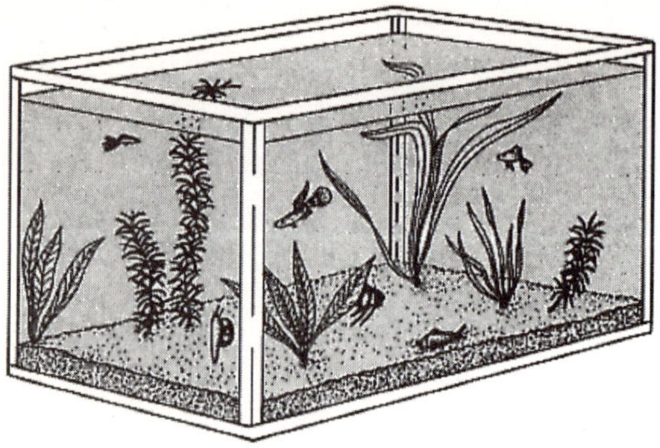

14. All animals need oxygen to live, even animals that live in the water. What is the main source of oxygen in this ecosystem?
    (1) the water (3) the snail
    (2) the fish (4) the green plants

15. Many important, nonliving factors within an ecosystem are recycled, although some are not. The survival of this community depends on a constant external supply of
    (1) energy (3) carbon dioxide
    (2) oxygen (4) plants

16. The original source of energy in our environment is
    (1) food (3) the sun
    (2) water (4) air

Base your answers to questions 17 and 18 on the diagram below, which represents a food pyramid.

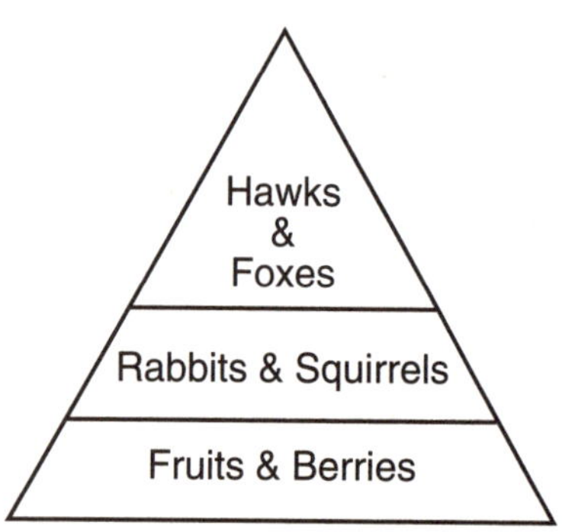

17. Based on this food pyramid, which would provide the greatest quantity of energy per pound when eaten?
    (1) apples (3) hawks
    (2) rabbits (4) foxes

18. Which organism can be considered a carnivore?
    (1) rabbits, because they eat fruit
    (2) berries, because they produce sugar from the sun's energy
    (3) squirrels, because they eat acorns
    (4) hawks, because they eat animals

Questions 19 and 20 refer to the following paragraph.

A student studies the animals in an environment and reports the following: There are trees and grasses. The rabbits and deer eat the grass; and the deer, squirrels, and small birds eat the fruits, berries, and leaves of the trees. Foxes hunt for the rabbits and squirrels. Large birds also eat the rabbits.

19. Copy the diagram below into your notebook. Fill in each of the organisms from the food web described above.

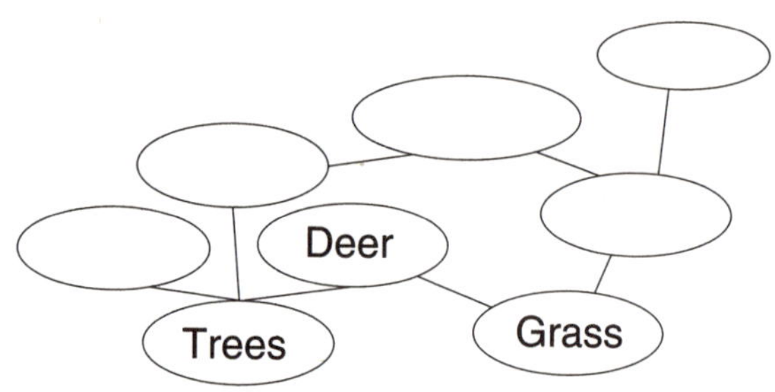

20. Copy the table below into your notebook. Based on the food web described above, identify each of the following as a producer, herbivore, or carnivore by checking the appropriate box. The first one has been done for you.

Organism	*Producer*	*Herbivore*	*Carnivore*
Grasses	✓		
Foxes			
Rabbits			
Deer			
Small birds			
Trees			
Large birds			
Squirrels			

Base your answers to questions 21–23 on the information given in Table 13-1, page 166.

21. Certain crocodiles allow a small bird called an Egyptian plover to sit inside their open mouths. The birds feed on harmful leeches and food particles found between the teeth of the crocodile. This relationship is best described as a type of
    (1) parasitism, because the crocodile is harmed
    (2) mutualism, because both organisms benefit
    (3) commensalism, because the crocodile is not affected
    (4) parasitism, because the bird is harmed

22. Which symbiotic relationship is an example of parasitism?
    (1) Bacteria live in our intestines and help us digest our food.
    (2) Ticks live on a dog's skin and suck its blood and transmit diseases.
    (3) Remora fish attach themselves to a shark and eat its leftover food bits.
    (4) Barnacles attach themselves to whales and are transported to new feeding grounds.

23. Small fish called clownfish are immune to the stinging cells of a sea anemone's tentacles. The clownfish live within these tentacles, where they are protected from larger fish (see figure, below). The sea anemone also benefits, because it can feed on scraps of food left over by the clownfish. This type of relationship is best described as

    (1) parasitism
    (2) commensalism
    (3) mutualism
    (4) a food web

24. Why are there fewer organisms at the top of a food pyramid than at the bottom?

25. What is the role of decomposer organisms in the environment? What might happen if there were no decomposers present on Earth?

14 Living Things Carry Out Life Processes

Points to Remember —

- All living things carry out life processes. These include nutrition, respiration, transport, excretion, and reproduction.
- All living things are composed of basic units called cells.
- Plant cells and animal cells have common structures. These include the nucleus, cytoplasm, and cell membrane.
- Plant cells have cell walls and chloroplasts, which are not found in animal cells.
- Both plants and animals must be able to adapt to changes in their environment.
- Plants and animals may use different strategies for their survival.

Characteristics of Living Things

In Chapter 13, you learned that all animals are ultimately dependent on plants for their survival. Likewise, many plants are dependent on animals. In this chapter, we will examine the similarities and differences between plants and animals.

All living things, or ***organisms***, share certain characteristics that set them apart from nonliving things. In particular, all organisms carry out ***life processes***, some of which are listed below in Table 14-1.

Table 14-1. Life Processes and Their Functions

Process	*Function*
Respiration	Releasing energy stored in food
Transport	Moving materials throughout the organism
hgestion	Taking food into the body
Digestion	Breaking down food into a form usable by cells
Excretion	Eliminating waste materials produced by the organism
Regulation	Responding to changes in the organism$ surroundings
Reproduction	Making more organisms of the same kind

The Cell

Living things are made up of basic units called ***cells***, each of which carries out life processes. Cells generally share some common structures, which perform these life processes. The ***nucleus*** controls cell activities. Surrounding the nucleus is a thick fluid called the ***cytoplasm***, which is where most life processes occur. The cytoplasm is contained within the ***cell membrane***—the "skin" of the cell—which controls the flow of materials into and out of the cell. Figure 14-1 shows a typical animal cell. Table 14-2 lists some structures within the cell and the functions they perform.

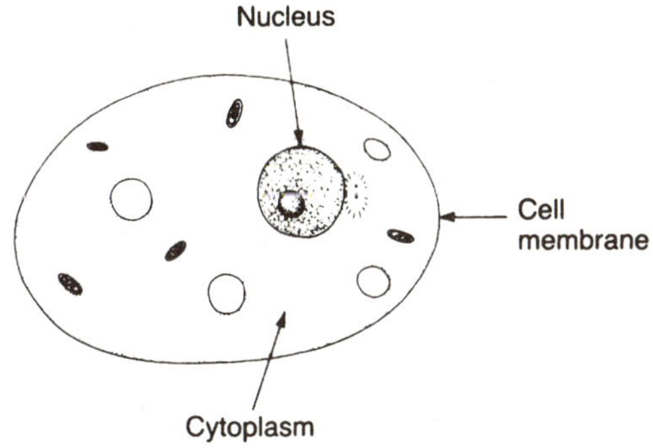

Figure 14-1. A typical animal cell.

Plant cells and animal cells have many structures in common, but they also have important differences (see Figure 14-2 on page 172).

1. ***Plant cells.*** In plant cells, a *cell wall* encloses the entire cell, including the cell membrane. The tough cell wall gives support to the plant's structure. Also found only in plant cells are structures called *chloroplasts*, which contain chlorophyll. Chloroplasts are the site of photosynthesis, the food-making process of plants.

Table 14-2. Some Cell Structures and Their Functions

Structure	Function
Mitochondria	Respiration—where food is ъurned"(combined with oxygen) to produce energy
Ribosomes	Synthesis—where proteins are made
ʎsosomes	Digestion—where digestive enʒmes are stored
Nucleus	Reproduction—where genetic material is stored
Vacuole	Digestion and excretion—where digestion occurs or where excess fluid is stored
Chloroplasts (in plant cells)	Photosynthesis—where glucose (sugar) is produced in green plants

2. ***Animal cells.*** Animal cells do not have cell walls; they are enclosed only by the cell membrane. Found only in animal cells are the *centrioles,* which participate in cell division.

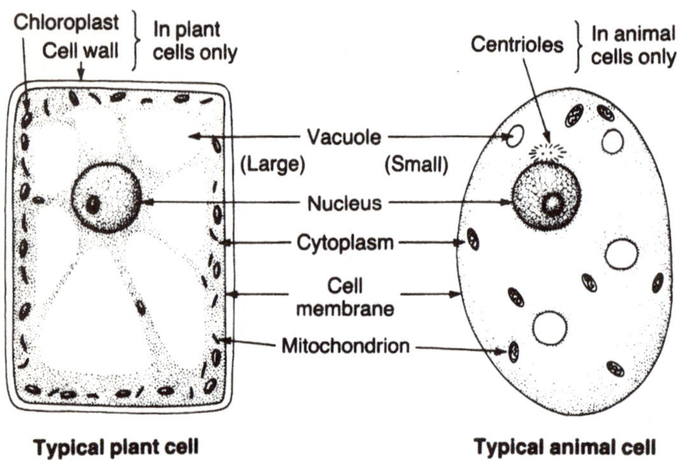

Figure 14-2. Comparison of plant and animal cells.

Nutrition

Because plants and animals must carry out the same life processes, there are many similarities between them. However, the methods by which these life processes are carried out often vary greatly. You have already learned in Chapter 13 that plants produce their own food through photosynthesis, whereas most animals rely on other organisms as a source of food.

As a result of these differences, plants and animals have developed different structures to help them obtain food. Leaves, for example, have a high concentration of chloroplasts, which are used to make food from sunlight. The broad, flat shape of a typical leaf helps it gather sunlight. The leaves are even capable of turning toward the light to maximize this process.

Animals, on the other hand, have developed very different methods for obtaining food. Plant-eating animals have a digestive system specifically adapted for breaking down plant cells. This includes flat teeth for grinding down the rigid cell walls of plant cells. Meat-eating animals may have sharp teeth, claws, or beaks for tearing apart flesh. Humans eat both meat and plants and have both types of teeth, as illustrated in Figure 14-3.

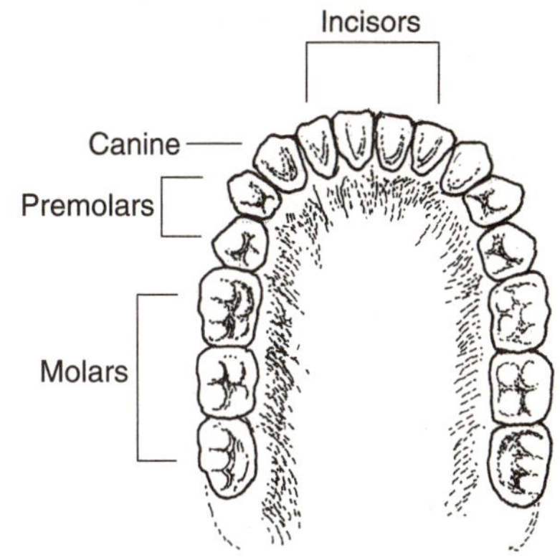

Figure 14-3. Human teeth (lower jaw).

Plants have also developed a root system for obtaining water. In wet environments, the root systems are not very large. However, in a dry environment, such as a desert, plant roots may run very deep in search of water.

Locomotion

Plants do not move from place to place. They don't need to. They do not have to search for food since they make their own. Animals, on the other hand, have developed various methods of locomotion—flying, crawling, swimming, and running—to enable them to find food and to avoid predators. Since animals reproduce sexually, they need to find mates. This also requires that animals be mobile.

Sexual Reproduction in Plants

Plants, like animals, may reproduce sexually. However, since plants are not mobile, they have developed different structures and techniques to accomplish this life process.

A flower is the reproductive organ of a plant. Pollen produced in the anther (see Figure 14-4) contains the male sperm cells. The female egg cells

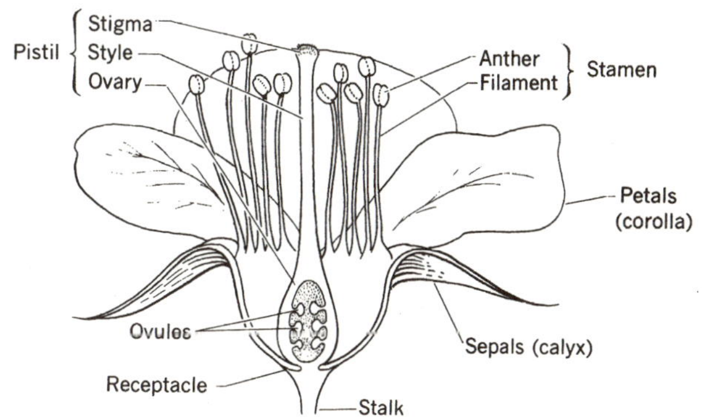

Figure 14-4. Reproductive parts of a flower.

Process Skills

INTERPRETING AND ANALYZING DATA IN A TABLE

In the United States, some birds migrate, while others do not. Our migrating birds fly south in the winter and north in the summer. The table below lists several bird species; their sizes; their diets; and the time of the year when they are most likely to be in Ohio. Base your answers to the following questions on the information in this paragraph and on the data in the table.

Bird Species	*Length (cm)*	*Diet*	*When Found in Ohio*
Yellowbilled cuckoo	305	Caterpillars	Summer
Blue jay	305	Acorns, corn, berries, seeds, insects	All year
Eastern kingbird	203	Flying insects	Summer
Goldencrowned kinglet	102	hsects and insect eggs	Winter
Baybreasted warbler	152	hsects and berries	May and October (during migration)
Cardinal	203	hsects, fruit, seeds, corn, oats, rice	All year

1. Which two bird species spend the whole summer north of Ohio?
   (1) blue jay and yellow-billed cuckoo
   (2) golden-crowned kinglet and bay-breasted warbler
   (3) cardinal and eastern kingbird
   (4) yellow-billed cuckoo and eastern kingbird

2. We can infer that birds that do not migrate are those that
   (1) are largest
   (2) are smallest
   (3) eat berries
   (4) have varied diets

3. Which summer and winter locations seem most likely for the bay-breasted warbler?
   (1) winter in South America, summer in Canada
   (2) winter in southern Canada, summer in northern Canada
   (3) winter in Mexico, summer in southern U.S.
   (4) winter and summer in northern U.S.

4. Some people put out sunflower seeds to feed birds. They most likely attract
   (1) yellow-billed cuckoos and blue jays
   (2) blue jays and golden-crowned kinglets
   (3) eastern kingbirds and bay-breasted warblers
   (4) blue jays and cardinals

are located in the ovary at the base of the pistil. In plants, as in animals, the male and female reproductive cells must join in order for fertilization to take place. In Chapter 1, you studied the way that several different types of animals accomplish this. A plant depends on wind, rain, insects, or birds and other small animals to transfer the pollen from the anther to the pistil. Some plants are completely dependent on bees as their means of fertilization.

The fertilized egg develops into a seed, as the ovary develops into a fruit. Animals that eat the fruit help distribute the seed to new locations. This is called *seed dispersal.* Wind is another source of seed dispersal. Different types of seeds have developed to take advantage of various methods of dispersal, as illustrated in Figure 14-5.

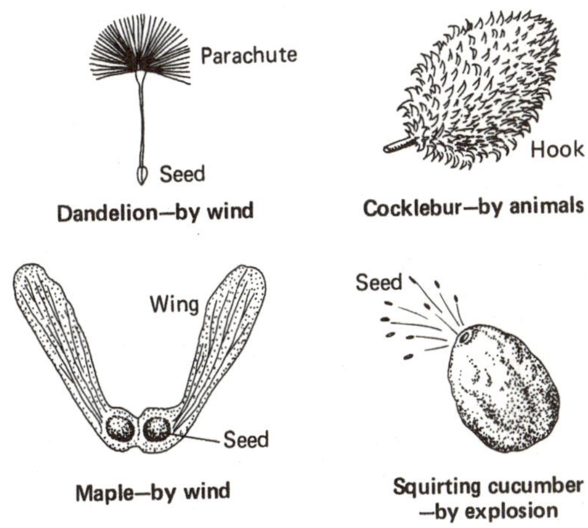

Figure 14-5. Methods of seed dispersal for different types of seeds.

Adapting to Changing Climates

For their survival, both plants and animals must be able to adapt to changes in their environments. One such environmental change is the change in seasons. In places that have large seasonal changes in temperature, amount of sunlight, and water supply, plants and animals adapt to these changes. For example, during the winter, many trees lose their leaves and go through a period of dormancy. During this dormant period, there is very little growth and a minimal amount of life activity.

Some animals, such as bears, have a very similar response in the winter. They ***hibernate,*** greatly reducing their life activities. During hibernation, the animal remains practically motionless, with its heartbeat and rate of respiration at a minimal level. Other animals ***migrate*** to a different climate when conditions are unfavorable. Migrating animals may travel long distances in search of food, water, or favorable temperatures.

Many plants, called annual plants, live for only one season. They go through their entire life cycle, including reproduction, during that season. Their seeds remain dormant until conditions are favorable the following year, when they grow and develop. Some insects, like the mosquito, also live for only one season. Their larvae survive the winter and hatch the next summer.

Changes in behavior that help an organism adjust to its environment are called *behavioral adaptations.* Structural differences, such as flat teeth, sharp claws, or long beaks, are called *physical adaptations.*

EXERCISE

1. Which life process is carried out by the flower in plants?
   (1) respiration (3) reproduction
   (2) digestion (4) excretion

2. The structure that is found in both plants and animals, and is used to control the flow of materials into and out of the cell, is the
   (1) cell membrane (3) chloroplast
   (2) nucleus (4) centrioles

Question 3 refers to the following short paragraph.

Did you ever wonder how trees can grow so tall without falling down? The trunk of a tree is made of a rigid material called *cellulose*. Cellulose is the main component of the cell walls of plant cells. The cellulose supports the tree in much the same way that bones support you.

3. Complete the following analogy: Bones are to animals as
   (1) cell walls are to animals
   (2) chloroplasts are to plants
   (3) cell membranes are to plants
   (4) cell walls are to plants

4. A bear and a maple tree are similar in that both
   (1) reduce their life processes during the winter
   (2) rely on other animals as a food source
   (3) use the sun's energy to make their own food
   (4) rely on cell walls for support

Questions 5–8 refer to the diagram below, which represents a beetle and a maple tree.

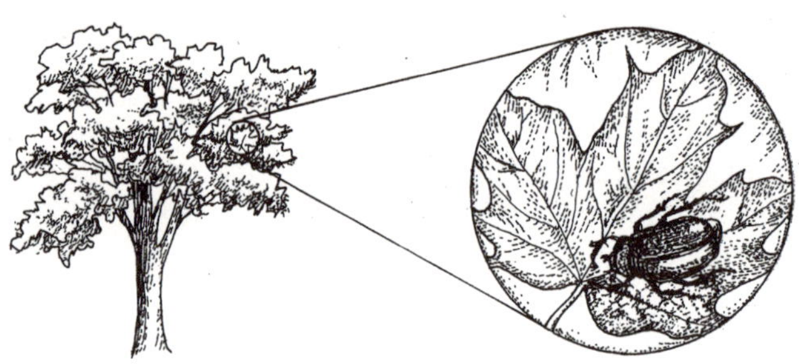

5. What can the beetle do that the tree cannot do?
   (1) make its own food
   (2) move from place to place
   (3) reproduce
   (4) take in oxygen to burn food for energy

6. What can the maple tree do that the beetle cannot do?
   (1) make its own food
   (2) move from place to place
   (3) reproduce
   (4) take in oxygen to burn the food for energy

7. What structures do both the maple tree and the beetle have in common?
   (1) cell walls
   (2) chloroplasts
   (3) cell membranes
   (4) centrioles

8. Which structure appears in the cells of a maple tree but does *not* appear in the cells of a beetle?
   (1) nucleus
   (2) vacuole
   (3) cell membrane
   (4) cell wall

15 Biological Diversity and Heredity

Points to Remember —

- Living things have developed special characteristics called adaptations, which enable an organism to live in a particular environment or habitat.
- Reproduction is the process by which an organism produces new individuals of its own kind.
- Asexual reproduction involves only one parent and produces offspring identical to the parent.
- Sexual reproduction involves two parents, male and female, and produces offspring not identical to either parent.
- All cells come from other cells through the process of cell division.
- Genetic information is passed from one generation to the next through chromosomes during reproduction.
- When two genes match for a trait, an individual is called pure for that trait. When two genes differ for a trait, the individual is called hybrid for that trait.
- In a hybrid, often only one trait is visible; this is called a dominant trait. The trait that is not exhibited in a hybrid is called a recessive trait.
- A change in a gene is called a mutation.

Diversity of Life

Why are there so many different types of animals and plants? Part of the diversity of life is due to changes that have occurred in living things over many generations. Different environments have promoted different changes.

Living things have special characteristics called ***adaptations*** that enable them to survive under a given set of conditions. Organisms may be adapted for life in water, soil, or air. For example, a fish has gills so that it can breathe underwater. An earthworm's body shape helps it move through the soil. A bird has wings and light, hollow bones so it can fly. Many adaptations help an organism obtain food or escape predators in its environment. Figure 15-1 shows how certain birds are adapted for survival.

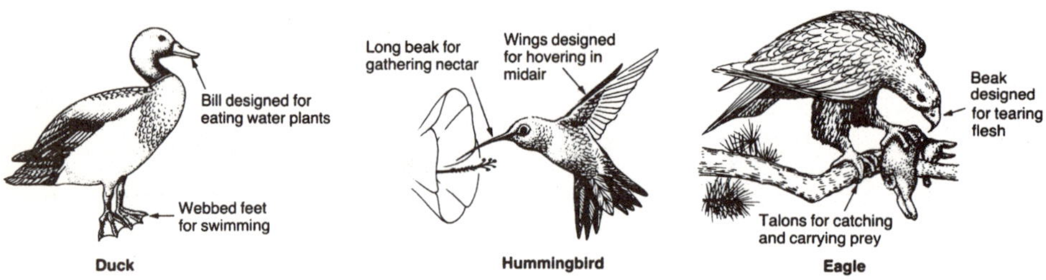

Figure 15-1. Some adaptations of birds for survival in different habitats.

Earth's many environments include oceans, deserts, tropical rain forests, and the frozen Arctic tundra. Adaptations permit an organism to live in its own particular environment, or ***habitat.*** Organisms living in a dry, desert environment have adaptations that enable them to obtain and conserve water. For example, the cactus plant has an extensive root system that helps it reach what little water there is in the desert.

Animals living in the icy Arctic have adaptations that help them to endure the region's very cold temperatures. For instance, polar bears have thick coats of fur; seals and whales have layers of protective fat called *blubber.* Table 15-1 lists organisms from various habitats and their specialized adaptations.

Table 15-1. Organisms and Their Adaptations

Organism	*Habitat*	*Adaptation*	*Function*
Arctic hare	Arctic	White fur in winter	Provides camouflage from predators
Monkey	Rain forest	Grasping tail	Acts as an extra hand to help movement through trees
Cactus	Desert	Waxy skin	Reduces water loss from evaporation

There are millions of different types of plants and animals. Keeping track of them can be a difficult task. So scientists classify organisms in various ways to better understand their similarities and differences. One way of classifying animals is based on the way they reproduce. You learned in Chapter 1 that methods of reproduction and development are important factors in distinguishing among amphibians, reptiles, birds, mammals, fish, and invertebrates such as insects.

Reproduction

All living things come from other living things. ***Reproduction*** is the process by which an organism produces *offspring*—new individuals of its own kind. Each particular kind of organism is called a ***species.*** Lions are a species. Tigers are a different species. Since every individual organism eventually dies, reproduction ensures the continuation of its species.

There are two types of reproduction: *asexual* and *sexual. Asexual reproduction* involves only one parent. The offspring created are identical to the parent. Figure 15-2 shows examples of asexual reproduction.

Sexual reproduction involves two parents and produces offspring that are not identical to either parent. The female parent produces an egg cell, and the male parent produces a sperm cell. The joining together of these cells is called ***fertilization.*** The fertilized egg grows into a new individual.

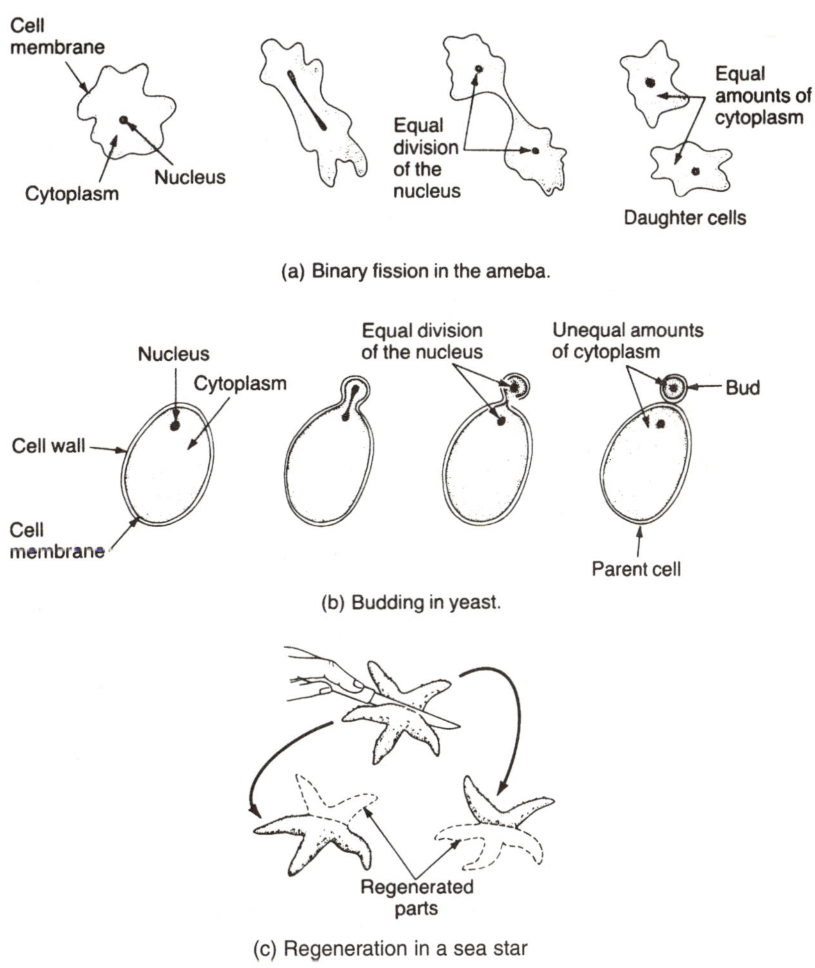

Figure 15-2. Asexual reproduction: (a) binary fission; (b) budding; and (c) regeneration.

Life Cycles

A puppy resembles an adult dog. A young elephant looks like a small version of its parents. However, this is not true of a frog or butterfly. Some organisms, like frogs and most insects, change so dramatically during their

Process Skill

INTERPRETING AN EXPERIMENT

About three hundred years ago, the Italian scientist Francesco Redi wondered where maggots—small, wormlike organisms—came from. The popular belief at the time was that rotting meat turns into maggots. This idea, that living things could come from nonliving material, was called *spontaneous generation*. Redi designed an experiment to test this belief. He placed meat into eight jars. Four jars were left open; four were tightly sealed. Diagram 1 shows what Redi observed.

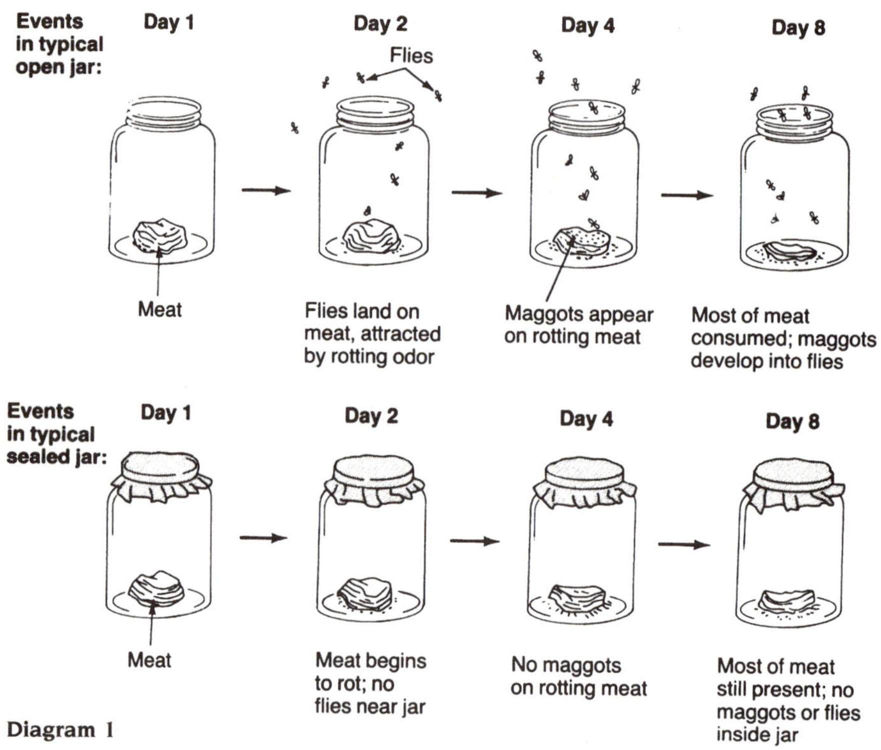

Diagram 1

As you can see, no maggots appeared on the rotting meat in the sealed jars. However, not everyone was convinced that Redi's experiment had disproved spontaneous generation. Some people claimed that fresh air was needed for spontaneous generation to occur. Therefore, Redi performed a second experiment. This time the jars were covered by fine netting, which allowed fresh air into the jars but prevented flies from entering and landing on the meat. Diagram 2 shows what Redi observed in his second experiment. Study both diagrams and then answer the following questions.

lives that the young may not resemble the adults at all. The changes an organism undergoes—whether from puppy to dog, or tadpole to frog—as it develops and then produces offspring make up its ***life cycle.*** Figure 15-3 illustrates the life cycles of a frog and a butterfly.

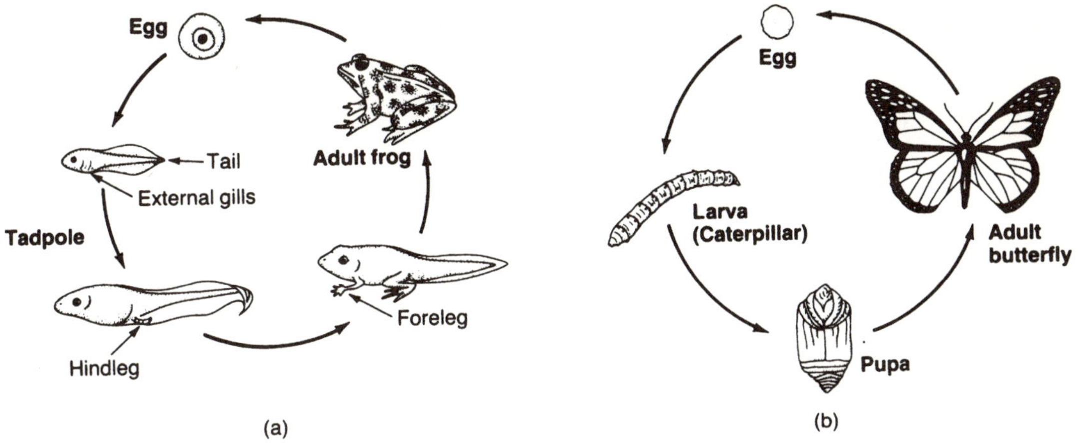

Figure 15-3. Life cycles: (a) the frog and (b) the butterfly.

Eventually, the young offspring of a frog or a butterfly do resemble their parents. Information was carried from the parent to the offspring, determining what they will be. This information is passed on from generation to generation during reproduction.

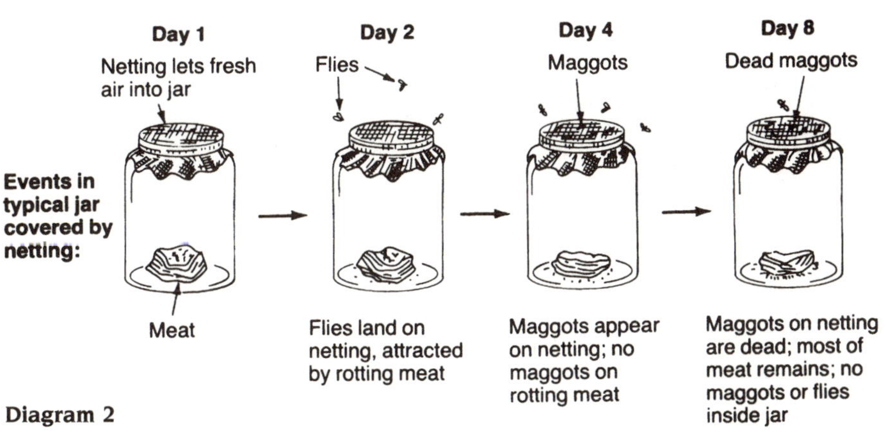

Diagram 2

1. Redi's second experiment provided clear evidence that the source of maggots was
   (1) air (2) flies (3) rotting meat (4) either flies or air

2. Redi's experiments
   (1) supported the concept of spontaneous generation
   (2) did not support the concept of spontaneous generation
   (3) provided no evidence for or against spontaneous generation
   (4) proved that living things come from both living and nonliving things

Cellular Reproduction

All cells come from other cells through the process of ***cell division.*** In this process, one "parent" cell divides into two new "daughter" cells. Every parent cell passes along to its daughter cells a set of "operating instructions" necessary for the cells to function properly. This *genetic information* is contained in threadlike structures called *chromosomes*, found in the cell nucleus.

The genetic information in the chromosomes also gives the cell, or the organism it belongs to, its individual characteristics or *traits*, such as size and shape. All members of a given species have the same number of chromosomes in each body cell. Chromosomes and their genetic information are passed on to the next generation during reproduction.

One-celled organisms reproduce through a kind of cell division called *mitosis*. In this process, a cell divides into two identical daughter cells, each of which contains the same number of chromosomes as the original parent cell (see Figure 15-4). When a one-celled organism undergoes mitosis, each new cell produced is a complete new organism, identical to the original cell.

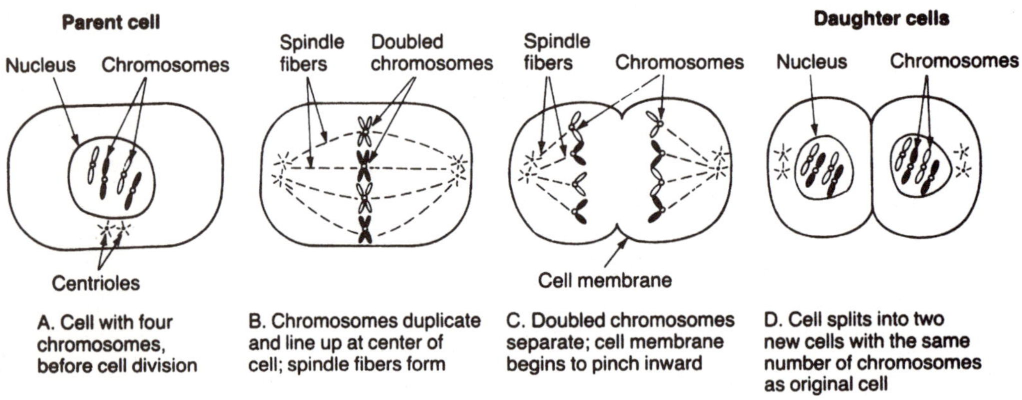

Figure 15-4. Cell division: Mitosis produces two new cells with the same number of chromosomes as the original parent cell.

In an organism made up of many cells, the cells must duplicate themselves to build new tissue for growth and to repair damaged tissue. They do this through mitosis, too. Some of these organisms also produce offspring by mitotic cell division. In fact, any organism that reproduces *asexually* (with just one parent) does so through mitosis. The offspring created are genetically identical to the parent.

Sexual Reproduction

Some organisms reproduce *sexually*, with two parents. Sexual reproduction involves the joining of two special reproductive cells, one from each parent. These sex cells (sperm cells and egg cells) have only half the number of chromosomes held by other body cells. Sex cells are formed by a type of cell division called *meiosis*.

During sexual reproduction, a sperm cell from the male parent joins with an egg cell from the female parent. This is called *fertilization*. Since each

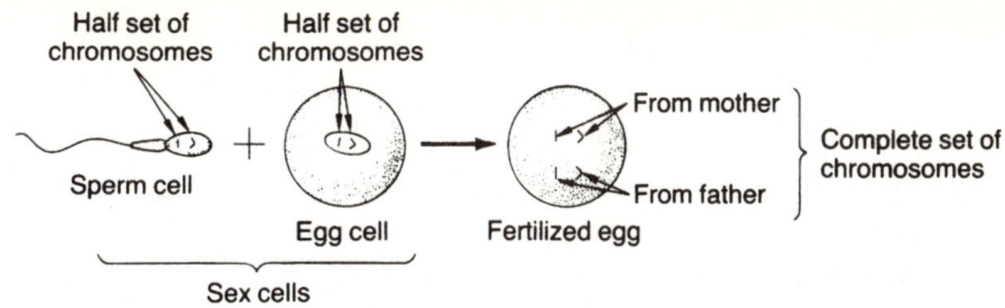

Figure 15-5. Fertilization: The fertilized egg has a complete set of chromosomes.

sex cell contains half the number of a normal set of chromosomes, when joined they form one cell with a complete set of chromosomes (Figure 15-5). This cell then develops into a new organism through mitosis. The new organism is not identical to either parent but has traits from both. In this way, sexual reproduction leads to variation in the next generation. Figure 15-6 shows a possible result of sexual reproduction in chickens.

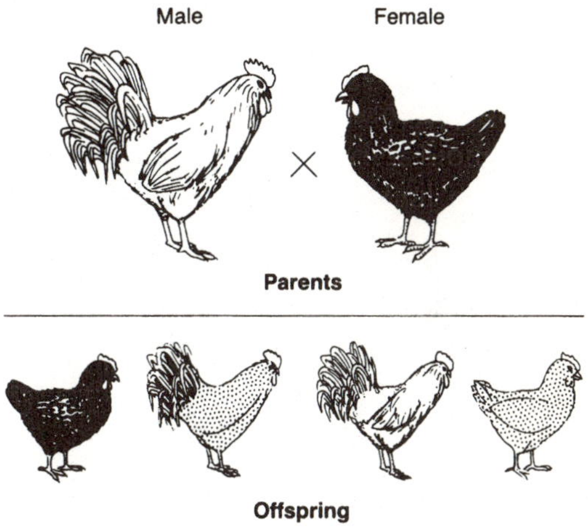

Figure 15-6. Sexual reproduction leads to variation in the offspring.

Inheritance of Traits

What type of earlobes do you have? Did you know that there are two types of earlobes, "attached" and "free" (see Figure 15-7)? The type of earlobes you have is determined by the genetic information you received from each parent. A piece of genetic information that influences a trait is called a ***gene.*** You received one gene from each of your parents containing information for the type of earlobes you have. These two genes can be the same or different.

If both genes match for a trait, that same trait will appear, and you are said to be "pure" for that trait. However, if you have differing genes for a trait, one that says "attached" and another that says "free," you are said to be "hybrid" for that trait.

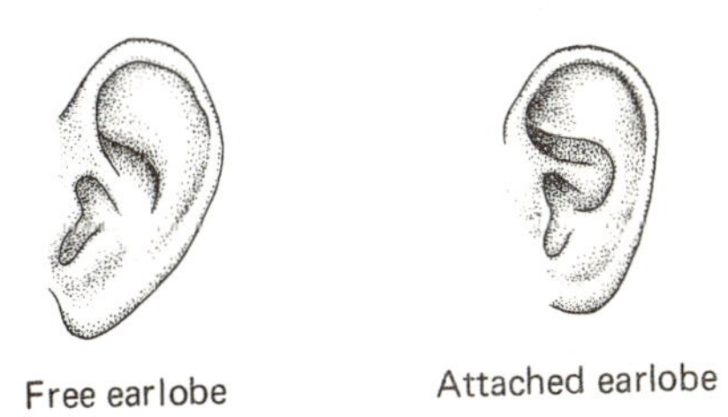

Figure 15-7. The two types of earlobes that can be inherited: "attached" and "free."

Mendelian Genetics

An Austrian monk named Gregor Mendel (1822–1884) performed experiments with pea plants to investigate how traits are inherited. He determined that for certain traits that are controlled by two differing genes (which he referred to as "factors"), one of these genes would be dominant over the other. An individual with both genes (a hybrid) will exhibit only the characteristics of the dominant gene. The gene for the trait that is not exhibited is called a recessive gene.

Mendel crossed pure tall pea plants with pure short pea plants. He observed that all the offspring were tall. He concluded that tall is the dominant trait. He also concluded that all of the offspring of this generation must be hybrids. When the hybrid tall pea plants were crossed with one another, most of the offspring were tall, but some were short. This proved that the hybrid tall pea plants still contained a gene (factor) for shortness. Only when the offspring received two short genes did they appear as short plants. Table 15-2 indicates some common dominant and recessive traits studied in genetics.

Table 15-2. Common Traits Studied in Genetics

Organism	*Trait*	*Dominant*	*Recessive*
Human	Eye color	Brown	Blue
Human	Earlobe	Free	Attached
Human	Blood type	A or B	O
Fruit fly	Wing	Normal	Vestigial
Fruit fly	Eye color	Red	White
Pea plant	Height	Tall	Short
Pea plant	Pea color	Yellow	Green
Pea plant	Seed shape	Round	Wrinkled

Using a Punnett Square

Fruit flies are often used to study genetics because they mature and reproduce quickly. Several generations can be observed in a short period of time.

One characteristic studied is eye color. Fruit flies can have red eyes or white eyes. Red is dominant, while white is recessive. We can predict the possibility of an offspring having red eyes or white eyes if we know the types of genes in the parents. To represent the gene, we use a capital letter for the dominant trait (R for red) and a lowercase of the same letter for the recessive trait (r for white). The possible gene combinations are outlined in Table 15-3.

Table 15-3. Possible Gene Combinations

Type of Genes	*Representation*	*Appearance*
Pure red	RR	Red
Hybrid red	Rr	Red
Pure white	rr	White

A diagram called a Punnett square can be used to predict the probability of an organism inheriting a given trait. The genes for one parent are placed at the top of the square and the genes for the other parent are placed at the side. Below is a Punnett square for two hybrid parents.

Each box is filled in with the letter appearing above it and to its left. These boxes represent the possible combinations of genes. We see that of the four boxes, one will be pure dominant (RR), one will be pure recessive (rr), and two will be hybrid (Rr). This can be summarized as shown in Table 15-4.

	R	r
R	RR	Rr
r	Rr	rr

Key:
R = gene for red eyes
r = gene for white eyes

Table 15-4. Probabilities of Offspring of Hybrid Red-Eyed Fruit Flies

Type of Genes	*Appearance*	*Probability*
RR—pure dominant	Red eyes	$\frac{1}{4}$ (25%)
Rr—hybrid dominant	Red eyes	$\frac{2}{4}$ (50%)
rr—pure recessive	White eyes	$\frac{1}{4}$ (25%)

EXERCISE 1

1. Based on Table 15-4, what percentage of the fruit fly offspring are likely to have red eyes?
   (1) 25% (2) 50% (3) 75% (4) 100%

2. What fraction of the red-eyed flies produced are pure for the trait?
   (1) $\frac{1}{2}$ (2) $\frac{1}{3}$ (3) $\frac{1}{4}$ (4) $\frac{2}{3}$

3. A farmer wanted to get rid of a recessive gene from a population of livestock. For generation after generation, she bred only animals that showed the dominant trait. Would she be successful? Explain.

Process Skill

INTERPRETING THE RESULTS OF AN EXPERIMENT

Indira studied the traits of an unusual insect. She noticed that some of the insects had long antennae and some had short antennae. She separated the insects into two groups based on their antenna size and allowed them to mate only within each group. After the eggs were laid, the parents were removed. The eggs were allowed to develop. In one group, all the offspring had short antennae. Offspring in the other group had both types of antennae.

1. Indira can tell from this information that
   (1) short is dominant
   (2) long is recessive
   (3) short is recessive
   (4) there were two different species of insects

2. In describing the parents, it would be safe to infer that
   (1) all those with long antennae were pure
   (2) all those with long antennae were hybrid
   (3) some of those with long antennae were hybrid
   (4) some of those with short antennae were hybrid

3. To improve this experiment, which of the following should *not* be done?
   (1) Separate the insects to be mated as soon as they are hatched.
   (2) Feed the long- and short-antennae groups different diets.
   (3) Maintain a constant environment for both groups.
   (4) Repeat the experiment several times.

4. Predict the results of a cross between a pure long-antennae insect and a pure short-antennae insect.
   (1) Some of the offspring will have short antennae.
   (2) All of the offspring will have short antennae.
   (3) Some of the offspring will have long antennae.
   (4) All of the offspring will have long antennae.

Mutations

If parents and offspring are so similar, why are there so many adaptations and such diversity among living things? Genetic accidents do occur. Sometimes genetic material does not reproduce properly. This may be caused by a natural "accident" or by something in the environment. Such a genetic accident is called a *mutation*. The new genetic information will cause a variation in the offspring. If this change is harmful to the organism, it will be less likely to survive and reproduce. If the change is beneficial to the organism, it will be better able to survive and reproduce. The new genetic information will then be passed on to each new generation. If these changes increase the likelihood that the organism will reproduce, the changes will become more prevalent within the population. *Natural selection* favors those organisms that are best able to survive and reproduce. After many generations, and many mutations, the organism may look and behave so differently from its ancestors that it has become a new species. This process, called ***evolution,*** may be very slow and take millions of years.

Evolution accounts for the great diversity among living things. The evolutionary process is dependent on the passing of genetic information from one generation to the next through chromosomes. This occurs during reproduction.

EXERCISE 2

1. Brown eyes are dominant over blue eyes in humans. What is the likelihood of two blue-eyed parents having a brown-eyed child?
(1) 0% (2) 25% (3) 50% (4) 100%

2. Examine the diagrams. What conclusion may be drawn based on the similarities of these structures?

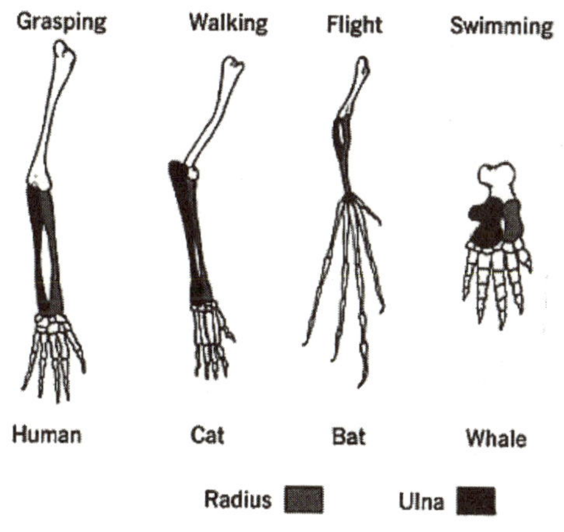

(1) All of these organisms are of the same species.
(2) These organisms live in the same environment.
(3) All living things have similar hands.
(4) All of these organisms may have had a common ancestor.

3. Penguins are birds that cannot fly. Instead, they are powerful swimmers that catch and eat fish. These birds have evolved in this way because they
(1) live in a cold climate
(2) live entirely on land
(3) use their wings to swim in search of food
(4) have no place to fly to

4. The diagram shows an example of

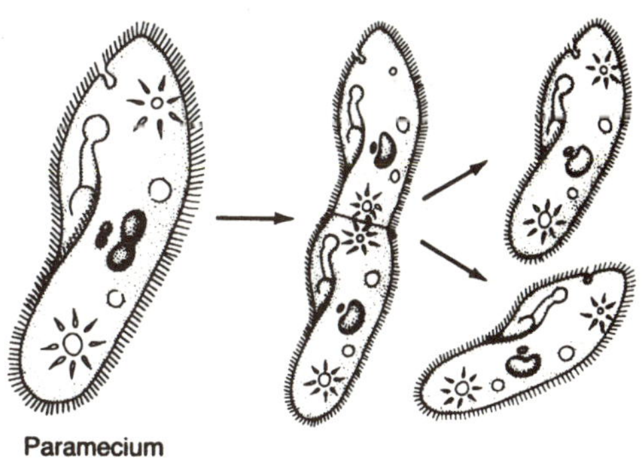

(1) spontaneous generation
(2) asexual reproduction
(3) photosynthesis
(4) respiration

5. Which statement is true of *all* living things?
   (1) They have two parents.
   (2) They are exact copies of their parents.
   (3) The young look like small adults.
   (4) They come from other living things.

6. What is shown in the diagram below?

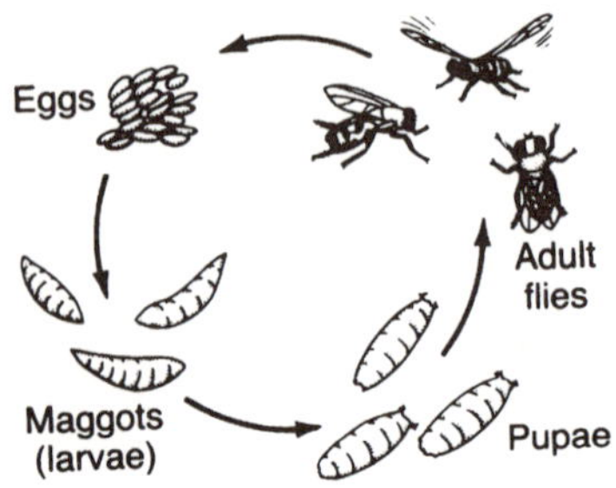

   (1) a life cycle (3) an ecosystem
   (2) a food cycle (4) a community

7. In which environment would you most likely find an animal with thick fur?
   (1) desert (3) tundra
   (2) tropical rain forest (4) grassland

8. Life in the desert is harsh because there is very little
   (1) sunshine (3) oxygen
   (2) water (4) sand

9. Birds that are adapted to live in an aquatic environment would most likely have the type of feet shown in

(1)

(3)

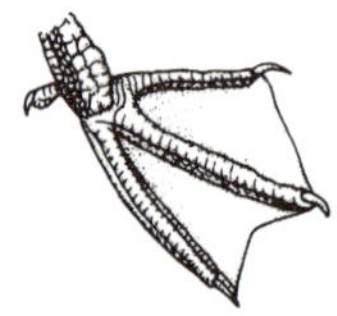
(2)

(4)

10. Genetic variation in a new generation of organisms is the result of
   (1) sexual reproduction involving one parent
   (2) sexual reproduction involving two parents
   (3) asexual reproduction involving one parent
   (4) asexual reproduction involving two parents

---

11. Which diagram shows sexual reproduction?

(1) 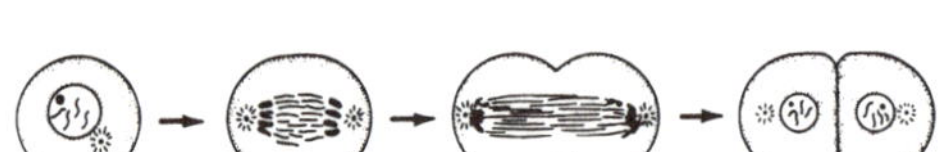

(3) 

(2) (2) 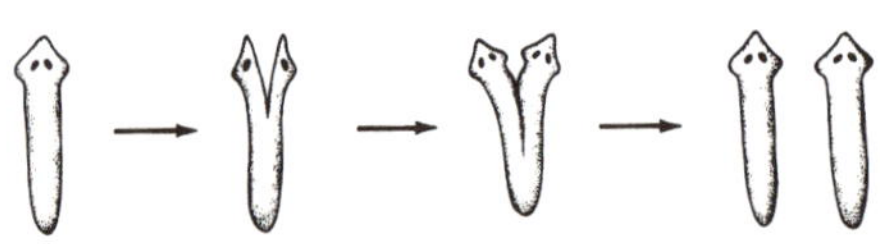

(4) 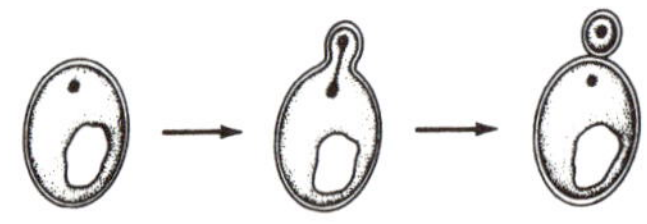

16 Living Systems: Human Body Systems

Points to Remember

- Complex organisms show several levels of body organization.
- Cells are the basic unit of life. Tissues are groups of similar cells that act together to carry out a life process.
- Organs are groups of tissues that act together to carry out a life process.
- Organ systems are groups of organs that act together to carry out a life process. Different organ systems act together to carry out life processes.
- The human body includes the following types of tissues: blood, bone, muscle, nerve, and skin.
- The human body includes the skeletal, muscular, nervous, endocrine, digestive, circulatory, respiratory, excretory, and reproductive systems.
- All body systems are interdependent; for example, the nervous system and endocrine system act together to regulate and control body activities.
- The circulatory system and respiratory system act together to bring oxygen in the blood to all body cells.

Organization, Support, and Movement of the Body

Human Body Systems Are Interdependent

A human being is a complex organism made up of a number of different *body systems.* Each system carries out a specific life process and thereby contributes to the operation of the body as a whole.

In addition, all body systems are interdependent and work with one another to keep a person alive. For instance, the respiratory system brings needed oxygen into the body; the oxygen is then transported throughout the body by the circulatory system.

Levels of Organization in the Human Body

1. *Cells.* Living things are made up of basic units called ***cells.*** The human body contains many types of cells, each designed to perform a different function. Figure 16-1 shows several kinds of cells.

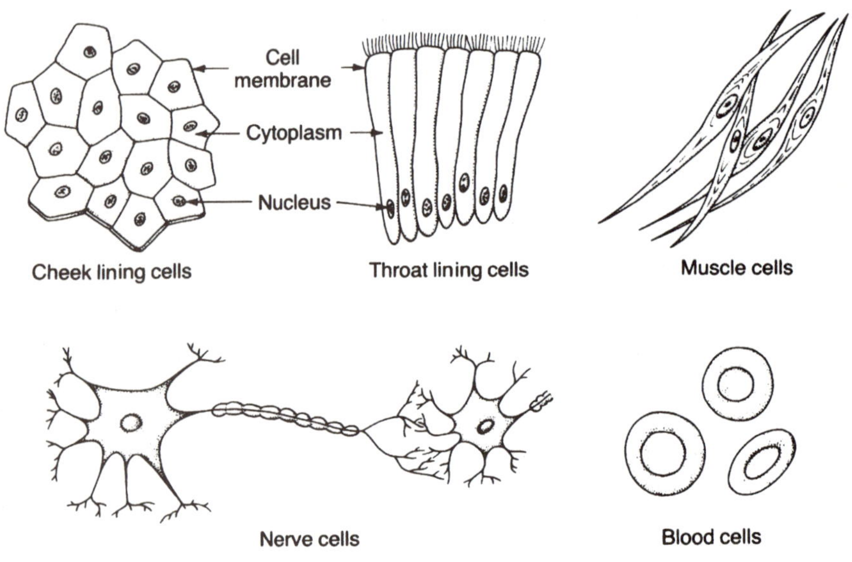

Figure 16-1. Different types of cells.

2. *Tissues.* A group of similar cells acting together to carry out a life process forms a ***tissue.*** For example, muscle tissue produces body movements. Table 16-1 lists some types of human tissues.

Table 16-1. Types of Human Tissues and Their Functions

Tissue	*Function*
Blood	Transports materials throughout the body
Bone	Supports and protects the body and organs; helps body move
Muscle	Helps body move; aids in circulation, digestion, and respiration
Nerve	Carries messages throughout the body
Skin	Covers and protects the body; excretes wastes

3. *Organs.* A group of tissues working together form an ***organ***. The heart is an organ that pumps blood throughout the body. It is composed mainly of muscle tissue, but it also contains blood tissue and nerve tissue. Table 16-2 lists some important organs.

Table 16-2. Important Organs and Their Functions

Organ	*Functions*
Heart	Pumps blood throughout the body
Kidney	Removes wastes from the blood
Lung	Exchanges gases with the environment
Stomach	Breaks down food by physical and chemical means
Brain	Controls thinking and voluntary actions
Skin	Covers and protects the body; excretes wastes

4. *Organ systems.* A group of organs acting together to carry out a specific life process makes up an ***organ system.*** For example, the circulatory system carries out the process of transport, moving materials throughout the body. Table 16-3 lists the human organ systems.

Table 16-3. Human Organ Systems

System	*Function*	*Example of Organs or Parts*
Skeletal	Supports body; protects internal organs	Skull, ribs
Muscular	Moves organs and body parts	Arm and leg muscles
Nervous	Controls body activities; carries and interprets messages	Brain, spinal cord
Endocrine	Regulates body activities with hormones	Adrenal glands
Digestive	Breaks down food into a usable form	Stomach, intestines
Circulatory	Carries needed materials to body cells and waste materials away from cells	Heart, arteries, veins
Respiratory	Exchanges gases with the environment	Lungs, bronchi
Excretory	Removes wastes from the body	Kidneys, skin
Reproductive	Produces offspring	Ovaries, testes

The Skeletal System

The human *skeletal system,* shown in Figure 16-2, supports and protects the body and its organs. The skeletal system includes the *skull, spinal column, breastbone, ribs,* the bones of the *limbs* (arms and legs), and *cartilage.*

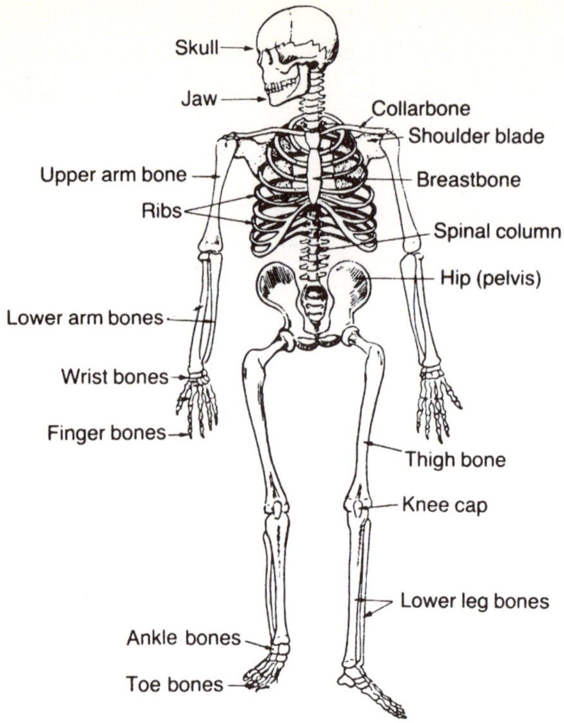

Figure 16-2. The human skeleton.

1. *Bones and cartilage. Bones* are made of hard, strong material. ***Cartilage*** is a softer, more flexible tissue. Cartilage acts as a cushion between bones and provides flexibility at the ends of bones. Disks of cartilage separate the bones of the spinal column, cushioning them from one another.

2. *Joints.* Where one bone meets another bone, a ***joint*** is formed. Most joints, such as the knee and elbow, allow the bones to move. However, some joints, like those in the skull, do not allow movement. Figure 16-3 shows three types of joints.

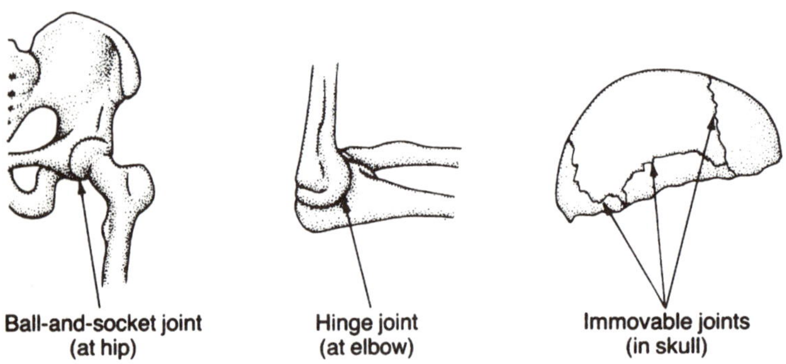

Figure 16-3. Three types of joints.

3. *Ligaments and tendons.* At movable joints, the bones are held together by strips of tissues called ***ligaments.*** Bones are moved by muscles, which are attached to bones by ***tendons,*** cordlike pieces of tissue. A common sports injury is a torn Achilles tendon in the back of the lower leg.

The Muscular System

Muscles are masses of tissue that contract to move bones or organs. The *muscular system* includes two main kinds of muscles: *voluntary* and *involuntary*.

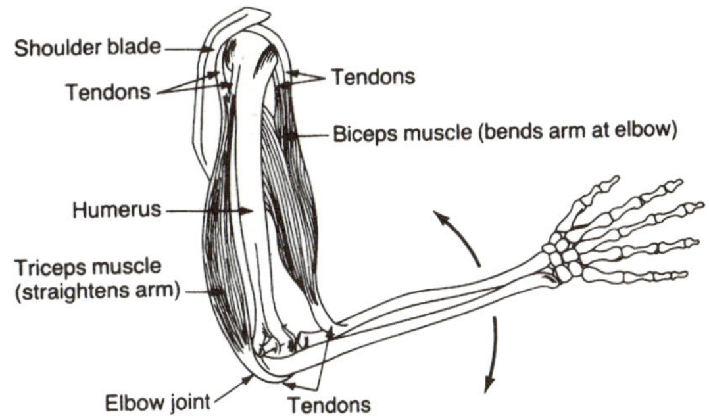

Figure 16-4. Muscles, tendons, and bones of the arm enable movement.

1. *Voluntary muscles.* The skeletal muscles, which move bones, are examples of ***voluntary muscles***—muscles that are controlled by our will. Skeletal muscles work with the skeleton to move body parts (see Figure 16-4), and thereby produce locomotion. ***Locomotion*** is the movement of the body from place to place. The muscles in the face and around the eyes are also voluntary muscles.

2. *Involuntary muscles.* ***Involuntary muscles*** are not under our conscious control. There are two types of involuntary muscle: cardiac and smooth. *Cardiac* muscle, present only in the heart, pumps blood throughout the body. *Smooth* muscle, found in the respiratory, circulatory, and digestive systems, aids in breathing, controlling blood flow, and movement of food.

EXERCISE 1

1. A group of organs working together to carry out a life process is called
   (1) a cell (3) an organ
   (2) a tissue (4) an organ system

2. A group of similar cells acting together makes up
   (1) a cell (3) an organ
   (2) a tissue (4) an organ system

3. A tissue designed to carry messages throughout the body is most likely to be
   (1) skin (3) nerve
   (2) muscle (4) bone

4. Going from the simple to the complex, which order correctly represents the organization of the human body?
   (1) organ system → organ → cell → tissue
   (2) cell → tissue → organ → organ system
   (3) tissue → cell → organ → organ system
   (4) organ → organ system → cell → tissue

5. Which body system supports and protects other body systems?
   (1) skeletal (3) reproductive
   (2) endocrine (4) digestive

6. Which body system provides for movement of the body?
   (1) digestive (3) muscular
   (2) circulatory (4) endocrine

7. Which group lists three parts of the skeletal system?
   (1) heart, stomach, brain
   (2) tendons, nerves, brain
   (3) bones, nerves, blood
   (4) cartilage, ligaments, bones

8. Which type of muscle is found only in the heart?
   (1) voluntary (3) cardiac
   (2) smooth (4) involuntary

9. Which activity is most likely to be controlled by a smooth muscle?
   (1) breathing (3) chewing
   (2) walking (4) thinking

10. Figure 16-4, page 193, best demonstrates that
   (1) the skeleton protects body organs
   (2) bones are held together at joints and ligaments
   (3) muscles and bones work together to move body parts
   (4) cartilage protects and cushions bones

Regulation, Digestion, and Circulation

Regulation

The *nervous system* and the *endocrine system* work together to regulate body processes and actions. They provide us with a way of detecting and responding to stimuli.

The nervous system (Figure 16-5) is made up of the *brain, spinal cord, nerves*, and parts of the *sense organs*.

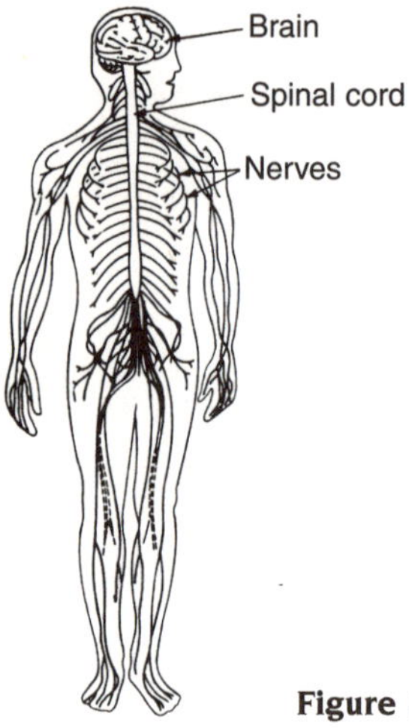

Figure 16-5. The human nervous system.

1. The ***brain*** receives and interprets *nerve impulses* ("messages") and controls thinking, voluntary action, and some involuntary actions, such as breathing and digestion.

2. The ***spinal cord*** channels nerve impulses to and from the brain and controls many automatic responses, or *reflexes*, such as pulling your hand away from a flame.

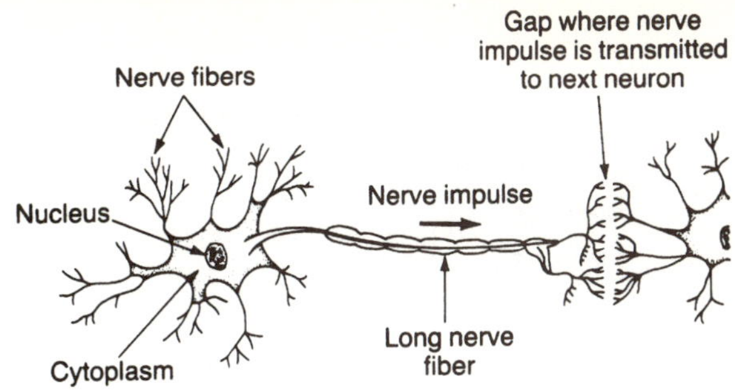

Figure 16-6. A typical neuron, or nerve cell.

3. ***Nerves*** provide a means of communication among the sense organs, the brain and spinal cord, and the muscles and glands.

4. The ***sense organs,*** which include the skin, eyes, ears, nose, and tongue, receive information from the environment.

Nerve cells, also called ***neurons,*** receive and transmit nerve impulses (Figure 16-6). There are two kinds of neurons that transmit messages. *Sensory neurons* carry information from the sense organs to the brain or spinal cord. *Motor neurons* carry messages from the brain or spinal cord to muscles and glands, which respond to the messages.

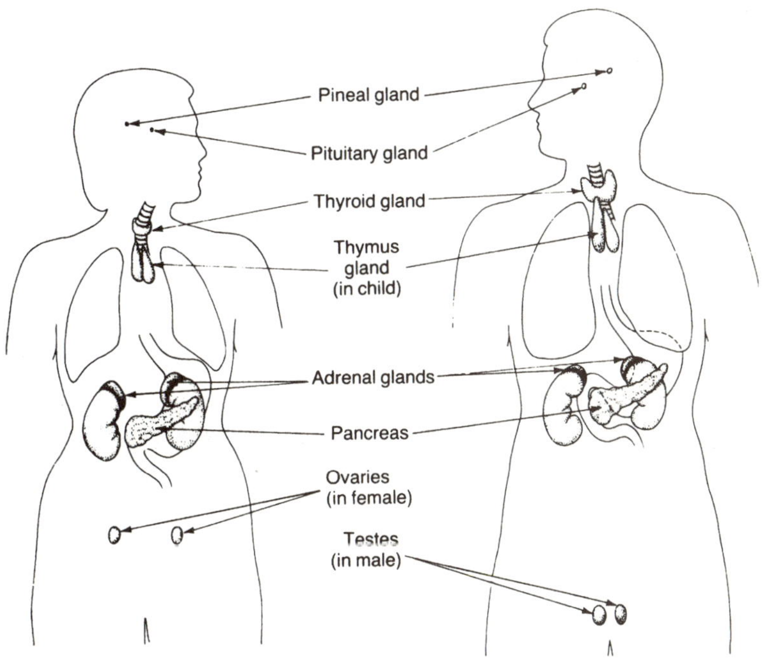

Figure 16-7. Some major glands of the human endocrine system.

The endocrine system is made up of glands. A ***gland*** is an organ that makes and *secretes* (releases) chemicals called ***hormones.*** Figure 16-7 shows some of the endocrine glands. When an endocrine gland secretes a hormone into the bloodstream, the blood carries the hormone to an organ, which responds in some way. For example, if you are suddenly faced with a danger, such as a snarling dog, the hormone *adrenaline* is released by your *adrenal gland.* The adrenaline makes your heart beat faster and your breathing more rapid. More sugar is released into your bloodstream to provide energy. These changes prepare your body to respond to the danger.

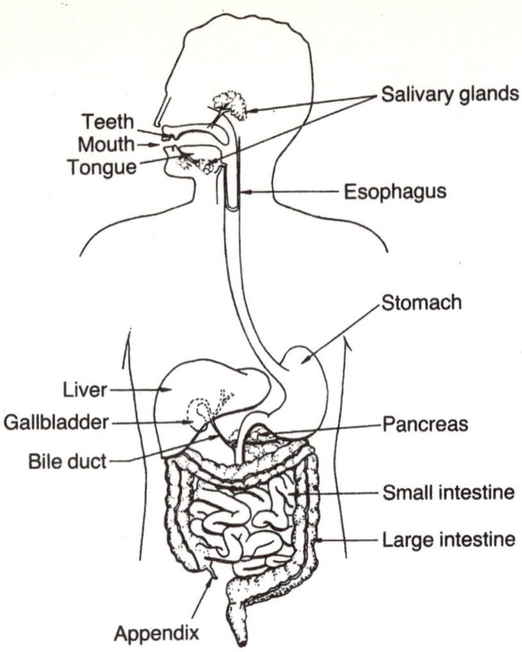

Figure 16-8. The human digestive system.

The Digestive System

Our cells need nutrients from food for energy, growth, and repair. The *digestive system* breaks down food into nutrients that can then be absorbed into the bloodstream and carried to the cells.

The digestive system, shown in Figure 16-8, consists of the digestive tract and the accessory organs.

1. The *digestive tract* is a tube in which food travels through the body. It begins at the mouth and continues through the *esophagus, stomach, small intestine,* and *large intestine.*

2. The *accessory organs* are the *pancreas, gallbladder,* and *liver.* They produce digestive juices that are released into the digestive tract. Table 16-4 lists the digestive juices, where they are produced, and what foods they digest.

Table 16-4. Digestive Juices

Organ	*Digestive Juice*	*Foods Acted On*
Mouth	Saliva	Starches
Stomach	Gastric juice	Proteins
Small intestine	Intestinal juices	Sugars, proteins
Pancreas	Pancreatic juice	Proteins, starches, fats
Liver	Bile	Fats

Note: Bile and pancreatic juice are secreted by the liver and pancreas into the small intestine, where digestion occurs.

The digestive system breaks down food by *physical* and *chemical* means.

1. *Physical.* Food is physically broken down into small bits by chewing and by the action of muscles in the digestive tract.

2. *Chemical.* This breakdown of food into nutrients that can be used by cells is accomplished by chemicals called *enzymes*, found in the digestive juices.

Digestion starts in the mouth and continues in the esophagus, stomach, and small intestine. When digestion has been completed, digested materials are absorbed into the bloodstream through the walls of the small intestine. Undigested materials, which make up the solid wastes called *feces*, pass on through the large intestine and are expelled from the body.

The Circulatory System

Nutrients absorbed into the blood must be transported to all body cells. This is the job of the *circulatory system*: to bring needed materials such as nutrients, water, and oxygen to the cells and to carry away wastes, like carbon dioxide, from the cells.

The components of the circulatory system are the *blood, heart, blood vessels* (*arteries, veins, and capillaries*), *lymph*, and *lymph vessels*.

1. *Blood.* The ***blood*** is a liquid tissue that contains red blood cells, white blood cells, and platelets. The blood also carries dissolved nutrients, wastes, and hormones.

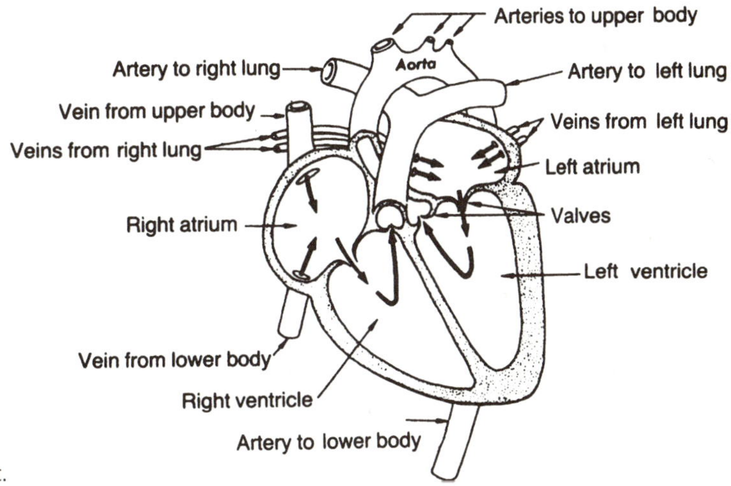

Figure 16-9. The human heart.

2. *Heart.* The ***heart*** (Figure 16-9) is a muscle that contracts regularly to pump blood throughout the body. The blood is pumped from the heart to the lungs, where it receives oxygen and gets rid of carbon dioxide. The blood then returns to the heart, to be pumped to the rest of the body, as shown in Figure 16-10.

3. *Blood vessels.* The blood flows through a network of tubes called ***blood vessels.*** There are three types of blood vessels. ***Arteries*** carry blood away from the heart, while ***veins*** return blood to the heart. Connecting arteries to veins are the ***capillaries.*** Through these extremely thin blood vessels, essential materials are exchanged between the blood and the body's cells. Dissolved nutrients, water, and oxygen pass from the blood into the cells, and some wastes (such as carbon dioxide) pass from the cells into the blood.

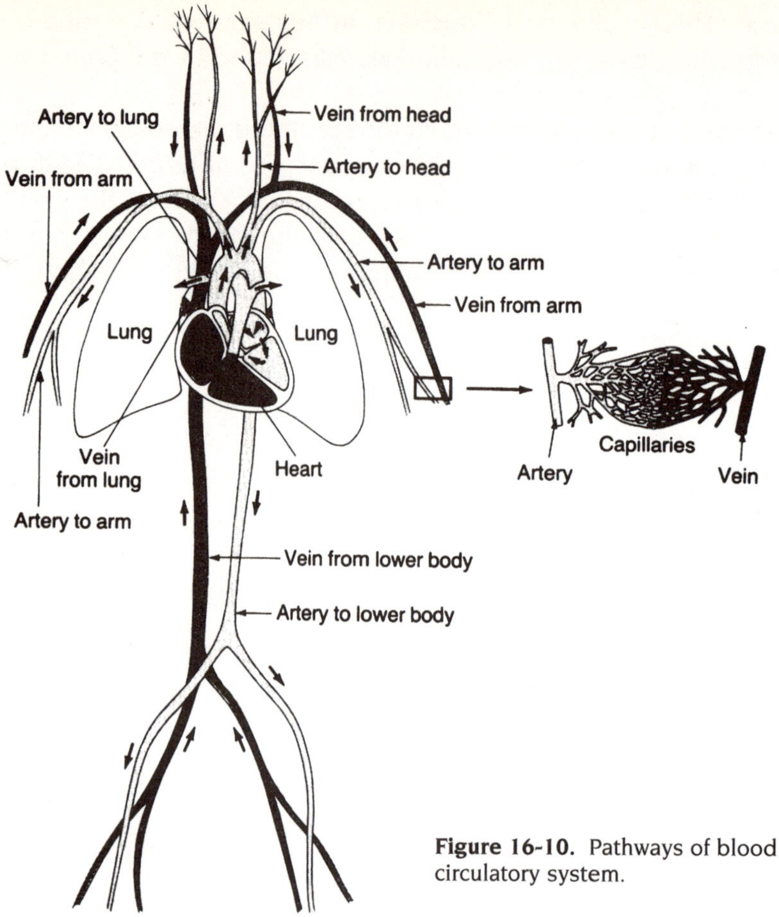

Figure 16-10. Pathways of blood through the human circulatory system.

4. *Lymph.* Some of the watery part of the blood filters out through the walls of the capillaries into the surrounding tissue. This fluid, called ***lymph,*** bathes all the cells of the body. Lymph acts as a go-between in the exchange of materials between the blood and the cells. After receiving wastes from the cells, lymph is collected and returned to the bloodstream through ***lymph vessels.***

EXERCISE 2

1. The brain, spinal cord, and sensory neurons are all part of the
   (1) nervous system
   (3) circulatory system
   (2) respiratory system
   (4) endocrine system

2. Hormones are chemicals that are secreted by the
   (1) gallbladder (3) endocrine glands
   (2) brain (4) small intestine

3. The endocrine system works with the nervous system to
   (1) digest nutrients
   (2) exchange gases with the environment
   (3) produce energy
   (4) regulate body activities

4. The human cell shown in the diagram functions to

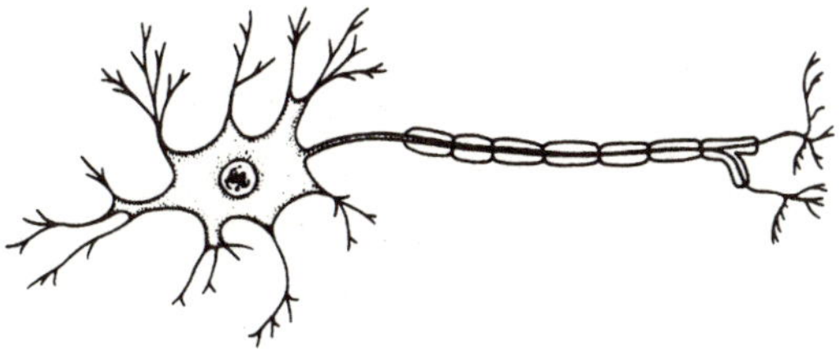

(1) store excess food
(2) send and receive nerve impulses
(3) cover and protect the body
(4) carry oxygen to other cells

5. Food is broken down into a usable form by the
(1) nervous system
(2) skeletal system
(3) digestive system
(4) circulatory system

6. Which group of structures lists parts of the digestive system?
(1) heart, lungs, pituitary gland
(2) adrenal gland, pituitary gland, thyroid gland
(3) skin, kidneys, lungs
(4) stomach, intestines, pancreas

7. Solid materials that are not digestible are eliminated from the body as
(1) urine (3) lymph
(2) perspiration (4) feces

8. The function of the circulatory system is to
(1) carry materials to and from the cells
(2) break down food into a usable form
(3) regulate body activities
(4) respond to stimuli

9. Which group of structures all belong to the circulatory system?
(1) heart, liver, lungs
(2) arteries, veins, capillaries
(3) arteries, kidneys, stomach
(4) bones, cartilage, ligaments

10. Which represents the correct pathway of the nutrients in an apple after you take a bite?
(1) circulatory system → cell → digestive system
(2) cell → digestive system → circulatory system
(3) digestive system → circulatory system → cell
(4) circulatory system → digestive system → cell

Questions 11 and 12 refer to the four diagrams of organ systems shown below.

11. Which diagram represents the digestive system?

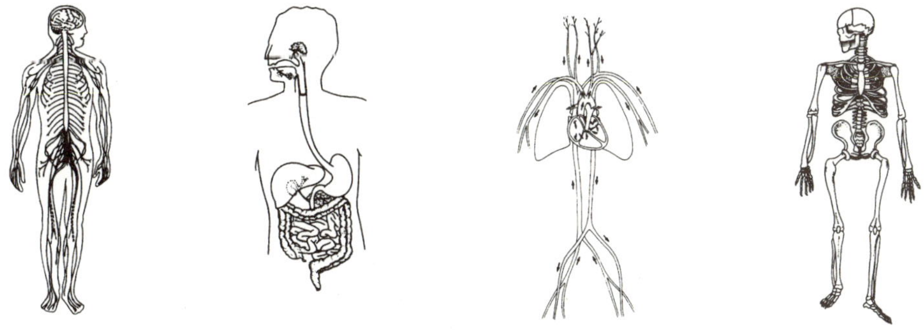

(1) (2) (3) (4)

12. Which system transports needed nutrients to all the cells of the body?
(1) 1 (2) 2 (3) 3 (4) 4

Respiration, Excretion, and Reproduction

The Respiratory System

The circulatory system provides oxygen to the cells. Cells use this oxygen in the process of ***cellular respiration,*** in which nutrients from digested food combine with the oxygen to release energy and produce the waste materials carbon dioxide and water. This chemical process takes place in all body cells.

The *respiratory system,* illustrated in Figure 16-11, brings oxygen from the air to the blood and returns carbon dioxide from the blood to the air. This process is called ***respiration.***

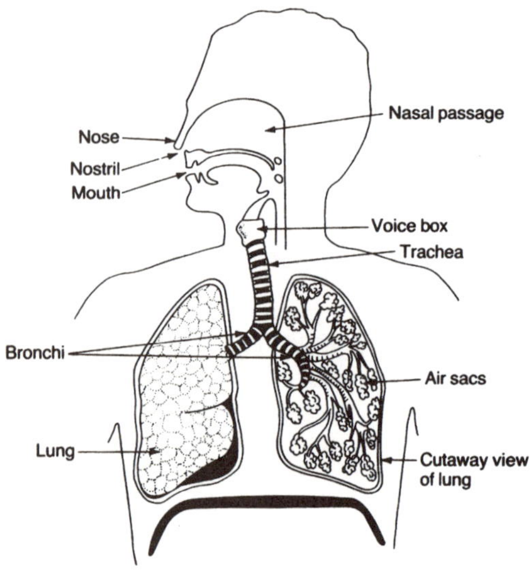

Figure 16-11. The human respiratory system.

When you breathe in (*inhale*), air enters the nose or mouth and passes through the ***trachea,*** or windpipe. The trachea branches off to each lung through tubes called ***bronchi.*** The lungs contain millions of tiny *air sacs,* surrounded by capillaries. Here, respiratory gases are exchanged—oxygen enters the blood while carbon dioxide leaves the blood and is breathed out (*exhaled*).

The oxygen that enters the blood is carried to the cells of the body, where an exchange of gases again takes place. This time, oxygen leaves the blood and enters the cells, while carbon dioxide leaves the cells and goes into the blood. The carbon dioxide is returned to the lungs to be exhaled. This process is repeated constantly.

Process Skill

INTERPRETING A DIAGRAM

The diagram below is a *schematic* representation of the circulatory system. In other words, it is not meant to be a realistic drawing of body parts, but only to show the basic scheme of the system—the relationships among its parts and the sequence of events that occur in the system.

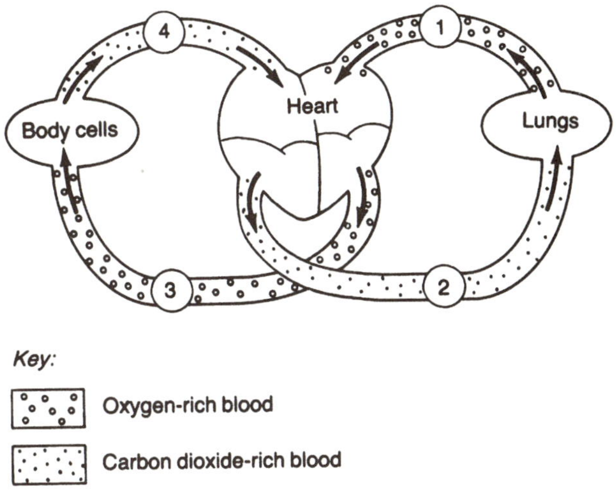

The circulation of blood is vital to the process of respiration, since the blood carries fresh oxygen to the cells of the body and returns carbon dioxide to the lungs to be expelled.

As you have learned, arteries are blood vessels that carry blood away from the heart. Which blood vessels in the diagram are arteries? The arrows indicate that blood vessels 2 and 3 carry blood away from the heart, so they are arteries. Blood vessels 1 and 4, which return blood to the heart, are veins. Study the diagram and then answer the following questions.

1. Blood rich in oxygen is found in blood vessels
   (1) 1 and 2 (2) 2 and 3 (3) 1 and 3 (4) 2 and 4

2. Compared with blood vessel 1, the amount of carbon dioxide in blood vessel 2 is
   (1) more (2) less (3) the same

3. Which statement is true?
   (1) All arteries carry oxygen-rich blood.
   (2) All veins carry oxygen-rich blood.
   (3) Arteries from the heart to the lungs carry oxygen-rich blood.
   (4) Veins from the lungs to the heart carry oxygen-rich blood.

The Excretory System

The activities of the body's cells produce waste materials that must be removed. These wastes are eliminated from the blood and, eventually, from the body by the *excretory system*.

The excretory system consists of the *lungs, skin, kidneys*, and *liver*.

1. The ***lungs*** expel the waste products carbon dioxide and water vapor from your body each time you exhale.

2. The ***skin*** expels wastes when you perspire. Microscopic sweat glands deep in the skin excrete *perspiration*, a liquid waste consisting mostly of water and salts. Perspiration leaves the body through the *pores*, which are tiny openings in the surface of the skin (Figure 16-12).

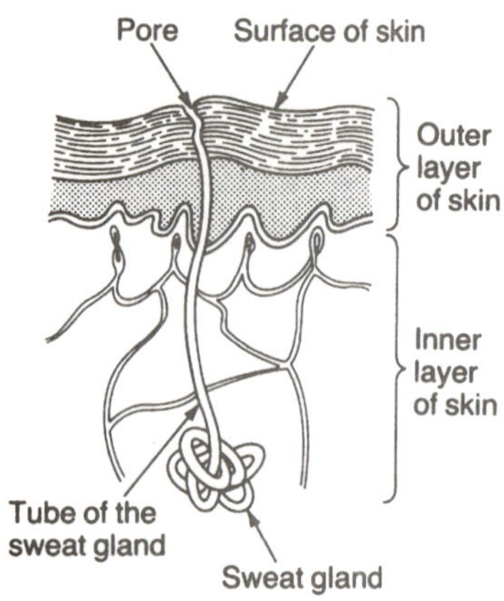

Figure 16-12. Sweat glands in the skin expel wastes through the pores.

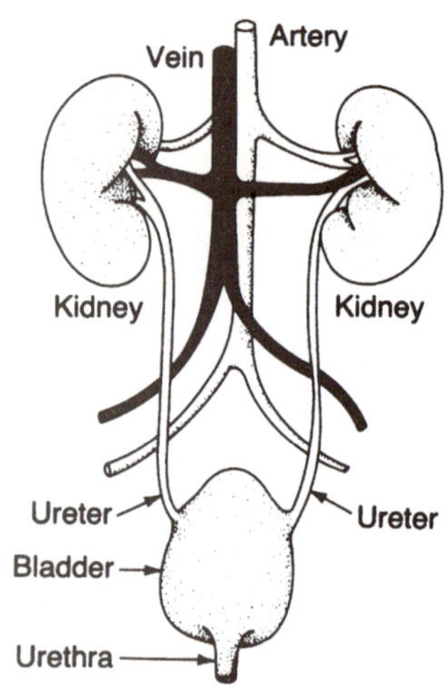

Figure 16-13. The human urinary system.

3. The two ***kidneys*** (Figure 16-13) help to maintain the proper balance of water and minerals in the body. As blood flows through the kidneys, the body's excess water, salts, urea, and other wastes are removed from the blood. These substances make up a fluid called *urine*. Urine is sent through a tube from each kidney to the *bladder*, where it is stored until excreted from the body.

4. The ***liver*** produces ***urea***, a waste that results from the breakdown of proteins. Urea is taken by the blood to the kidneys; it is later expelled from the body in urine. The liver also removes harmful substances from the blood.

The Reproductive System

The reproductive system is responsible for the production of offspring. There are two human reproductive systems—the male and the female—as shown in Figure 16-14.

1. *Male*. The male reproductive system consists of the *testes, penis*, and *sperm ducts*. The ***testes*** produce *sperm cells*, the male sex cells. During reproduction, these cells pass through tubes called ***sperm ducts***, where they mix with a fluid to form *semen*. The semen is delivered through the *penis* into the female's reproductive system.

2. *Female*. Making up the female reproductive system are the *ovaries, oviducts, uterus,* and *vagina*. The ***ovaries*** produce *egg cells*, the female reproductive cells. Once a month, an egg cell leaves an ovary and travels through one of the ***oviducts*** to the ***uterus***, or womb. If sperm cells are present in the oviduct, fertilization may take place.

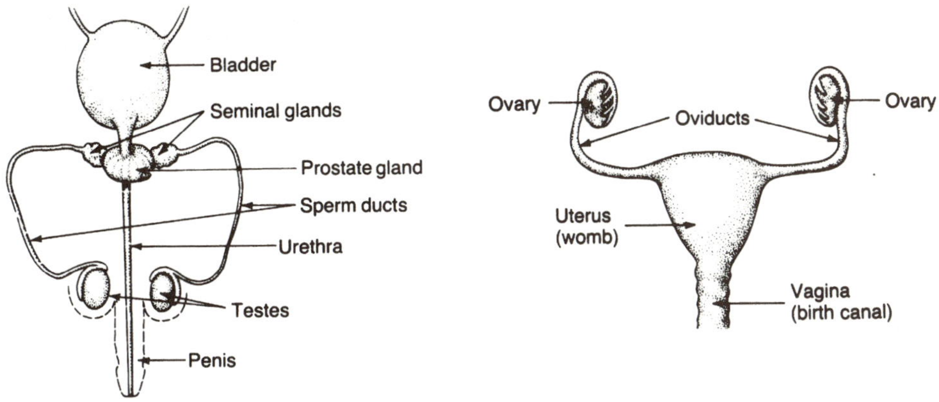

Figure 16-14. The human reproductive systems: male (left) and female (right).

After fertilization has occurred, the fertilized egg attaches itself to the inner wall of the uterus. There it develops into a new offspring over a period of about nine months. At the end of this time birth takes place, and the offspring emerges through the ***vagina***, or birth canal. The newborn baby may be fed milk produced by the mother's ***mammary glands***, or breasts.

EXERCISE 3

1. The process by which energy is released from nutrients is called
   (1) cellular respiration (3) digestion
   (2) excretion (4) circulation

2. Cellular respiration takes place in
   (1) the blood only
   (2) the lungs only
   (3) the heart only
   (4) all body cells

3. The respiratory system includes the
   (1) heart, liver, lungs
   (2) lungs, trachea, nose
   (3) stomach, esophagus, liver
   (4) heart, arteries, veins

4. Which represents the correct order in which oxygen enters the body?
   (1) nose, trachea, bronchi, lungs
   (2) bronchi, nose, trachea, lungs
   (3) lungs, bronchi, trachea, nose
   (4) nose, bronchi, trachea, lungs

5. The exchange of gases between the air and the blood takes place in the
   (1) nose (3) bronchi
   (2) trachea (4) lungs

6. Which organ belongs to both the excretory system and the respiratory system?
   (1) heart (3) lungs
   (2) kidney (4) liver

7. At each body cell,
   (1) carbon dioxide enters the blood, and oxygen leaves the blood
   (2) both carbon dioxide and oxygen enter the blood
   (3) both carbon dioxide and oxygen leave the blood
   (4) oxygen enters the blood and carbon dioxide leaves the blood

8. The excretory system includes the
   (1) kidneys, liver, lungs
   (2) lungs, trachea, nose
   (3) stomach, esophagus, liver
   (4) heart, arteries, veins

9. Which helps remove wastes from the body?
   (1) skull (3) spinal cord
   (2) skin (4) stomach

10. Male is to sperm as female is to
   (1) egg (3) oviduct
   (2) testes (4) uterus

Use the diagrams below to answer question 11.

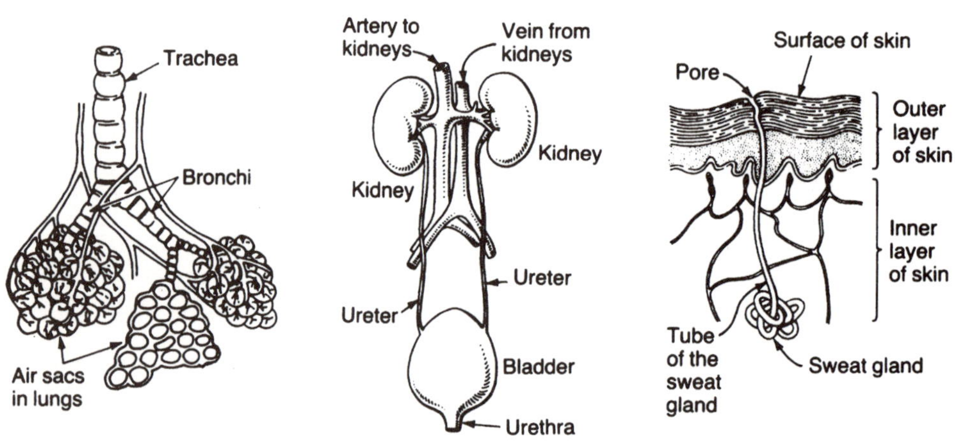

11. To which system do all three structures belong?
   (1) excretory (2) circulatory (3) respiratory (4) skeletal

12. All veins in the body carry oxygen-poor blood to the heart, except one. Which vein carries oxygen-rich blood? Where is the oxygen-rich blood pumped next?

17 Scientific Discovery

Points to Remember —

- Science is the search for the truth about the universe. The processes of science include questioning, exploring, experimenting, observing, measuring, concluding, and communicating. Science also includes a body of knowledge consisting of facts, theories, principles, laws, and so on.
- The scientific method is a procedure by which scientific discovery can be accomplished. However, scientific discovery is not a single process; it can be accomplished in many different ways.
- Factors that influence scientific discovery may be social, economic, and technological.
- The ways scientific discovery occurs vary greatly in time and the number of scientists contributing to the discovery. It often takes many years and many individuals to arrive at the truth about some aspect of the universe.

Science Defined

Science, briefly defined, is the search for the truth about the universe. The search extends from the smallest part of the atom to the whole of space and its contents. It is a process by which information about the universe is gathered and organized into bodies of knowledge such as biology, physics, chemistry, geology, and astronomy.

Most important, science is a process that seeks answers about the world around us. Why do objects fall? How do plants produce food? What causes the flu? These are just a few of the many questions that scientists have devoted lifetimes to in seeking an answer. In addition to questioning, science also involves exploring, experimenting, observing, measuring, concluding, and communicating.

Science is also a body of knowledge consisting of facts, laws, principles, and theories. This body of knowledge contains the known truths of the universe and is based on evidence gathered by inquiring scientists for more than 2000 years. Newton's laws of motion, the movements of air masses, and the dependence of green plants on sunlight are all part of this body of knowledge.

The Scientific Method

A fundamental part of the process of science is the *scientific method* (see Figure 17-1), and although it is not the only procedure used by scientists, it does provide a model of how science adds to the body of knowledge. Many scientific discoveries have been made using this step-by-step procedure; however, other discoveries have been made purely by accident or without any logical process.

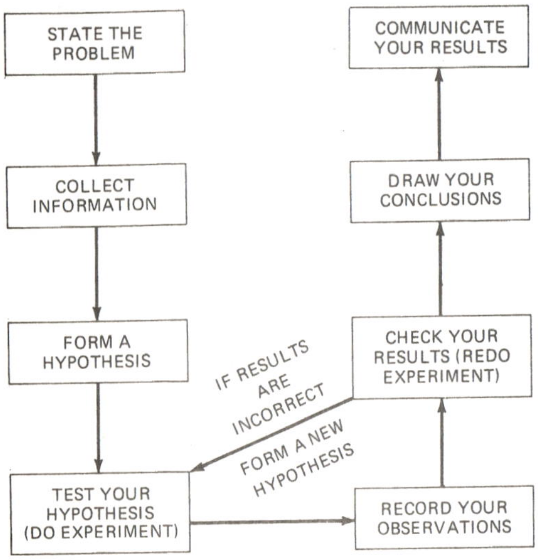

Figure 17-1. The scientific method.

People involved in the study of any scientific topic commonly start with a question about some aspect of that topic. History has shown that such people need not be scientists; they can be any people with a strong desire to understand the world around them. In an effort to answer the question, they research the related body of knowledge to learn what others have observed and what previous evidence has been gathered. A hypothesis, or plausible guess based on the known facts, is proposed to answer the questions, and further independent experimentation and observation are carried out to test the hypothesis. If all evidence gathered supports the hypothesis and appears to lead to new truths about the universe, a theory may be presented to the scientific community, usually in the form of a published report or presentation. Thus, other scientists will be able to review the evidence for the theory and judge its worth. The theory, in turn, will raise new questions that will lead to new hypotheses, new experiments, and new studies. (See Figure 17-2.)

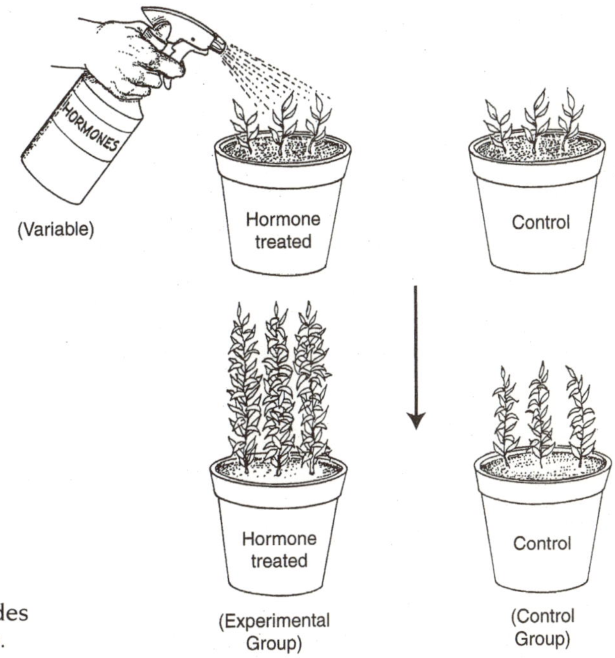

Figure 17-2. A scientific experiment includes a control group and an experimental group.

EXERCISE 1

1. Science is best described as
   (1) a body of knowledge
   (2) a process to study the universe
   (3) both a process to study the universe and a body of knowledge
   (4) neither a process of study nor a body of knowledge

2. The scientific method should be thought of as
   (1) the method by which scientific information is discovered
   (2) the method by which all experiments are conducted
   (3) the method used by all scientists to make discoveries
   (4) a model of a process that can be used to find answers to scientific problems

Questions 3–5 refer to the following article, which appeared on the Internet at http://www.eurekalert.org/E-ert/current/public_releases/mars/mrel.html.

> Washington, D.C.—Ever since scientists learned that water once flowed on Mars, they've wondered whether life might also have flourished on the apparently now-dead planet. In the 16 August 1996 issue of *Science*, McKay et al. report the first identification of organic compounds in a Martian meteorite. The authors further suggest that these compounds, in conjunction with a number of mineralogical features observed in the rock, may be evidence of ancient Martian microorganisms.

3. The science question addressed by this article is,
   (1) Do meteorites come from Mars?
   (2) Was there life on Mars?
   (3) Is there water on Mars?
   (4) What compounds are found in meteorites?

4. The printing of this article in the 16 August 1996 issue of *Science* represents what step in the scientific process?
   (1) presentation of a theory to the scientific world and general public
   (2) development of an initial question to be studied
   (3) experiments and observations to test a hypothesis
   (4) development of a hypothesis

5. After reading the brief article above, which question would you expect to find answered in the complete article?
   (1) Does water flow on Mars?
   (2) Can humans live on Mars?
   (3) What other forms of life are on Mars?
   (4) What mineralogical features did the authors find in the meteorite?

6. A scientist may attempt to answer the question, "How can we efficiently harness the energy of the sun to meet humanity's energy needs?" To begin solving this problem, the scientist would first most likely
   (1) develop a theory
   (2) develop a hypothesis
   (3) research what other scientists have done on this topic
   (4) publish an article about the question

7. To be accepted by the scientific community, a theory must be
   (1) voted on by scientists
   (2) supported by evidence
   (3) determined by the scientific method
   (4) agreed on by all scientists

Example of Major Scientific Discoveries

Many of the great discoveries in science have come about over a span of many years, with contributions by many individuals. Throughout history, scientific theories are constantly challenged, changed, and refined by new observations and new evidence.

A major scientific theory may take decades or even centuries to gain acceptance. Typically, such theories are the result of observations and evidence gathered by many individuals. The following three examples briefly describe the chronological events leading to a major scientific theory.

Plate Tectonics

Question 1: How did the continents get their present shapes and positions?

Question 2: What process could cause the continents to shift on Earth's surface?

1912: Alfred Wegener proposed that the continents were drifting across Earth's surface, based on the shapes of the continents and on some matching features of rocks and fossils on the continents on either side of the Atlantic Ocean. He thought that the continents were once together in a single landmass that broke apart. Thus, the theory of continental drift was proposed with some limited supporting evidence.

1912-1960: Continental drift was discussed by the scientific community for about a decade, but eventually interest declined because there was no known mechanism within Earth that could account for how the continents moved. Although the evidence presented by Wegener was intriguing, his theory lacked a satisfactory answer to this important question.

1960: Oceanographers discovered an underwater mountain ridge (the Mid-Atlantic Ridge) running north-south down the middle of the Atlantic Ocean. Harry Hess suggested that new rock material was upwelling along the ridge and moving away from the ridge in a conveyor beltlike fashion, pushing out in opposite directions. (See Figure 17-3a.)

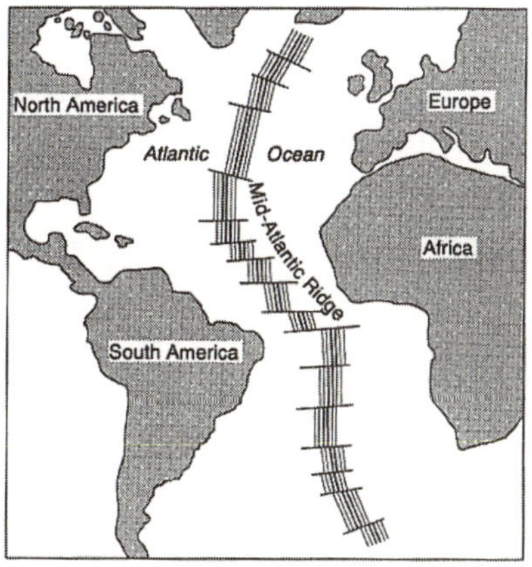

Figure 17-3a. New rock material upwells along the ridge and moves away from it in opposite directions.

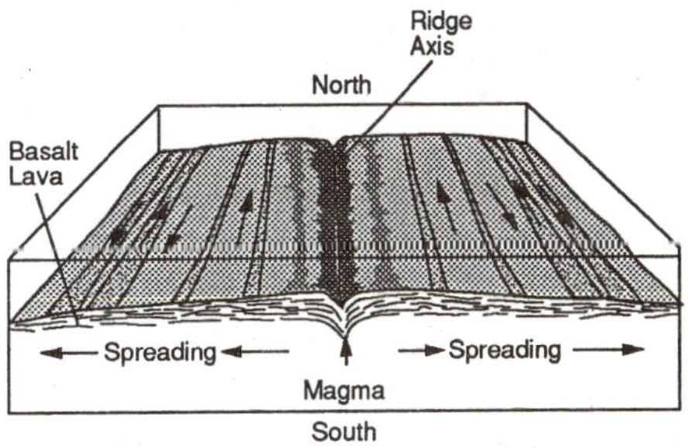

Figure 17-3b. Identical magnetic bands are seen on each side of the ridge, proof of seafloor spreading.

1969: Fred Vine and Drummond Matthews showed that strips of ocean floor with matching magnetic polarity found on opposite sides of the Mid-Atlantic Ridge constituted evidence that the process proposed by Hess, called seafloor spreading, was occurring. (See Figure 17-3b.) Seafloor spreading was the mechanism needed to support the theory of continental drift.

1970s: Detailed mapping of the ocean floor revealed other features that supported seafloor spreading as the mechanism for continental drift.

When Alfred Wegener proposed the theory of continental drift, other scientists were skeptical and raised important questions about the evidence. Although many geologists were intrigued with the concept, most were unsatisfied with the theory. They investigated other possibilities to account for Wegener's observations. Today, after much has been learned about the ocean floor, the continental drift theory has been thoroughly tested and is accepted by almost all scientists.

Heliocentric Solar System

Question: What is Earth's position in the solar system and in the universe?

6th century B.C.: Early Greek scholars believed that fire (the sun) occupied the center of the universe and that Earth moved around the fire. The concept was correct, but scientific evidence was lacking.

About 350 B.C.: Aristotle favored a geocentric (earth-centered) solar system, with the sun and planets revolving around Earth.

About 150 A.D.: Ptolemy reinforced the notion of an earth-centered solar system by accounting for the observed reverse, or "retrograde," motion of the visible planets which didn't seem to fit Aristotle's model. Although his explanation was incorrect, it was accepted because it seemed to fit the data, and it upheld the idea that Earth was central.

1543: Nicholas Copernicus proposed a *heliocentric* (sun-centered) theory of the solar system based on the observed motions of planets and stars. Copernicus claimed that the planets, including Earth, orbit the sun in perfect circles. (See Figure 17-4.)

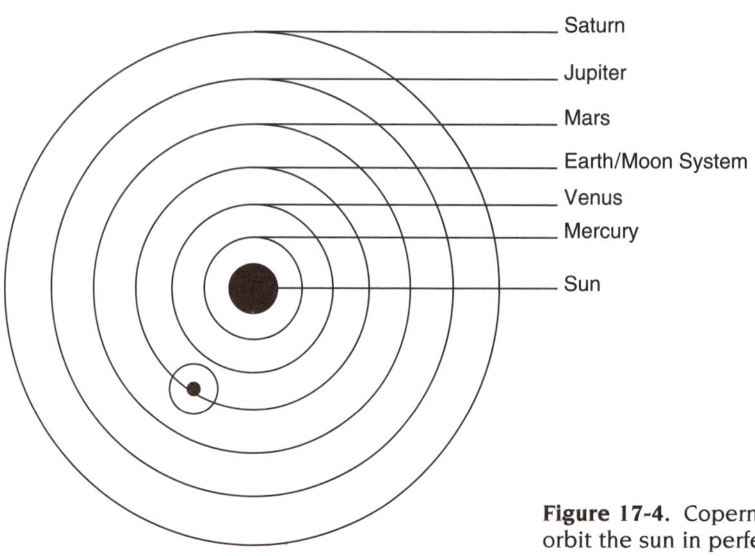

Figure 17-4. Copernicus claimed that the planets orbit the sun in perfect circles.

1632: Galileo Galilei presented arguments in support of the heliocentric theory. Using a telescope, which had recently been invented, he observed the changing phases of the planet Venus. He showed that this could best be explained by assuming that Venus orbits the sun and sometimes passes between Earth and the sun. Thus, new technology contributed evidence to reinforce the heliocentric theory.

1618: Johannes Kepler, using extensive observations of the planet Mars, was able to derive mathematical laws of planetary motion, and he calculated the orbits of the known planets. Kepler found that their orbits were slightly *elliptical* (oval), rather than circular, refining the heliocentric theory.

1784: William Herschel concluded from the distribution of stars in the sky that our solar system was part of a large disk-shaped cluster of stars, which he called the Milky Way Galaxy. Recognizing that the visible stars were more densely clustered in one area of the Milky Way and less densely grouped in other directions, he pictured the sun in the center of this great mass of stars. Although his general concept of the Milky Way was correct, he was later proved wrong about locating the sun in the center of the galaxy.

It took about 2500 years for scientists to fully understand Earth's place in the universe. Each scientist contributed new evidence to change or refine earlier thinking. Most formally published their ideas with supporting evidence that could be evaluated by other scientists. Galileo also demonstrated how new technology contributed to science by providing better observational data. Even today, astronomers are gathering new evidence in efforts to further refine the answer to the original question (see Figure 17-5).

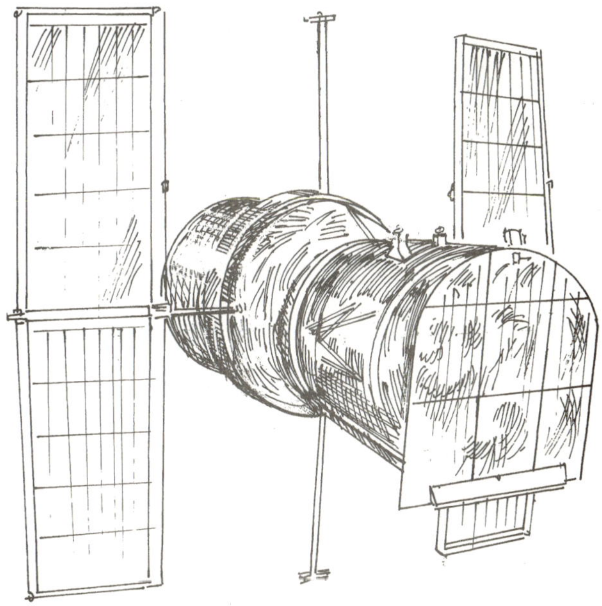

Figure 17-5. The Hubble telescope helps astronomers gather more information about the universe.

Germ Theory

Question: What causes disease?

1800: Benjamin Rush, a well-known "man of medicine" in his time, believed that an imbalance of body fluids caused disease. He felt that diseases of the human body could be cured by bleeding, blistering, and vomiting, and that these acts restored the natural balance of the fluids.

1820–1900: The American public became wary of physicians and believed that anyone could cure ailments by applying commonsense techniques. Scientific medicine was replaced by quackery, and diseases were attributed to all sort of fanciful causes. The scientific causes of disease were generally neither sought nor accepted. Unscientific approaches, such as hydrotherapy (external and internal water purging), electrical treatment, and patented "cure-all" medical potions, were practiced until the turn of the century. As scientific discoveries regarding the causes of disease were made, these methods became less popular.

1840–1870s: Some European doctors conducted studies into the causes and occurrence of diseases and found that increased cleanliness and use of purified water were decreasing the number of childbed fever and cholera cases.

1860s: Civil War casualties forced the medical profession to study the causes of disease and infection more scientifically.

1870: Louis Pasteur developed the germ theory of disease. The germ theory stated that microorganisms were the cause of some diseases. It took several decades for the germ theory to replace quackery medicine in the United States. (See Figure 17-6.)

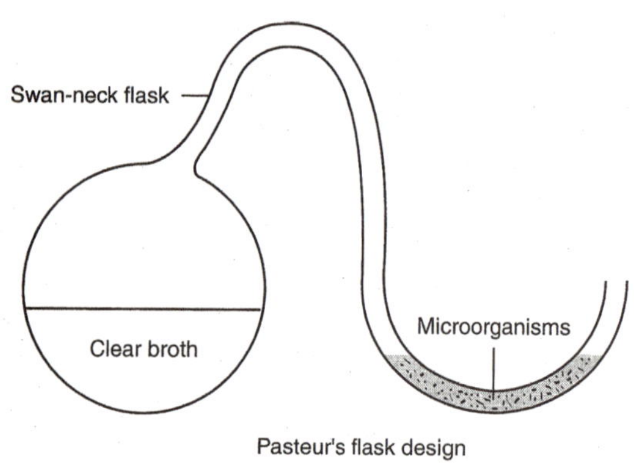

Figure 17-6. Pasteur's experiment proved his germ theory of disease.

Scientific medical treatment of disease was accepted slowly because the public lacked confidence in aristocratic physicians and because of the emotional appeal of quacks and their claims of commonsense cures for diseases. These medicines lacked a scientific basis for the cause of disease, and many of these practices were damaging to patients.

Atomic Theory

Question: What is the structure of the atom?

The atom has long been proposed to be the smallest particle of an element. But the exact structure of the atom was not always known. An outline of the major discoveries associated with the development of the atomic theory is presented below.

Early 1800s: John Dalton stated that all matter is composed of extremely small, indivisible particles called atoms. He believed that all atoms were similar (hard, round, solid spheres), but that each elementary substance had it own atoms, and that different atoms could unite to form compound substances.

1897: Joseph J. Thomson discovered that atoms contain negatively charged particles, called electrons.

1903: Philip Lenard reasoned that the atom was not a solid sphere but consisted mostly of empty space.

1911: Ernest Rutherford concluded that the atom consisted of a relatively massive central nucleus surrounded by mostly empty space and electrons.

1913: Niels Bohr suggested that electrons traveled in circular orbits around the nucleus of the atom.

Many new scientific discoveries have been made in recent years, and it appears that scientific discovery is accelerating at an ever-increasing rate. Each new discovery leads to more questions and eventually to more research in an effort to answer the questions. The accelerated pace of scientific discovery can be attributed in part to the use of powerful computers with the ability to analyze vast amounts of data quickly and accurately.

EXERCISE 2

1. A scientific theory may not be accepted because
   (1) the public is not ready to accept a change in thinking
   (2) the evidence is incomplete
   (3) the theory has not been presented or published
   (4) all of the above are possible

2. Which sentence best describes the rate of scientific discovery over the past 2000 years?
   (1) Scientific discovery is a slow process.
   (2) Scientific discovery is a fast process.
   (3) Scientific discovery occurs at variable rates.
   (4) Scientific discovery occurs at a steady rate.

3. Major scientific discoveries are often credited to one person; for example, James Chadwick is credited with the discovery of the neutron in the atom. However, such a discovery was possible because
   (1) many other discoveries about the atom already had been made
   (2) Chadwick used the scientific method
   (3) Chadwick was the first person to study the atom
   (4) Chadwick was the first person to develop a hypothesis about the atom

Questions 4 and 5 refer to the following paragraph.

Sir William Ramsay and Lord Rayleigh (John William Strutt) discovered the element argon in 1895. While making measurements on the density of nitrogen, Lord Rayleigh found what appeared to be errors in his calculations. Upon investigating, he discovered the existence of a new atmospheric gas, later called argon. Over the next five years, Ramsay refined the original calculations even further and thereby discovered neon, krypton, and xenon in the atmosphere. Both scientists were awarded Nobel Prizes in 1904 for their discoveries.

4. Lord Rayleigh discovered argon
   (1) because he was seeking an unknown gas
   (2) by accident
   (3) because his hypothesis was incorrect
   (4) because his theory was correct

5. Ramsay's discovery of neon, krypton, and xenon are an indication that
   (1) a single scientist is not capable of making scientific discoveries
   (2) scientists always use the scientific method to make discoveries
   (3) the work of previous scientists often makes possible new scientific discoveries
   (4) a pair of scientists can always accomplish more than a single scientist

6. Major scientific theories are generally developed
   (1) by one person over a short period of time
   (2) by many people over a short period of time
   (3) by one person over a long period of time
   (4) by many people over a long period of time

Use the information in the outline of major discoveries on page 213 and in the table below to answer questions 7–9.

Number of Elements Discovered, by Decade

Pre-1700	13	1810s	3	1900s	2
1730s	2	1820s	4	1910s	1
1740s	1	1830s	2	1920s	3
1750s	1	1840s	3	1930s	2
1760s	1	1850s	0	1940s	7
1770s	5	1860s	4	1950s	5
1780s	4	1870s	6	1960s	2
1790s	5	1880s	5	1970s	3
1800s	12	1890s	9	1980s	2
				1990s	2

7. The outline of major discoveries associated with the development of the atomic theory lists
   (1) all the scientists who contributed to the atomic theory
   (2) most of the scientists who contributed to the atomic theory
   (3) a few of the scientists who contributed to the atomic theory
   (4) a complete picture of the atomic theory

8. According to the table and the outline above, the number of elements discovered by decade appears to
   (1) correlate well with John Dalton's and Joseph Thomson's contributions to atomic theory
   (2) correlate well with Joseph Thomson's and Niels Bohr's contributions
   (3) correlate well with John Dalton's and Niels Bohr's contributions
   (4) not correlate at all with major discoveries in atomic theory

9. The outline of major discoveries associated with the development of the atomic theory
   (1) provides only the major contributions to the atomic theory
   (2) is lacking evidence for each of the contributions
   (3) is incomplete because it is only for 1800 to 1913
   (4) all of the above

10. In one or two sentences, tell the differences between a scientific hypothesis and a scientific theory.

11. A scientific theory can change over time. In one or two sentences, explain why this is so.

18 Energy and Resources

Points to Remember —

- Fossil fuels are our main sources of energy.
- Fossil fuels come from the remains of dead plants and animals. Oil, coal, and natural gas are fossil fuels. Fossil fuels are nonrenewable resources.
- Burning fossil fuels produces gases such as carbon dioxide, which may affect Earth's atmosphere.
- Other sources of energy include hydroelectric, nuclear, solar, and wind.
- Each energy source has both advantages and disadvantages.
- Decisions about energy sources must weigh many factors. These include supply, cost, and the effect on the environment.
- It is important to conserve our energy supply. Methods of conserving energy include reducing consumption, recycling materials, and reusing materials.
- The products we buy and the ways we use energy affect our energy supply and our environment.

Sources of Energy

Everything that occurs in the universe involves energy. As you have learned, energy is the capacity to do ***work***—the ability to make something move. Heat, light, sound, and electricity are all forms of energy. Humans have learned to describe, explain, and measure energy and to harness it for their use.

Measuring Energy

To compare the amounts of energy stored in various substances, we need to describe energy with some unit of measurement. The energy in foods and fuels can be measured and compared using a unit called ***calorie.*** One calorie is the amount of heat energy needed to raise the temperature of one gram of water by one degree Celsius.

When describing the energy in food, we use the word *Calorie* spelled with a capital C. This "food Calorie" is equal to 1000 ordinary calories, or 1 *kilocalorie* (*kilo* means "one thousand"). Calories indicate how much energy you can obtain from various foods. Digested food containing energy that is not needed by the body is stored, usually as fat. When you go on a diet, you count Calories to make sure you don't eat more food than your body needs for energy.

The *rate* at which energy is used (the amount of energy used over a certain time interval) can also be measured. The rate at which electrical energy is used is measured in a unit called the ***watt.*** We can use watts to compare the rates at which different electrical devices use energy. For instance, a 100-watt lightbulb uses twice as much electricity each second as a 50-watt bulb. Table 18-1 lists some common electrical devices and their wattages.

Table 18-1. Some Electrical Devices and Their Wattages

Electrical Device	*Wattage*
Hair dryer	1200 watts
Lightbulb	100 watts
Electric shaver	7 watts
Small air conditioner	860 watts
Microwave oven	750 watts
Stereo	240 watts
Toaster oven	1400 watts

Energy Consumption

Humanity's consumption of energy is constantly increasing. Our growing populations and economies require more and more energy. *Fuels* are sources of energy. As our demand for energy increases, so does our demand for fuel. We need more fuel to cook meals, heat homes, run industries, and power cars, ships, trains, and airplanes. We also use more fuel to produce electricity.

Electricity has become essential to our society. It is used for many purposes, such as heating and cooling buildings, running machines and appliances, and providing lighting. Electrical energy can be transmitted easily over conductors such as metal wires. However, electrical energy must itself be produced from other energy sources.

Fossil Fuels

The main energy sources used to produce electricity are ***fossil fuels.*** They are called fossil fuels because they were formed from the remains of plants and animals that lived and died long ago. Over time, these organic remains were changed into energy-rich substances. The most commonly used fossil fuels are oil (also called *petroleum*), coal, and natural gas.

Oil is a sticky black liquid usually found trapped within rock layers deep underground. ***Coal*** is a black rock that occurs in layers, or seams, between other rock layers. ***Natural gas*** is commonly found underground with oil deposits. Each of these fossil fuels can be burned to provide energy for the production of electricity. Figure 18-1 shows the relative amounts of oil, coal, natural gas, and other energy sources used to produce electricity in the United States.

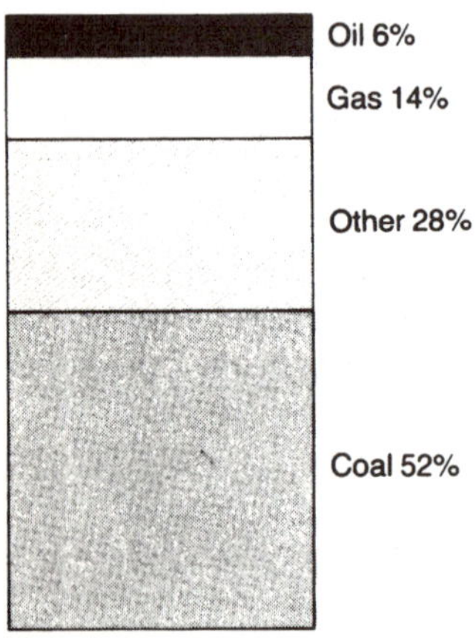

Figure 18-1. Relative percentages of energy sources used to produce electricity in the United States.

Other Uses of Fossil Fuels

Fossil fuels have many uses besides producing electricity. Gasoline is used to power automobiles. Heating fuel is used to heat homes and industries. Both gasoline and heating fuel are obtained from oil and are its main products used for energy. Natural gas is used to heat homes and industries and for cooking. (Gas stoves use natural gas.) Coal, although mainly used to produce electricity, is also used to provide heat for industrial processes, such as the making of steel.

Fossil fuels are used to make many other important substances. Oil, in particular, has many such uses. Plastics, fertilizers, certain drugs, and synthetic fabrics like nylon and polyester are all products made from petroleum.

Process Skill

INTERPRETING A GRAPH

A *pictograph* represents numbers, or quantities, by using pictures. In the accompanying diagram, each picture of a barrel represents one million barrels of oil. For example, the category "Industry and electricity" is represented by three barrels, which means that three million barrels of oil are used each day for these purposes.

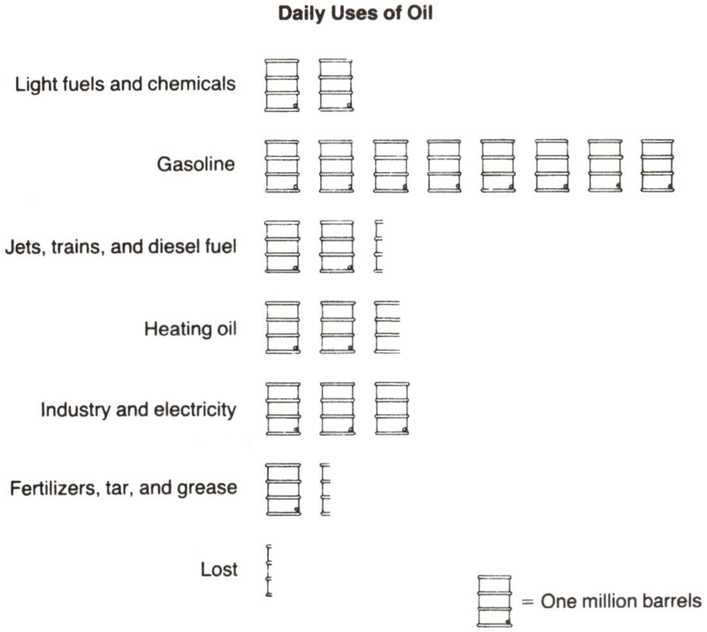

How much oil is used daily for "Jets, trains, and diesel fuel"? Looking at the row of barrels for that category, you see there are two complete barrels plus part of a third barrel. This part is about one-quarter of a complete barrel, so it represents one-quarter of a million barrels of oil. This makes the total daily use of oil for "Jets, trains, and diesel fuel" equal to $2\frac{1}{4}$ million barrels. Use the diagram to help you answer the following questions.

1. Which products use the least amount of oil each day?
   (1) light fuels and chemicals
   (2) gasoline
   (3) heating oil
   (4) fertilizers, tar, grease

2. About how much oil is lost every day?
   (1) half a million barrels
   (2) less than one-quarter of a million barrels
   (3) one million barrels
   (4) more than one-quarter of a million barrels

Other Sources of Energy

While most of the energy we consume comes from fossil fuels, there are several other energy sources currently in use. The most important of these are *hydroelectric energy* and *nuclear energy.*

1. ***Hydroelectric energy*** is electricity produced by the power of flowing water. Water in motion has kinetic energy, which can be transformed into electrical energy. When water flows steeply downhill at a waterfall or a dam on a river, it can be used to operate a *generator,* which produces electricity. A generator contains a *turbine,* a device similar to a paddle wheel on an old-fashioned steamboat (Figure 18-2). The moving water turns the turbine, which spins a coil of wire inside an electromagnet in the generator. This creates an electrical current.

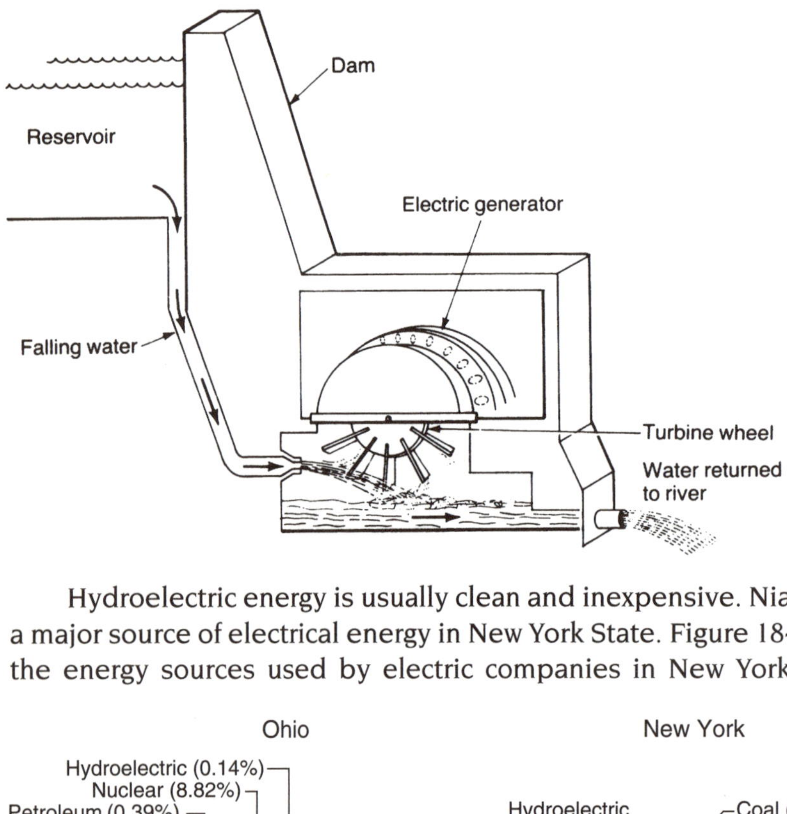

Figure 18-2. A hydroelectric dam is used to produce electricity.

Hydroelectric energy is usually clean and inexpensive. Niagara Falls is a major source of electrical energy in New York State. Figure 18-3 compares the energy sources used by electric companies in New York with those

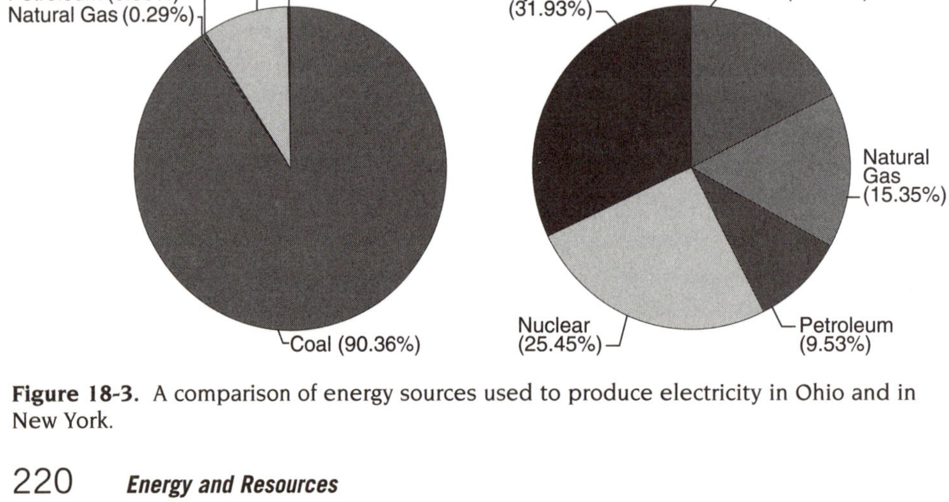

Figure 18-3. A comparison of energy sources used to produce electricity in Ohio and in New York.

used in Ohio. Notice that New York uses a much greater percentage of hydroelectric energy than does Ohio, while the percentage of coal used in New York is much smaller. The choice of energy sources varies from one state to the next, as different states have different sources available to them.

2. ***Nuclear energy*** is the energy stored in the nucleus of an atom. When this energy is released, it creates heat, which can be used to produce electricity. Nuclear power plants use ***uranium*** (a radioactive element found in certain rocks) as their fuel source. Uranium atoms are naturally unstable and can be readily split apart to release heat energy. The heat created by the uranium fuel in a nuclear reactor is used to boil water, thereby producing steam. The steam turns turbines that generate electricity. This is illustrated in Figure 18-4.

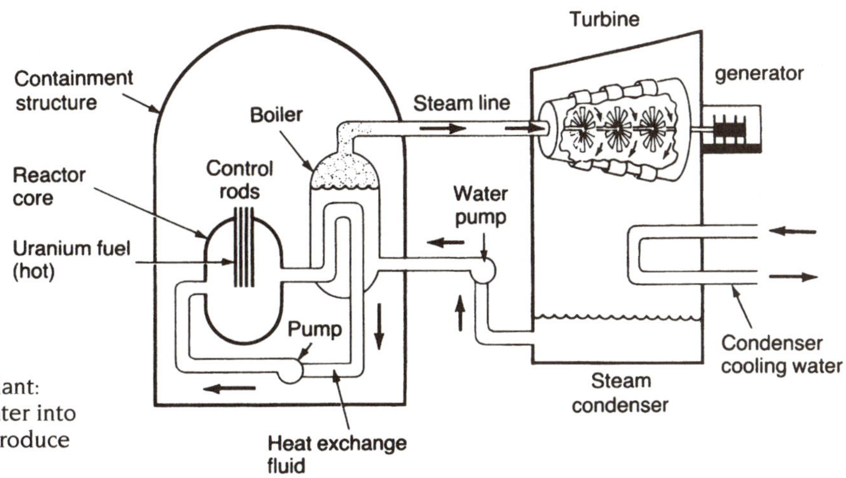

Figure 18-4. A nuclear power plant: Heat from a reaction changes water into steam, which turns turbines to produce electricity.

EXERCISE 1

1. Which is *not* a form of energy?
   (1) heat (3) gasoline
   (2) electricity (4) light

2. The energy obtained from foods is generally measured in
   (1) Calories (3) volts
   (2) watts (4) degrees

3. A unit that measures the rate at which electrical energy is used is a
   (1) volt (3) calorie
   (2) watt (4) degree

4. Which is *not* a fossil fuel?
   (1) uranium (3) oil
   (2) coal (4) natural gas

5. Plastics, fertilizers, and synthetic fabrics are products commonly made from
   (1) coal (3) gasoline
   (2) oil (4) uranium

6. Fossil fuels were formed from
   (1) rocks and minerals
   (2) uranium deposits
   (3) remains of dead plants and animals
   (4) moving water

Base your answers to questions 7–10 on the graphs below, which compare energy used in the United States with energy used in Ohio and New York.

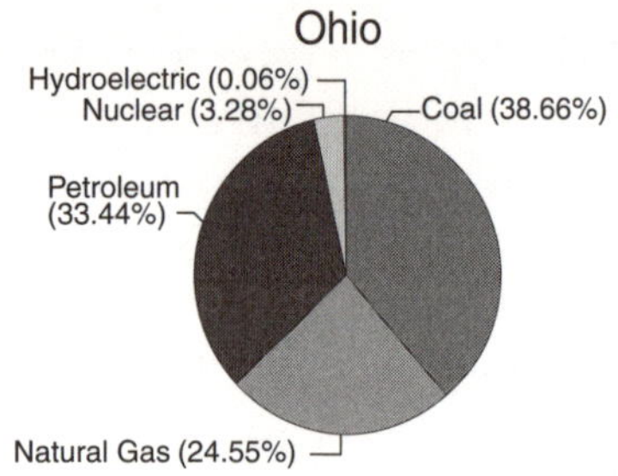

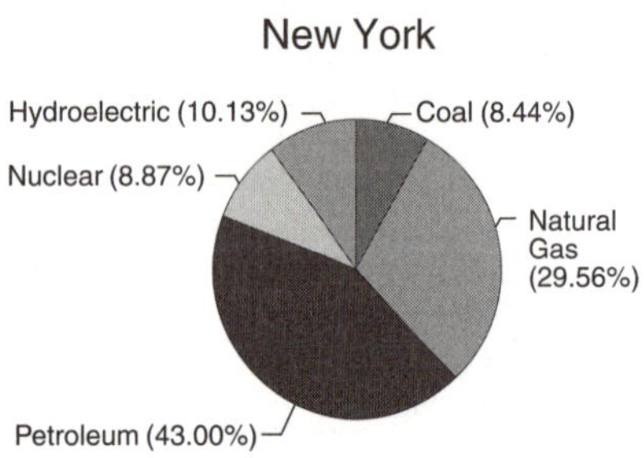

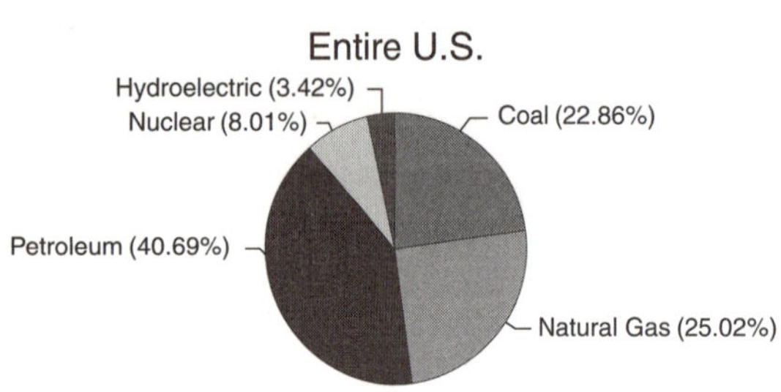

7. Approximately what fraction of Ohio's energy comes from petroleum?
   (1) one-half (3) one-third
   (2) one-quarter (4) two-thirds

8. Compared to the entire United States, Ohio uses
   (1) a larger percentage of oil and a larger percentage of coal
   (2) a smaller percentage of oil and a larger percentage of coal
   (3) a smaller percentage of oil and a smaller percentage of coal
   (4) a larger percentage of oil and a smaller percentage of coal

9. To the nearest whole number, what percentage of Ohio's energy comes from fossil fuel?
   (1) 33% (2) 72% (3) 63% (4) 97%

10. From the graphs of New York and Ohio, we could reasonably infer that
    (1) Ohio has more nuclear power plants than New York
    (2) New York has more available sources of moving water than Ohio
    (3) Coal is more available in New York than in Ohio
    (4) Ohioans use their cars more than New Yorkers do

11. Higher gasoline prices would most likely result from a shortage of
    (1) coal (3) natural gas
    (2) oil (4) heating fuel

12. What are three ways that you can conserve energy in your home or in your school?

Problems With Energy Sources

Problems With Fossil Fuels

In the United States, most of the demand for energy is met by the fossil fuels: oil, coal, and natural gas. However, the burning of fossil fuels creates air pollution. When fossil fuels burn, chemicals are released into the air that pose dangers to living things and their environment. This is especially true of coal.

1. *Coal.* The supply of coal found in the United States is much greater than that of oil or natural gas. Because of its abundance, coal is relatively inexpensive. However, there are serious environmental and health problems involved with its use. The burning of coal contributes greatly to air pollution. Smoke from coal-burning power plants is the main cause of ***acid rain,*** which is harmful to the ecology of lakes and forests.

In addition, coal mining is dangerous for people who work in the mines. Breathing air that contains coal dust is unhealthful and can lead to *black lung disease*. Certain coal-mining techniques are also damaging to the environment. Sometimes large areas of land are dug up to reach the coal. This practice, called *strip mining*, destroys topsoil and scars the landscape.

Mining companies are now required by law to restore the land they have damaged. Advances in technology have made coal mining safer and reduced the amount of pollution caused by burning coal. Nevertheless, these measures have only begun to solve the problems with using coal as a fuel.

2. *Oil and natural gas.* Although oil and natural gas burn cleaner than coal, they cause other environmental problems. Offshore drilling for oil and transporting oil by ship can lead to accidental oil spills that kill marine wildlife and cause severe pollution of land and sea. Pipelines built to transport oil and gas over land may alter the ecology of areas they cross.

Problems With Hydroelectric and Nuclear Energy

The production of electricity using moving water or nuclear reactions is generally much "cleaner" than energy production with fossil fuels. This is because hydroelectric and nuclear power do not involve burning anything. However, even these "clean" energy sources have environment costs.

1. *Hydroelectric energy.* Building a dam on a river to produce hydroelectric power changes the surrounding area, as you can see in Figure 18-5 on page 224. The area upriver is flooded, creating a large lake over land that may have once provided a habitat for wildlife or been used for farming. The area downriver from the dam receives a diminished flow of water. These changes greatly affect the ecology of the area around a dam.

2. *Nuclear energy.* Although nuclear power plants do not cause air pollution, they use water from nearby lakes or rivers to cool their nuclear

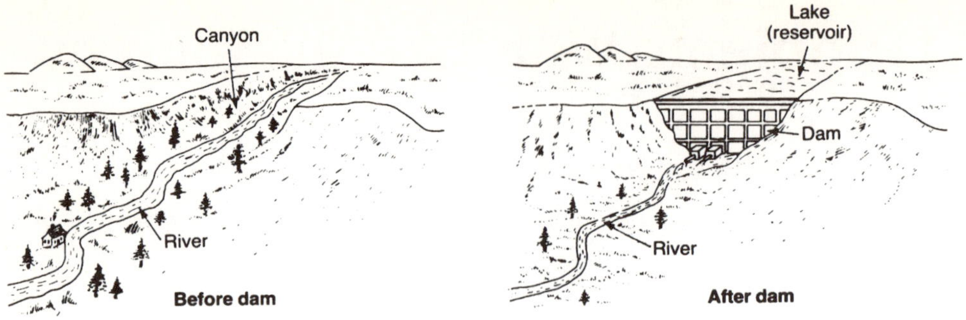

Figure 18-5. The effect of a hydroelectric dam on the environment.

reactors. The water is then returned to the environment several degrees warmer. This increase in the temperature of the environment, called ***thermal pollution,*** can be harmful to organisms living in the water.

An even more serious problem with nuclear power is how to dispose safely of the used-up uranium material fuel, known as ***nuclear waste.*** This poisonous, radioactive material must be stored where it will never leak into the environment. Most people do not want nuclear waste stored in, or even transported through, their communities. Disposal of nuclear waste is a difficult problem, and scientists continue to disagree on whether any of the proposed solutions are adequate.

EXERCISE 2

1. Most of the energy consumed in the United States comes from
   (1) nuclear energy
   (2) fossil fuels
   (3) moving water
   (4) solar energy

2. Killing of marine wildlife and pollution of shorelines result from accidents involved in
   (1) coal mining
   (2) storing nuclear waste
   (3) building dams
   (4) transporting oil by ship

3. Acid rain is mainly caused by
   (1) burning natural gas
   (2) nuclear reactors
   (3) drilling for oil
   (4) burning coal

4. The threat of global warming is mainly associated with
   (1) radioactive wastes
   (2) use of hydroelectric energy
   (3) burning of fossil fuels
   (4) thermal pollution of lakes and rivers

5. Which energy source produces the most air pollution?
   (1) coal
   (2) uranium
   (3) moving water
   (4) natural gas

6. Thermal pollution from nuclear power plants involves
   (1) release of toxic chemicals into the environment
   (2) storage of nuclear wastes
   (3) an increase in temperature of the environment
   (4) a decrease in temperature of the environment

7. The safe storage of hazardous wastes is a problem involved mainly with
   (1) transporting oil
   (2) strip mining
   (3) hydroelectric power
   (4) nuclear power

8. A problem with dams and pipelines is that they
   (1) contribute to air pollution
   (2) produce toxic wastes
   (3) cause thermal pollution
   (4) alter the ecology of surrounding areas

9. The energy source being used to make electricity in the diagram below is

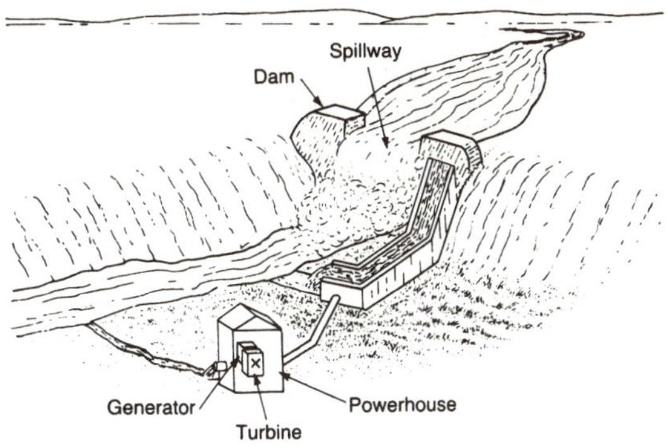

   (1) nuclear power
   (2) moving water
   (3) a fossil fuel
   (4) wind

Energy for the Future

Energy Conservation

Most energy resources do not exist in unlimited amounts. To guarantee an adequate supply of energy for the future, we must practice conservation. ***Conservation*** is the saving of natural resources through wise use. This means using resources more efficiently and eliminating unnecessary waste. The methods used in conserving our resources are often summarized as the three R's of conservation—reduce, reuse, and recycle.

Reducing Energy Consumption

1. *High-efficiency appliances.* We can contribute to conservation efforts by purchasing high-efficiency appliances. These appliances consume less energy than do less efficient appliances while doing the same job. For instance, a car that can travel 48 kilometers on a gallon of gasoline is more efficient than a car that gets only 24 kilometers per gallon. An air conditioner with a high "energy efficiency rating" uses less electricity than one with a low rating, but it cools a room just as well.

Although a high-efficiency appliance may cost more to buy, it costs less to use. This means that it will save money in the long run, while helping to conserve energy. Many appliances carry an "energy efficiency rating" label that shows how they compare with other models. Figure 18-6 on page 226 shows an example of such a label.

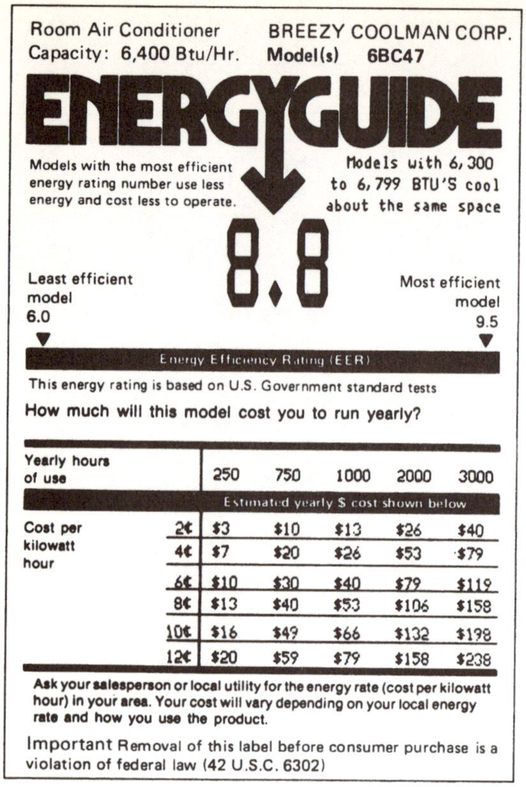

Room Air Conditioner
Capacity: 6,400 Btu/Hr.

BREEZY COOLMAN CORP.
Model(s) 6BC47

ENERGYGUIDE

Models with the most efficient energy rating number use less energy and cost less to operate.

Models with 6,300 to 6,799 BTU'S cool about the same space

8.8

Least efficient model 6.0

Most efficient model 9.5

Energy Efficiency Rating (EER)

This energy rating is based on U.S. Government standard tests

How much will this model cost you to run yearly?

Yearly hours of use		250	750	1000	2000	3000
		Estimated yearly $ cost shown below				
Cost per kilowatt hour	2¢	$3	$10	$13	$26	$40
	4¢	$7	$20	$26	$53	$79
	6¢	$10	$30	$40	$79	$119
	8¢	$13	$40	$53	$106	$158
	10¢	$16	$49	$66	$132	$198
	12¢	$20	$59	$79	$158	$238

Ask your salesperson or local utility for the energy rate (cost per kilowatt hour) in your area. Your cost will vary depending on your local energy rate and how you use the product.

Important Removal of this label before consumer purchase is a violation of federal law (42 U.S.C. 6302)

Figure 18-6. Example of an appliance's "energy efficiency rating" label.

There are some disadvantages of high-efficiency appliances. They often require the use of more expensive materials. Many people will not buy more expensive appliances unless they can see that they will save money in the long run. Scientists and engineers are constantly working to develop machines that combine high efficiency with low cost.

Sometimes increasing the energy efficiency of a machine affects its performance. Lighter-weight automobiles burn less gasoline than do heavier ones but may not be as safe in the event of a collision. Cars with larger engines, which can produce greater acceleration, usually burn more gasoline than do cars with smaller engines. Yet many people prefer big, fast cars despite their greater cost and possible harm to the environment.

2. *Insulated buildings.* Energy use at home and at work can be reduced through improved *insulation*. A well-insulated building prevents heat loss in winter and keeps heat out in summer. These benefits can be achieved by constructing walls in two layers, with insulating material in between. Cracks around doors and windows can be sealed with weather stripping for further insulation. With these improvements, less energy is needed to maintain a comfortable indoor temperature year-round.

There are some disadvantages to insulation as well. Some materials that have been used as insulators emit toxic fumes when burned. Safe, efficient insulating materials can be expensive. In some cases, insulation can affect the performance of an appliance. For example, better insulation can greatly increase the efficiency of a refrigerator, but the space occupied by the insulating material may decrease the amount of food the refrigerator can hold.

Reusing Materials

Every time you go to the supermarket you receive one or more bags in which to carry your purchases. Many of these bags could be used more than once, yet we usually throw them away. Today, in an effort to save energy and improve the environment, some supermarkets are offering a slight discount to shoppers who bring their own bags. Reusing resources is efficient in terms of energy and waste management. Other reusable products include glass and plastic bottles, cans, and even worn clothing. Some disadvantages of reusing material are the time and effort required in saving and collecting them.

Recycling

This practice also helps conserve energy resources. The graph in Figure 18-7 shows that making bottles and cans from recycled materials consumes less energy than does making those products from raw materials.

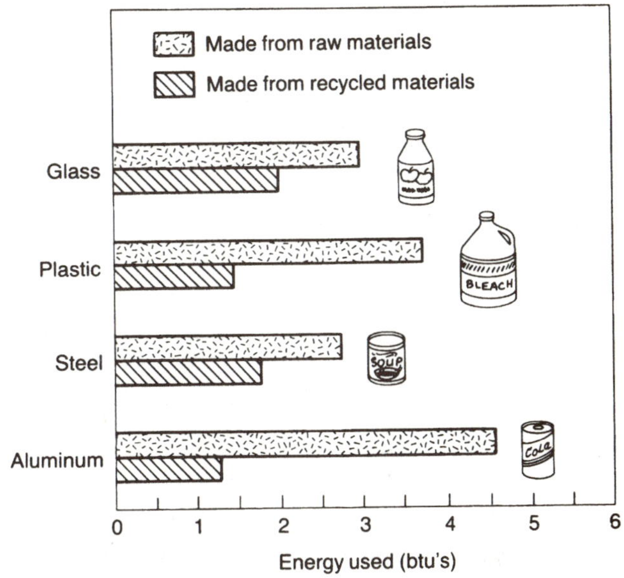

Figure 18-7. Less energy is consumed when products are made from recycled materials than when they are made from raw materials.

As you can see in the figure, less energy is used to make glass containers from recycled materials than from raw materials. However, the collection of glass for recycling requires the cooperation of large numbers of people, who must separate glass containers from other waste. These must be cleaned, collected, and transported to a recycling plant. Sometimes, the cost of collection and transportation is so high that the process no longer saves money. However, recycling is important even when money is not the issue. When materials such as glass and paper are recycled, they no longer have to be disposed of. Recycling benefits the environment by reducing the amount of waste that must be dumped in landfills or burned in incinerators.

It's Your Decision

In many places in the United States, recycling is the law. People are required to sort their garbage so that recyclable materials can be collected. This requires extra time and effort, but they have decided, through their elected representatives, that the time and effort are worth it. You will be making decisions like these in the future. Let us examine some of the issues you should consider in making your decisions.

Have you ever noticed that a greeting card printed on recycled paper costs more than one printed on new paper? Why should we recycle paper when it saves us neither energy nor money? Paper is made from wood from our forests. By using recycled paper we are cutting fewer trees and benefiting the environment. Table 18-2 lists some arguments for and against the recycling of paper.

Table 18-2. The Cases For and Against Recycling Paper

For	*Against*
Preserves forests, which soak up carbon dioxide and provide habitat for wildlife	Trees can be replaced by new trees, minimizing damage to forests; may cause job layoffs in timber industry
Decreases amount of solid waste in landfills	Expensive
Decreases air pollution in areas where paper is burned in incinerators	Separation of used paper from other waste is time-consuming

The decision of whether to recycle affects people's jobs and businesses as well as the environment. Environmentalists and businesspeople will try to convince you of their points of view. Listen carefully to both sides before you make your own decisions.

Renewable and Nonrenewable Resources

Earth's supply of fossil fuels is rapidly being used up. We continually remove these resources from the earth, but we cannot replace them. Nature does not create new deposits of oil, coal, and natural gas within the time span of human history. For this reason, fossil fuels are considered ***nonrenewable resources.*** Uranium is also a nonrenewable resource.

Renewable resources are those that can be replenished by nature within a relatively short time span. Moving water, wind, plants, and sunshine do not run out as we use them because they are constantly being replaced by natural processes. Table 18-3 lists some renewable and nonrenewable energy resources.

Table 18-3. Energy Resources

Renewable	*Nonrenewable*
Hydroelectric	Oil
Solar	Coal
Wood	Natural gas
Wind	Nuclear

Using Renewable Resources

Even if we practice conservation, our supply of nonrenewable energy resources may not be sufficient to meet the energy demands of the future. Renewable resources offer alternatives to fossil fuels and radioactive minerals. Unlike nonrenewable resources such as oil, coal, and uranium, renewable energy resources cannot run out. They also cause fewer environmental problems than fossil fuels and nuclear energy do. For these reasons, scientists and engineers are seeking more and better ways to use renewable resources for our growing energy needs.

As you know, moving water can be used to run generators and produce electricity. The natural water cycle of evaporation, condensation, and precipitation renews the water supply that feeds the rivers used for this purpose. However, not all areas have rivers suitable for producing hydroelectric energy.

The wind can be used to generate electricity by turning the blades of a *wind turbine* (see Figure 18-8 on page 230). In windy areas, wind turbines can provide safe, clean electricity. But the wind is not as constant and reliable as a flowing river. When there is only a slight wind or the air is calm, little or no electricity is produced.

Plant matter and animal wastes can be burned to produce heat, or they can be changed to other fuels. For instance, decaying plant matter

Table 18-4. Advantages and Disadvantages of Energy Sources

Energy Source	*Advantages*	*Disadvantages*
Oil	Efficient; can be converted into different types of fuel	Causes air pollution; risk of spills while drilling or transporting; limited reserves in U.S.; nonrenewable
Natural gas	Available in U.S.; clean	Difficult to store and transport; mostly nonrenewable
Coal	Abundant in U.S.; inexpensive	Causes air pollution and acid rain; mining practices may be harmful to miners' health and destructive to the environment
Nuclear	Abundant fuel in U.S.; does not cause air pollution; can meet long-term energy needs	Causes thermal pollution; creates radioactive waste; risk of accidents releasing radioactivity into environment; uranium mining harmful to miners' health
Hydroelectric	Does not cause air pollution; inexpensive; renewable	Not available in all areas; affects local ecology
Wind	Does not cause pollution; clean; inexpensive; renewable	Not practical for large-scale generation; not always reliable (wind is not constant)
Solar	Does not cause pollution; clean; renewable	Expensive to convert into usable form; not always reliable (depends on the weather)
Plant matter and animal wastes	Renewable	Expensive to convert into usable form; inefficient

Figure 18-8. Wind turbines use the renewable energy of the wind to produce electricity.

and animal wastes produce *methane*, the main component of natural gas. Methane produced in this way is a renewable resource, unlike natural gas found underground. At present, however, converting these materials into fuel on a large scale is too expensive to be practical. Unfortunately, *every* energy source has both advantages and disadvantages, as outlined in Table 18-4.

Solar Energy

The primary source of energy on Earth is the sun. Energy from the sun is called ***solar energy.*** The energy in fossil fuels came originally from sunlight absorbed by plants during photosynthesis million of years ago. The moving water used for hydroelectric energy is replenished by the water cycle, which is powered by the sun's energy. Wind, which can be used to make electricity, is caused by the sun's heating of the atmosphere.

People have found ways to use the sun's energy directly to provide heat and hot water for homes, offices, and factories. For example, a device called a *solar collector* absorbs solar energy and converts it into heat energy. The heat is transferred to water circulating through the collector. This hot water can be used to run a home heating system, as shown in Figure 18-9.

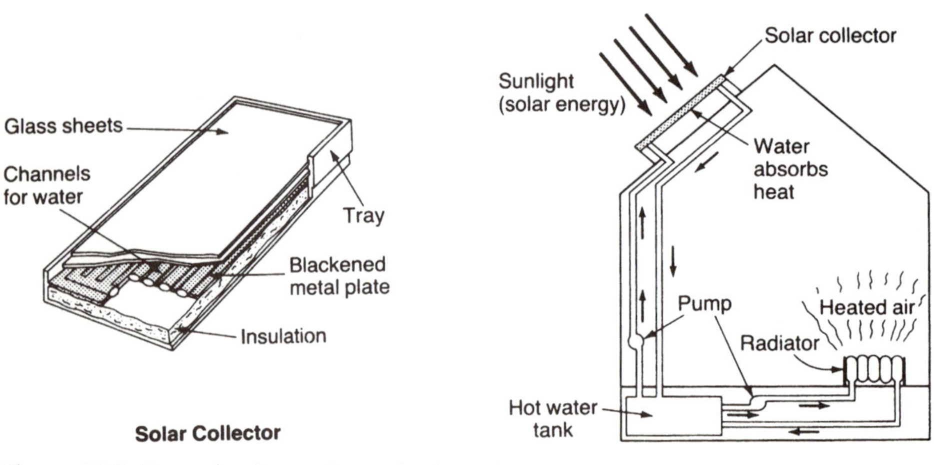

Figure 18-9. Example of capturing and using solar energy to run a home heating system.

People have also developed ways to transform solar energy into electrical energy. For instance, a *solar cell* is a device that converts light directly into electricity. Some calculators and light meters in cameras use solar cells. However, to generate large amounts of electricity this way requires a huge number of these cells, which are very expensive.

Electricity can also be produced by heating water with solar energy. Water is heated to a boil by using mirrors to focus and concentrate sunlight (see Figure 18-10). The boiling water changes into steam, which turns turbines to produce electricity. This method is more economical than using solar cells to make electricity, but it is still more expensive than using fossil fuels. However, as fossil fuels become more scarce, this situation may change.

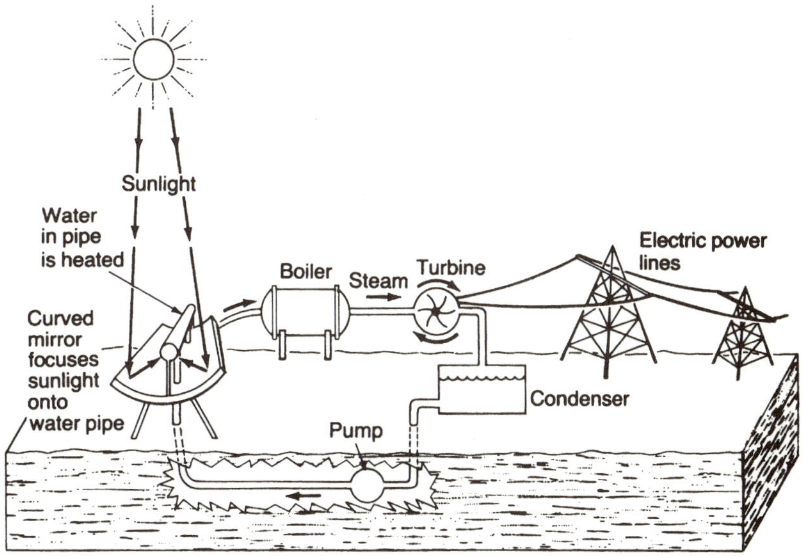

Figure 18-10. Solar energy changes water into steam, which turns turbines to produce electricity.

EXERCISE 3

1. Energy resources can be conserved by all of the following means *except*
   (1) better insulation
   (2) using high-efficiency appliances
   (3) increased mining of coal
   (4) increased use of solar energy

2. Which is a renewable energy resource?
   (1) moving water (3) coal
   (2) uranium (4) oil

3. The primary source of most energy on Earth is
   (1) moving water (3) coal
   (2) the sun (4) wind

4. *Gasohol* is a combination of gasoline and alcohol. Gasoline comes from oil, and alcohol is made from plant matter. Gasohol was developed to decrease our use of oil. However, gasohol costs more than ordinary gasoline and is therefore not commonly used. This paragraph suggests that
   (1) gasohol will soon replace gasoline
   (2) gasohol causes less pollution than gasoline
   (3) renewable resources are less efficient than nonrenewable resources
   (4) alternative energy sources may have both advantages and disadvantages

5. In the diagram below,

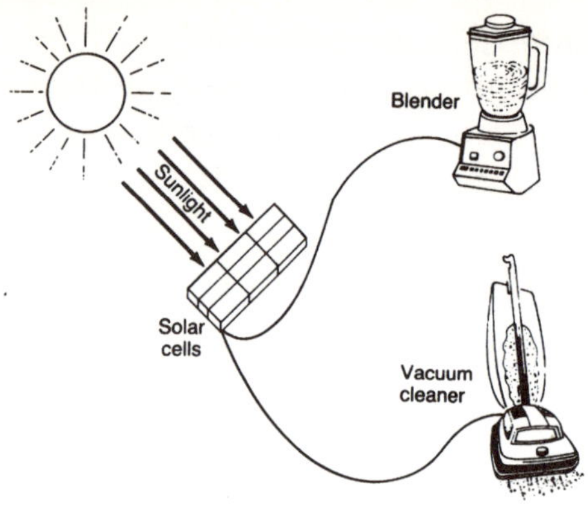

(1) electrical energy is being changed into solar energy
(2) solar energy is being changed into electrical energy
(3) chemical energy is being changed into solar energy
(4) solar energy is being changed into chemical energy

6. Because natural gas taken from underground is not quickly replaced by nature, it is considered a
(1) renewable resource
(2) nonrenewable resource
(3) pollutant
(4) solar energy source

7. Which are arguments *for* recycling paper?
A. Burning paper causes air pollution.
B. Paper comes from trees, which are a renewable resource.
C. Collection of paper for recycling is time consuming.
D. Used paper adds to the waste that is stored in landfills.
(1) A only (3) A, B, and C only
(2) A and D only (4) A and B only

Questions 8–11 are based on the graph below, which shows the monthly gas and electric bills of a family for one year.

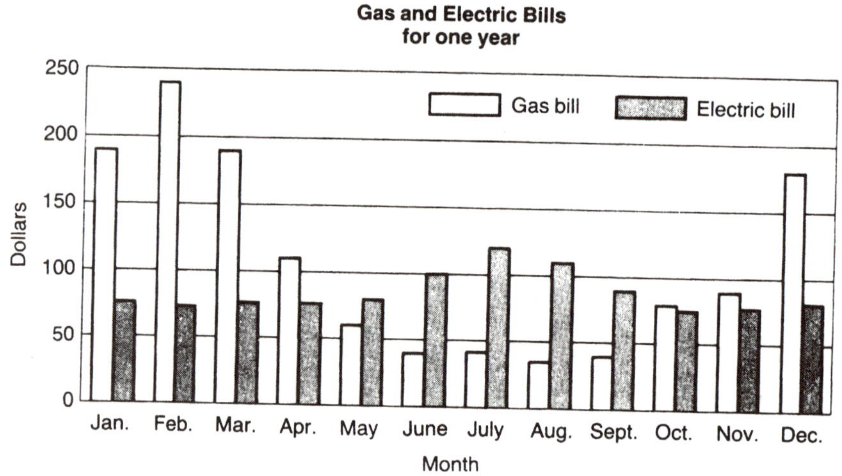

8. For which month was the electric bill highest?
(1) February (2) June (3) July (4) December

9. For which month were the gas and electric bills about the same?
(1) March (2) June (3) October (4) December

10. These bills are probably for a home that is heated by
(1) oil (2) electricity (3) wood (4) gas

11. Which would most likely result from improved insulation of this home?
    (1) Gas bills would decrease, and electric bills would increase.
    (2) Gas bills would increase, and electric bills would decrease.
    (3) Both gas and electric bills would decrease.
    (4) Both gas and electric bills would increase.

Questions 12 and 13 refer to the following paragraph.

Suppose that you need a new air conditioner. Two models are available that produce the same amount of cooling. A low-efficiency model costs $180, while a high-efficiency model costs $250.

12. Which argument(s) favor purchasing the high-efficiency model?
    A. Lower initial cost
    B. Lower energy costs
    C. Conservation of energy resources
    (1) A only
    (2) B only
    (3) B and C only
    (4) A, B, and C

13. If you use the air conditioner for 10 years, which will end up costing more?
    (1) the high-efficiency model
    (2) the low-efficiency model
    (3) they will cost the same amount after ten years
    (4) the total costs depend on the amount of use and the price of electricity in your area

14. What is a renewable resource? How does it differ from a nonrenewable resource? What are two examples of each type?

19 Science, Technology, and Society

Points to Remember —

- While the emphasis in science is on gaining knowledge of the natural world by asking questions, the emphasis in technology is on finding practical ways to apply that knowledge to solve problems.
- Science and technology frequently help to advance each other.
- We all interact with the products of technology in almost everything we do. Technology is used to extend or improve our abilities, and many products of technology affect the environment in some way.
- Science, technology, and society are constantly interacting. Often a change in one of these areas affects the other two.
- People have more choices in their everyday lives because of the products and processes of technology.
- Every technological process or device has advantages and disadvantages associated with its use, providing both benefits and burdens for people and the environment.

Relationship of Science and Technology

Science and Technology

Science and technology affect the lives of people all over the world. ***Science*** is the process of asking questions and seeking their answers to gain an understanding of the natural world. By providing insight into the workings of nature, science helps us predict the outcome of physical events.

Some questions that science attempts to answer include:

- What is the nature of matter?
- How did the universe come into being?
- How did life evolve on Earth?

Technology is the process of using scientific knowledge and other resources to develop new products and processes. These products and processes help solve the problems and meet the needs of the individual or society. Problems that technology attempts to solve include how to:

- increase the gas mileage of cars
- increase the productivity of farmland
- control industrial pollution

While the emphasis in science is on gaining knowledge of the natural world, the emphasis in technology is on finding practical ways to apply that knowledge to solve problems.

There are three major fields of science: life science, earth science, and physical science. Each of these fields contains a number of more specific sciences (see Figure 19-1).

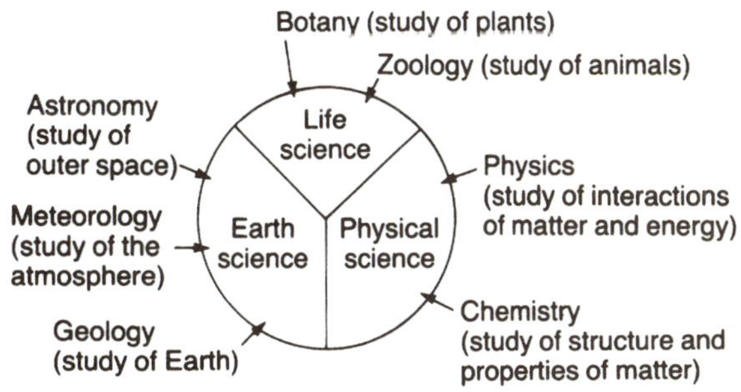

Figure 19-1. The major fields of science.

Biologists, chemists, and geologists are some types of scientists. Engineers, computer programmers, and medical technicians are examples of workers in the fields of technology.

Science and Technology Advance Each Other

Science and technology frequently help to advance each other. Scientific discoveries often lead to the development of new or better technological devices and processes. These technologies may, in turn, lead to new discoveries or to a better understanding of scientific principles.

For instance, scientists discovered various properties of light, such as how light is bent when it passes through different types of lenses. This knowledge led to the invention of the telescope and the microscope. Using these technological devices, scientists have made many more discoveries about the natural world.

Every technological device or process is based in some way on scientific principles, as the examples in Table 19-1 suggest.

Table 19-1. Relation of Scientific Principles and Technology

Scientific Principle	*Technological Device Process*
Cold temperatures kill or reduce growth of microorganisms.	Refrigerators and freezers
Sunlight contains energy.	Solar heating systems and solar cells
Splitting atoms of radioactive elements produces heat.	Nuclear power plants
Every action produces an equal and opposite reaction.	Rocket engines and jet engines

In fact, much technology involves knowledge from more than one field of science. For example, the artificial heart shown in Figure 19-2 involves knowledge from both life science (the structure of the human heart) and physical science (the mechanical principles of how the heart works).

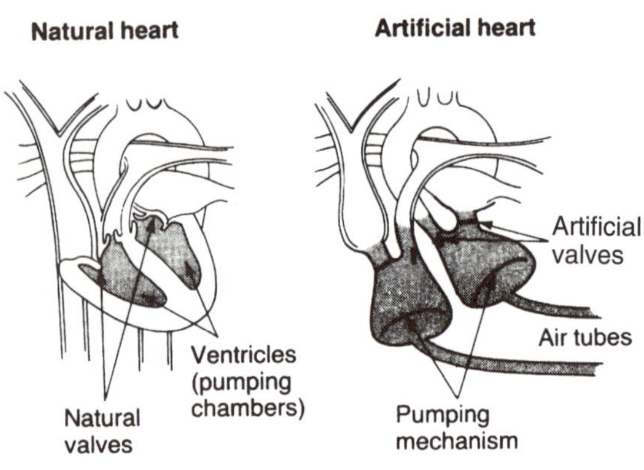

Figure 19-2. Scientific knowledge enabled development of the artificial heart.

Technology as a System

Some technological processes and devices can be viewed as systems. A *system* is a group of related elements or parts that work together for a common purpose. The parts of a system act in a series of steps, consisting of *input, comparison and control, processing, output,* and *feedback.*

A home-heating system with a furnace and a thermostat can be viewed in terms of these steps (Figure 19-3). Setting the thermostat to the desired temperature is the *input*. The thermostat *compares* the actual room temperature to the set temperature and *controls* the furnace, turning it on if the room's temperature is too low. The burning of fuel in the furnace is the process that produces the *output*—heat. The changing room temperature provides *feedback* to the thermostat, which turns the furnace off when the desired temperature is reached. In this way, the system maintains a constant indoor temperature.

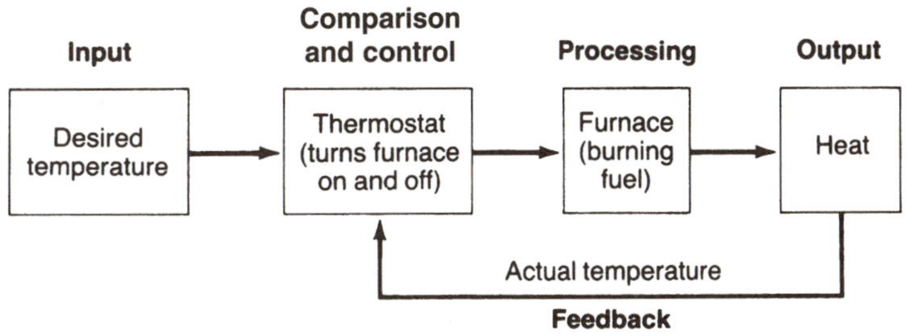

Figure 19-3. A home-heating system is a technological example of a feedback system.

Technology in Use

Everyone interacts with the products of technology. You do so when you wear clothing, sleep on a bed, watch television, eat with a knife and fork, ride a school bus—in almost everything you do.

People use technology for a number of reasons (Figure 19-4). To extend or improve our abilities, we use radios and telephones, calculators and computers, binoculars and telescopes, and other devices. Machines and appliances help us do work that requires more than human strength and at faster speeds than are humanly possible. To overcome physical disabilities, people use devices like eyeglasses, hearing aids, and heart pacemakers.

Figure 19-4. Some technological devices are used to extend human abilities or to overcome disabilities.

Many products of technology are used to change our environment. For example, we use electric lights so that our activities can continue after nightfall. Every technological process or device affects the environment in some way. Some of these effects may be harmful. Lightbulbs and many other appliances require electricity. The production of electricity may use precious natural resources or cause pollution of air and water. However, technology can also be used to protect the environment, as with sewage-treatment plants and pollution-control devices in cars and factories. Table 19-2 gives some examples of beneficial and harmful effects of technology on the environment.

Table 19-2. Technology and Our Environment

	Technology	*How It Affects the Environment*
Beneficial Effects	Electric lights	Extend light into nighttime
	Dams	Store water, generate electricity, create lakes for recreation
	Sewage treatment plants	Reduce pollution of rivers and lakes
Harmful Effects	Electrical generators	Consume natural resources and cause pollution
	Cars, boats, and airplanes	Consume natural resources and cause pollution

EXERCISE 1

Questions 1–3 refer to the following paragraph and the accompanying table and graph. The table below lists the numbers of different word-producing machines used in a newsroom from 1955 to 1990, at five-year intervals. These numbers show how the usage of manual typewriters, electric typewriters, and word processors has changed over time.

Numbers of Different Word-Producing Machines in a Newsroom, 1955–1990

Year	*Manual Typewriters*	*Electric Typewriters*	*Word Processors*
1955	45	0	–
1960	35	10	–
1965	25	20	–
1970	10	40	–
1975	0	50	0
1980	–	40	10
1985	–	15	40
1990	–	5	50

The relationships among the three sets of data may not be immediately clear from the table. However, if the same data are presented in a graph, these relationships become much easier to interpret.

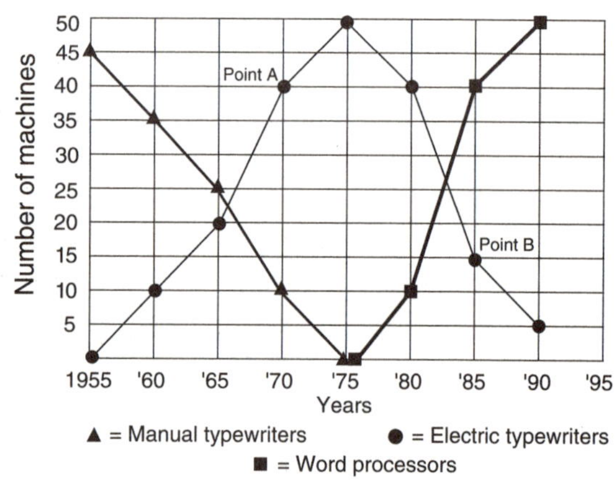

1. Point A on the graph represents the number of
   (1) manual typewriters in use in 1970
   (2) electric typewriters in use in 1980
   (3) electric typewriters in use in 1970
   (4) word processors in use in 1970

2. Point B represents the number of
   (1) electric typewriters in use in 1985
   (2) word processors in use in 1980
   (3) word processors in use in 1986
   (4) none of the above

3. If the trend shown for electric typewriters continued, in 1995 the number of electric typewriters in use would have been
   (1) less than in 1990
   (2) more than in 1990
   (3) the same as in 1990
   (4) impossible to determine

4. The process of using scientific knowledge to develop new products or processes is called
   (1) science
   (2) industry
   (3) technology
   (4) renewing resources

5. Using scientific knowledge about magnetism and electricity to build an electromagnet is
   (1) a scientific discovery
   (2) a technological development
   (3) predicting future physical events
   (4) observing the natural world

6. Using scientific knowledge, engineers built a space probe and sent it to the planet Jupiter. The probe sends data about Jupiter back to Earth, adding to our scientific knowledge. Which statement does this best demonstrate?
   (1) New technology sometimes builds on past technology.
   (2) Advances in technology cause some devices to become obsolete.
   (3) Technology affects our environment.
   (4) Science and technology help to advance each other.

7. Which technological device affects the environment in some way?
   (1) coffeepot
   (2) washing machine
   (3) air conditioner
   (4) all of these

8. Which statement is most correct?
   (1) Few people interact with the products of technology.
   (2) Only people in the United States interact with the products of technology.
   (3) Only adults interact with the products of technology.
   (4) Everyone interacts with the products of technology.

9. In this air-conditioning system, the output is the

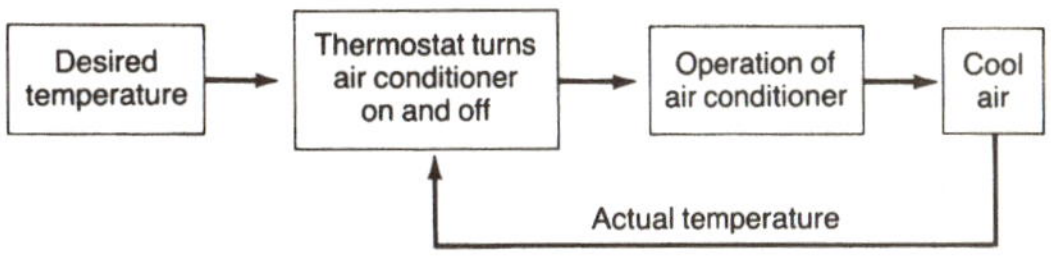

   (1) desired temperature
   (2) cool air
   (3) actual temperature
   (4) thermostat

10. The table below lists three general uses of technology. Examples of the first two uses are given in the table. Which three examples would best fit in the third column?
   (1) heart pacemaker, furnace, traffic light
   (2) binoculars, telephone, calculator
   (3) hammer, furnace, videocassette recorder
   (4) airplane, plumbing, coal mining

Uses of Technology

Affect Our Environment	*Overcome Disabilities*	*Extend Our Abilities*
Air conditioner Space heater Dam on a river	Hearing aid Wheelchair Eyeglasses	

11. The difference between the two heating systems shown below is that system A is lacking

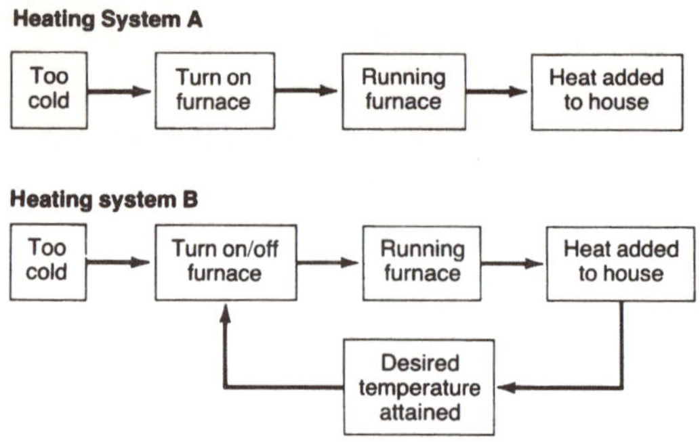

(1) input
(2) output
(3) feedback
(4) processing

12. A technological device or process may become outdated, or obsolete, when a new device or process is developed that does the job better. For example, the slide rule that was used to do mathematical calculations has been replaced by the hand calculator. Another example of an older device that has become obsolete and the new device that replaced it is
(1) buses and cars
(2) fountain pens and ball point pens
(3) eyeglasses and contact lenses
(4) shoes and sneakers

Interaction of Science, Technology, and Society

Effects of Science and Technology on Society

Science, technology, and society are constantly interacting (Figure 19-5). Often a change in one of these areas affects the other two. For example, scientific discoveries about the structure of matter led to many technological developments, including the production of microprocessors on tiny silicon chips. These microchips made possible many new products that have affected society by improving health care, communications, and transportation.

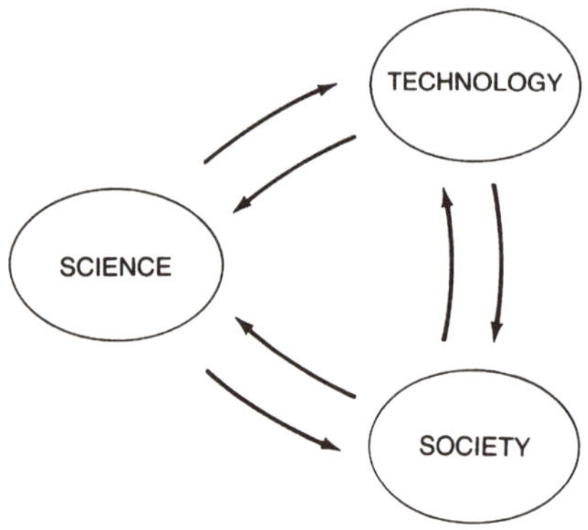

Figure 19-5. Science, technology, and society are constantly interacting and affecting one another.

Our culture, economy, and social systems are often affected by developments in science and technology. During the 1800s, the United States was transformed from a mainly agricultural society to a highly industrialized society. This period of cultural, economic, and social change was caused by the development of industrial machinery and new ways to power it. Other ways that science and technology have affected society are shown in Table 19-3.

Table 19-3. Ways Science and Technology Have Affected Society

New or Improved Products Raised Standard of Living
Health-care products allow us to live longer, healthier lives.
Work-saving home appliances provide more leisure time.
Created New Businesses and Industries
Film processing developed into the motion picture industry.
Music recordings changed from phonograph records to tapes to CDs.
Eliminated Businesses and Industries
Refrigerators and freezers eliminated the need for the ice-cutting industry.
Home-heating oil eliminated the need for coal company delivery.
Career Choices and Job Opportunities Eliminated, Modified, and Created
Telephone operators replaced by computers.
Typists had to learn word processing.
Computer systems needed computer analysts for implementation.
Solved Society's Problems
Vaccines eliminated certain diseases.
Communications systems allow people to work collaboratively.

While science and technology have solved many problems, they have also created new problems. Pollution of the environment and disposal of garbage and hazardous waste are problems caused, in part, by science and technology (Figure 19-6). Solving such problems requires the help of people working in government, industry, science, and technology.

Figure 19-6. Disposal of garbage is a social problem. Much of the garbage consists of products of science and technology.

Effects of Society on Science and Technology

Society also affects science and technology in many ways. New technology is often developed in response to the needs of individuals or society. For example, the need to help people overcome diseases and disabilities has encouraged the development of new medical procedures, such as chemotherapy and laser surgery, and new devices, like artificial organs and limbs.

The attitudes of people in a society may influence the direction of scientific research and technological development. In our society, public opinion has encouraged research to find a cure for AIDS. In contrast, public attitudes have largely discouraged the use of animals to test the safety of new cosmetic products.

Acceptance and use of an existing technology can also depend on people's attitudes. An example is nuclear energy. Most people agree that nuclear energy has both benefits and drawbacks. However, people disagree about whether its benefits outweigh its dangers. Public attitudes against nuclear energy have led some countries to ban its use. Other countries, on the other hand, generate most of their electricity with nuclear energy. Public opinion will undoubtedly influence the future of nuclear energy (Figure 19-7).

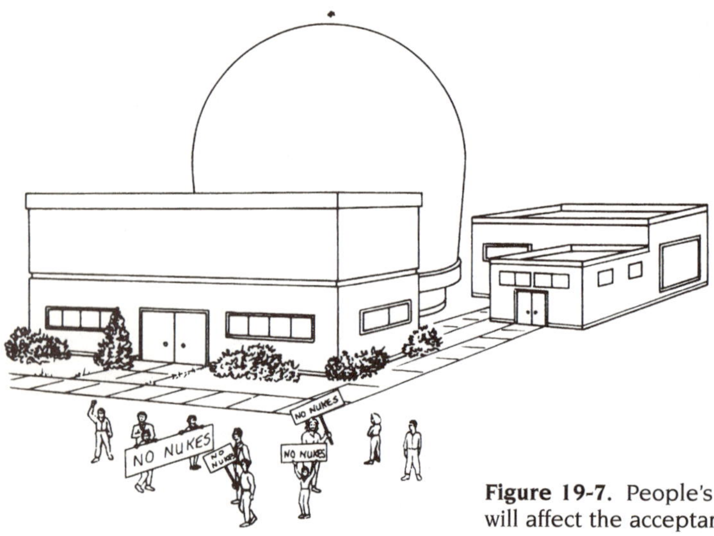

Figure 19-7. People's attitudes toward nuclear power will affect the acceptance and use of this technology.

Global Effects of Technology

Technology used in one country may have an international or global impact. For instance, in 1985, an accident at the Chernobyl nuclear power plant in the former Soviet Union released radiation that affected several neighboring countries. The radiation contaminated livestock, crops, and water. Another example is acid rain. Industries in the Midwest sometime create air pollution that drifts eastward with the prevailing winds. This causes acid rain to fall in New York, New England, and parts of Canada.

A possible global effect of technology is the destruction of Earth's ozone layer. Certain chemicals used in refrigerators, air conditioners, and spray cans are destroying the layer of the ozone gas that exists high in the atmosphere. The ozone layer screens out much of the sun's dangerous ultraviolet radiation. Its destruction could cause an increase in skin cancer and may have harmful effects on Earth's ecology.

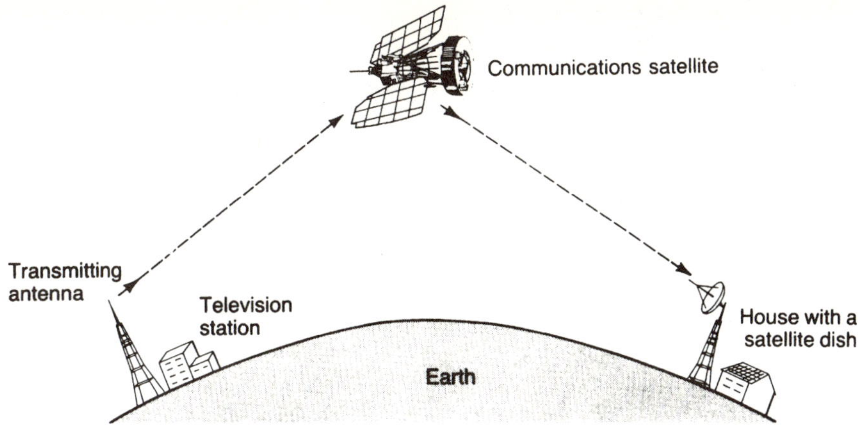

Figure 19-8. Communication satellites enable us to view distant events on television.

On the positive side, people around the world interact more frequently because of technological advances. Communications satellites let us make phone calls to people on other continents. We can also view distant events on television with the help of satellites (Figure 19-8). Technology helps inform us immediately of natural disasters like earthquakes and hurricanes, and it provides the means of sending aid to victims of these disasters.

EXERCISE 2

1. Society affects technology by
   (1) having problems that need to be solved
   (2) providing funds for research and development
   (3) its attitudes toward new research or products
   (4) all of the above

2. Carbon dioxide released into the atmosphere by industry and cars may cause a warming of Earth's climate, called the greenhouse effect. This is an example of
   (1) a problem solved by technology
   (2) people's attitudes affecting the use of technology
   (3) new industries being created by technology
   (4) a global impact of technology

3. An example of a job that has been created by recent developments in technology is
   (1) farmer (3) schoolteacher
   (2) astronaut (4) postal worker

4. Developments in microelectronics and computer science are changing the United States from a largely industrial nation into one more dependent on information services. This is an example of
   (1) science helping technology to advance
   (2) society affecting science and technology
   (3) science and technology affecting society
   (4) science and technology solving society's problems

5. In Europe, many people use irradiated milk (milk subjected to radiation that kills microorganisms). Containers of such milk can be stored at room temperature for a long time if unopened. In the United States, however, some people regard irradiated milk with suspicion, so its use is only slowly becoming accepted. This paragraph illustrates that
   (1) technology has affected society by raising our standard of living
   (2) people's attitudes can affect acceptance and use of technological devices or processes
   (3) products of technology may have an international impact
   (4) technology has caused some industries to become obsolete

6. The table below shows how the percentage of the workforce in three fields has changed over the years.

Year	*Agriculture*	*Information*	*Industry*
1800	75%	5%	20%
1850	60%	5%	35%
1900	40%	5%	55%
1950	10%	20%	70%
1990	5%	75%	20%

   From 1800 to 1990, the percent of the workforce in industry has
   (1) increased steadily
   (2) decreased steadily
   (3) first increased, then decreased
   (4) first decreased, then increased

7. In the past, lumberjacks chopped down trees using axes and handsaws. Today, they mostly use motorized chain saws. This example demonstrates that technology has
   (1) created new jobs
   (2) modified some jobs
   (3) eliminated some jobs
   (4) not affected lumberjacks

Changes in Average Life Expectancy from 1910–1970

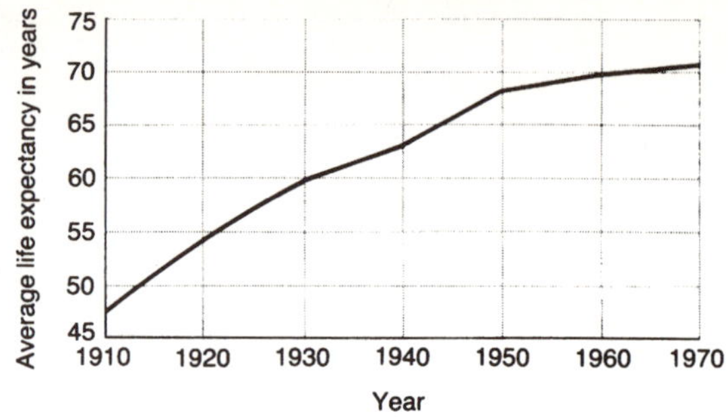

8. The graph above shows how average life expectancy has changed over time. This change is most likely a result of
   (1) advances in medical technology
   (2) harmful effects of technology on the environment
   (3) advances in educational technology
   (4) advances in communication technology

9. Which is the best example of the relationship shown in the diagram below?

   (1) principles of electricity → development of electrical appliances → more leisure time for people
   (2) effect of cold on microorganisms → people eat fresher, healthier food → development of refrigerators and freezers
   (3) principles of nuclear energy → development of nuclear power plants → development of nuclear weapons
   (4) development of radio and television → principles of electromagnetic waves → home entertainment for people

Making Decisions About Technology

Technology Increases Our Choices

People have more choices in their everyday lives because of the products of technology. For example, cable television and videocassette recorders have increased our choices in home entertainment. Figure 19-9 shows some leisure activities that are outgrowths of technology.

Technology has given us more forms of transportation to choose from. For instance, people may travel from New York City to Chicago by car, bus, train, or airplane. Shoppers may choose from a wide selection of home appliances produced by technology. Coffeemakers, microwave ovens, dishwashers, and vacuum cleaners are just a few of these products.

Figure 19-9. Many leisure-time activities are outgrowths of technology.

Assessing Technology

Every technological process or device has advantages and disadvantages associated with its use, providing both benefits and burdens for people and the environment. For instance, the automobile has given people greater mobility and contributed to our nation's prosperity. However, cars and trucks contribute to air pollution and lead to deaths and injuries in traffic accidents. Table 19-4, page 246, lists benefits and burdens of some technological devices and processes.

Technological processes and devices should be assessed by their advantages and disadvantages. When a device or process is adopted for use, information on its short-term and long-term effects should be continuously collected and evaluated. This helps us to identify and compare the benefits and possible adverse consequences of the technology for people and the environment, both for present and future generations.

Our society monitors the effects of many technological devices and processes, including medical treatments, food additives, industrial chemicals, and processes for generating electricity. This task is performed by various government agencies and public-interest groups.

Table 19-4. Benefits and Burdens of Technology

Technological Process or Device	Benefits	Burdens
Nuclear energy	Additional clean electricity	Risk of accidents; radioactive wastes
Painkilling drugs	Treat diseases, relieve pain	Addiction through abuse
Computer	Increased ability to process data	Loss of jobs; health problems from computer keyboards
Space travel	Increased knowledge	High financial cost
Life-sustaining medical devices	Keep people alive	Decisions about when to use or remove them
Automobile	Increased mobility	Increased pollution; deaths and injuries
Chemical fertilizers	Increased agricultural yields	Upset ecology of lakes and streams
Artificial sweeteners	Convenience for diabetics and dieters	Increased risk of cancer

Technology and Decision Making

Decisions about the use of technology must be made almost constantly. To make these decisions wisely, both short-term and long-term consequences should be considered. Sometimes the short-term benefits of a technology outweigh its long-term burdens. For example, dentists agree that the benefits of using X rays to find cavities in your teeth outweigh the possible long-term dangers of brief exposure to the radiation (Figure 19-10).

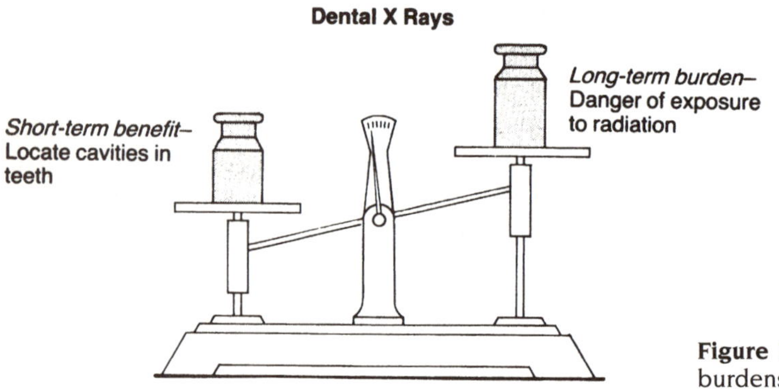

Figure 19-10. "Weighing" the benefits and burdens of a technology.

In other cases, long-term benefits may outweigh short-term burdens. Wearing a seat belt in a car may be a momentary discomfort. Over time, however, the use of seat belts reduces deaths and injuries from car accidents. Society's consideration of short-term and long-term effects has led to using unleaded gasoline for cars (which is less polluting) and the recycling of cans, bottles, and newspapers (to reduce waste).

EXERCISE 3

1. Should government regulate the use of technological devices and processes for the good of society? Some social problems can be solved or lessened by laws that regulate technology. But besides providing benefits, such laws may burden us in some ways. Some laws may infringe on our right to privacy or freedom of choice. Other laws may be costly to put into action and enforce. Examples of laws that regulate technology are listed in the table below—with benefits and burdens. Which law has its benefit and burden given in the wrong order?

Problem	*Law*	*Benefit*	*Burden*
Water pollution from industries	Effluent controls law	Cleaner water	Higher industrial costs
Head injuries from motorcycle accidents	Mandatory motorcycle helmet law	Reduces freedom of choice; uncomfortable	Fewer head injuries in motorcycle accidents
Crimes committed with easily obtained handguns	Handgun control law	Less crime; harder for criminals to get guns	Harder for people to get guns for protection
Bottle and can litter	Bottle deposit bill	Cleaner streets and sidewalks	Inconvenience of returning bottles and cans

(1) effluent controls law
(2) mandatory motorcycle helmet law
(3) handgun control law
(4) bottle deposit bill

---

2. The development of radio and television opened up many new career options for people. This is an example of
   (1) technology developing in response to society's needs
   (2) people's attitudes affecting the acceptance and use of new technology
   (3) increased choices brought about by technology
   (4) technology affecting the global environment

3. Which answer lists both a benefit and a burden of television?
   (1) provides entertainment and provides up-to-date news
   (2) provides entertainment and causes higher electric bills
   (3) causes eyestrain and discourages reading books
   (4) provides useful educational tool and provides entertainment

4. Most oil tankers have single hulls that break open fairly easily if the ship runs aground, causing oil spills. However, some tanker ships are now being built with double hulls that are more resistant to breaking. This best illustrates that
   (1) technology has increased our choices in life
   (2) a technology may be modified to reduce or eliminate its disadvantages
   (3) government should constantly monitor technology to determine possible adverse consequences
   (4) decisions about technology often involve trade-offs between benefits and burdens

5. The artificial sweetener saccharin can be safely used by people with diabetes, who cannot eat sugar. It also has fewer calories than sugar and causes less tooth decay. However, experiments have shown that using saccharin increases the risk of getting cancer. Saccharin provides
(1) a benefit only
(2) a burden only
(3) both a benefit and a burden
(4) neither a benefit nor a burden

6. In 1984, the bottle deposit bill was passed. This law requires people to pay an extra five cents for every bottle or can of beverage. The deposit is refunded when the empty bottles and cans are returned. The graph indicates that the bottle deposit bill has

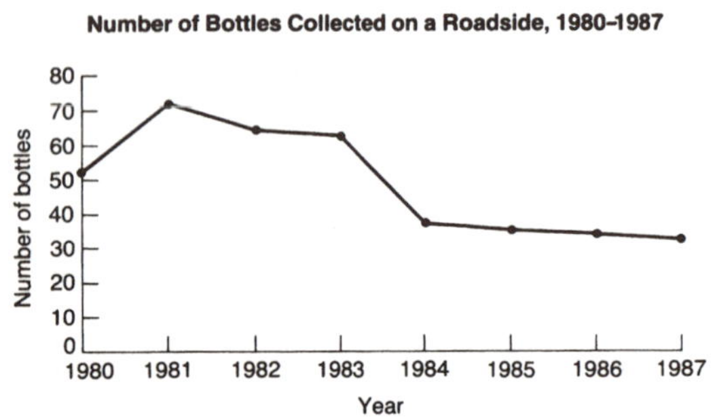

(1) increased roadside litter
(2) decreased roadside litter
(3) had no effect on roadside litter
(4) its effect cannot be determined from the graph

7. Big cars get poor gas mileage, but they are safer than small cars in the event of a traffic accident. Small cars get very good gas mileage but offer little protection in an accident. For these reasons, many people buy medium-size cars, which get fairly good gas mileage and offer some protection in accidents. This is an example of
(1) a burden of technology on the environment
(2) monitoring short-term and long-term effects of technology
(3) a decision about technology that involves a trade-off between benefits and burdens
(4) modifying technology to reduce or eliminate its drawbacks

8. Although the pesticide DDT was effective in killing insects that damage crops, its use was banned when it was found to be harmful to humans and wildlife. This shows that
(1) technological products have only disadvantages
(2) technological products have increased our choices
(3) use of a technological product may be terminated if its disadvantages outweigh its advantages
(4) use of a technological product may depend on people's attitudes

9. What are three benefits of modern technology that you enjoy? What are some possible harmful effects of these benefits?

20 Environmental Changes in an Ecosystem

Points to Remember

- Plants and animals within a community are dependent on each other and tend to develop a stable community in balance with their environment.
- Ecological succession is a natural process by which one community is replaced by another community in an orderly, predictable sequence.
- Certain natural processes, such as volcanic eruptions, severe weather changes, and forest fires, put stress on ecosystems and may disrupt or destroy them.
- Humans affect the environment by causing pollution, killing animal populations, destroying forests, and so on. These actions put stress on ecosystems and may disrupt or destroy them.
- Both natural and human processes that cause environmental disturbances can endanger species populations and even cause species extinction.

Changing Ecosystems

Biological Balance in Nature

An ecosystem consists of all factors, living and nonliving, necessary to sustain a living community (see Chapter 13). As long as these factors remain relatively constant, the biologic community will be stable and will change gradually through natural ecological processes. Animal and plant populations, and species variety, tend to remain constant within a narrow range as long as food, water, oxygen, proper temperature, and proper environment are available.

A biologic ***community*** can contain many types of plants and animals living in association with one another. The community contains many interrelationships in which the organisms depend on one another for food, environmental conditions, and maintaining a balanced population.

However, changes in the environment can put stress on an ecosystem and affect the living community. Environmental changes may be slow, such as ecological succession and climate change, or sudden, such as volcanic eruptions or forest fires. Table 20-1 gives examples of events or processes that cause change in ecosystems.

Table 20-1. Causes of Change in Ecosystems

Gradual Changes	*Sudden Changes*
Succession	Volcanic eruptions
Climate change	Forest fires
Human population growth	Human actions
Continental drift	Floods, meteorite impacts

Succession

Succession is a natural process by which one living community is replaced with a new community. An area of forest land that has been completely destroyed by fire provides us with an opportunity to see how an ecosystem develops by succession (see Figure 20-1).

After a forest fire has destroyed an ecosystem, the soil becomes enriched with minerals from the decaying remains of the plants and animals that once lived there. Soon after, mosses, lichens, and other small, new plants sprout. These provide shelter and food for insects and small animals. Eventually, these plants die and are replaced by larger plants like grasses, shrubs, and small trees. Large animals, such as foxes, deer, and rabbits, begin to move into the area. As the trees become larger, a forest develops, with squirrels, deer, and many kinds of birds in it.

Each new community changes the environment, making it more suitable for the next community. Finally, a community emerges that is not replaced. This stable community is called the *climax community*.

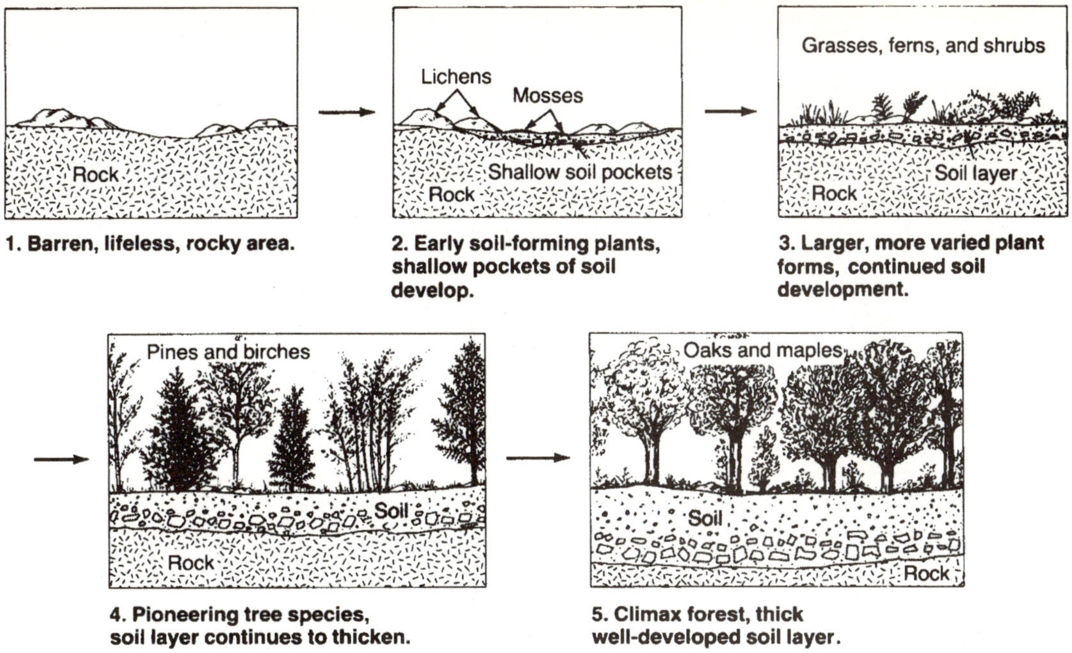

Figure 20-1. Ecological succession on land.

Competition

Competition for food and space is an important part of the relationships among living things in an ecosystem. Both the moose and the snowshoe hare live in the same habitat and compete for food from the birch tree. The moose, by far the larger of the two animals, has an advantage over the hare in obtaining food. In winter, when food is scarce, the hare is more likely to die of starvation because of competition for the limited food available. Plants also compete, for space and for sunlight (see Figure 20-2).

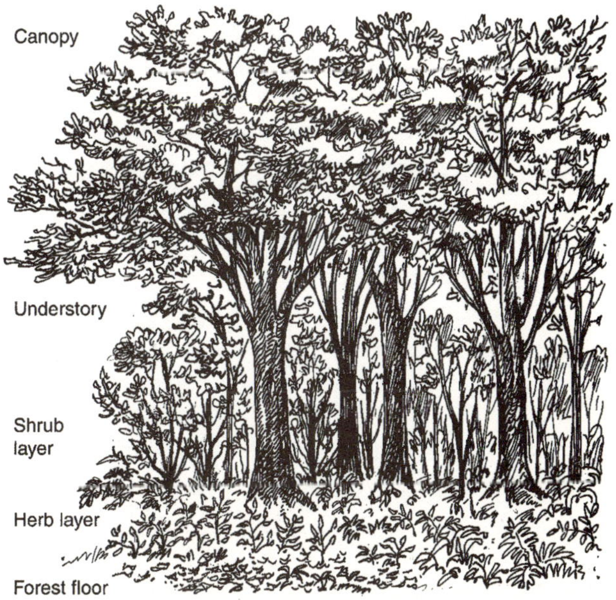

Figure 20-2. Plants compete for resources, such as growing space and sunlight.

EXERCISE 1

1. An example of an ocean food chain is shown below. If pollution destroys the shrimp population, the most likely result would be

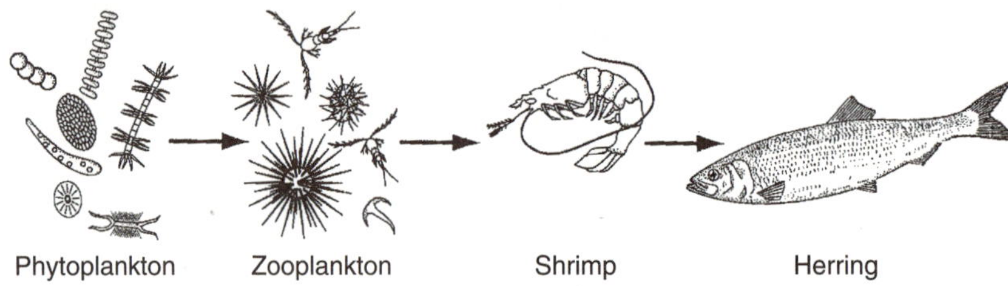

(1) a decrease in phytoplankton
(2) a decrease in zooplankton
(3) a decrease in fish
(4) an increase in fish

---

2. Tracy observed that the forest behind her house had many large trees, with some small trees growing in the shade of the larger trees. The large trees were blocking out the sunlight and preventing the sun's energy from reaching the small trees. The small trees will eventually die. This is an example of
(1) a climax forest
(2) a balanced community
(3) competition
(4) an unbalanced community

3. All ponds change over time. Soil and decaying plants and animals tend to slowly fill in a pond. Fish, frogs, and swamp grasses die as the water depth decreases and soil forms. Grass and small land plants grow in the soil. This process may take tens or hundreds of years, depending on climate, the size of the pond, and other environmental conditions. This process is called
(1) competition
(2) succession
(3) the balance of nature
(4) a food chain

Questions 4 and 5 refer to the following short paragraph and the accompanying graph.

A decrease in sea otters can lead to an increase in sea urchins. An increase in the sea urchin population results in less kelp (seaweed). Kelp "forests" provide protection to many fish, which will then be exposed and may not survive.

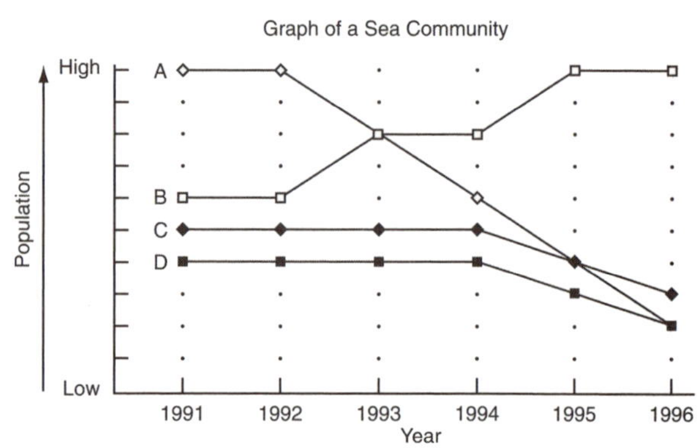

4. This chain of events demonstrates how
(1) living things depend on nonliving factors
(2) living things grow
(3) living things do not depend on each other
(4) living things depend on each other

5. The graph represents the increase and decrease in population for each of four species (sea otters, sea urchins, kelp, and fish). The line that represents the sea urchin population is
(1) line A
(2) line B
(3) line C
(4) line D

6. Succession is the process by which one living community is replaced by a new community. Recently cooled lava in Hawaii forms a rock surface devoid of living organisms. The order of succession of living organisms on the rock surface will most likely be
(1) trees, shrubs, grasses, mosses
(2) grasses, shrubs, mosses, trees
(3) mosses, shrubs, trees, grasses
(4) mosses, grasses, shrubs, trees

7. The area that looks most like a climax community is

(1)

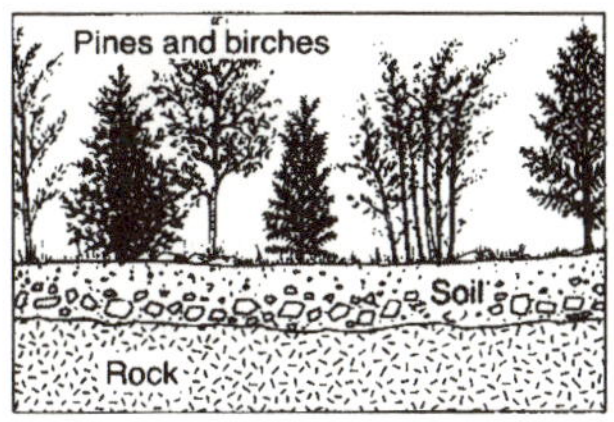

(2)

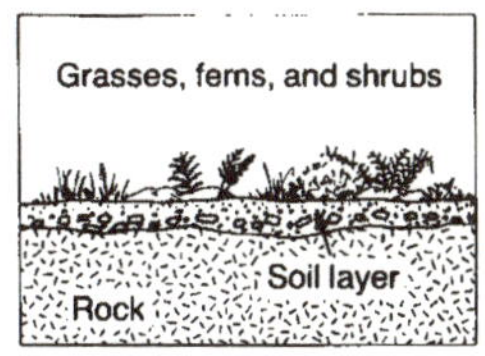

(3)

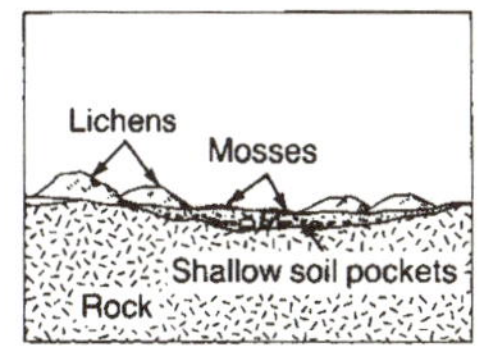

(4)

8. Nothing succeeds like succession. Explain this statement.

Environmental Threats That Upset the Ecosystem

Geologic Events

Some geologic events, such as volcanic eruptions, earthquakes, and tsunami waves, can completely devastate an ecosystem, or they can disrupt the environment and cause an ecosystem to change.

In 1980, Mount St. Helens, a volcano in the state of Washington, erupted violently. The explosion destroyed almost 100,000 acres of forest. In addition, this event caused mud slides, flooding, and the blanketing of land for many kilometers downwind by volcanic ash. Living communities on land and in the water were destroyed or disrupted. Many larger animals such as deer, elk, and bear perished, and so did many birds and smaller animals. However, some animals, like rodents, frogs, and insects, survived because they lived underground. Soon after the eruption, life started to return to the mountain. Plant roots began to sprout, and seeds that were spared destruction started to germinate. Living communities, in all stages of succession, reappeared around the mountain.

Climatic Changes

Climate is the average condition of the atmosphere over a large region for a period of many years. Life on Earth is well adapted to the varied climatic conditions of the planet; thus, a wide variety of ecosystems exist. However, climate changes can affect communities of living organisms within an ecosystem.

About 18,000 years ago, glacial ice covered a large portion of North America. Fossil evidence indicates that during this glacial episode, organisms such as the mastodon, woolly mammoth, and saber-toothed cat lived near the ice sheet (see Figure 20-3). Today, these animals are gone and appear to be victims of a major extinction event that took place across North America about 11,000 years ago. The warming climate seriously disrupted ecosystems and is probably partially responsible for this period of extinction.

Figure 20-3. Climate affects organisms: The woolly mammoth and the saber-toothed cat were adapted to ice age conditions.

Floods and Droughts

Flooding is a condition of excessive water on land. Heavy rainfall and rapidly melting snow and ice are the most common causes of flooding. Low-lying land areas near the ocean are commonly flooded by storm surges and exceptionally high tides. Ecosystems containing fish, birds, and invertebrates near rivers and streams may also be upset by floodings.

Droughts are caused by a lack of precipitation for extended periods of time. Droughts commonly cause many plants to die, removing the primary food source of animals in a community. The lack of water can also upset lake and stream ecosystems.

Storms

Thunderstorms, hurricanes, and *tornadoes* can seriously affect ecosystems. ***Thunderstorms*** may ignite fires, destroying large areas of forest (see Figure 20-4), and sometimes produce hail large enough to destroy crops and other plant life. ***Hurricane*** storm surges along the coast destroy beach ecosystems for nesting birds and spawning fish. ***Tornadoes*** are capable of severely disrupting life in a narrow strip of land.

In addition, blizzards and blizzardlike conditions, consisting of blowing snow, severe cold, and high winds, can cause animals to starve and many living things to freeze to death.

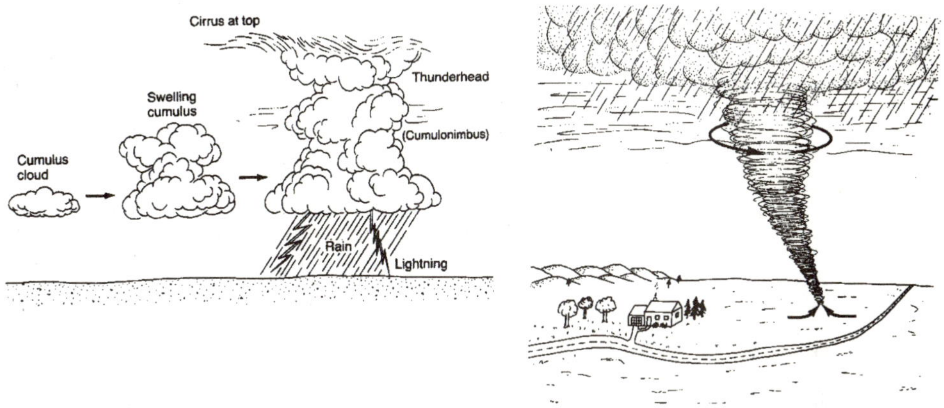

Figure 20-4. Thunderstorms and tornadoes can have a devastating impact on ecosystems.

Forest Fires

Forest fires destroy large areas of timber each year. In addition to the trees that are destroyed, many other living things in the ecosystem die. Forest fires may be started naturally by lightning strikes or by humans. Table 20-2 lists common causes of forest fires.

Table 20-2. Causes of Wildfires in the U.S. (1984-1990)

Cause	*Percent of Total Number*
Arson	30
Debris burning	25
Lightning	11
Miscellaneous	11
Cigarette smoking	6
Children	5
Equipment	5
Campfires	3
Railroads	3

Although forest fires can take lives and destroy property, natural forest fires play an important environmental role. Ecosystems are renewed by the carbon and nutrients returned to the soil, and processes such as succession and the introduction of new plants and animals replenish the living community.

Human Interference

1. *Pollution.* For hundreds of years humans showed little or no concern for the environment. Human activities have put stress on many ecosystems and sometimes destroyed them.

Factories, power plants, cars, and airplanes produce harmful substances called ***pollutants.*** Pollutants in smoke from factories and vehicles can increase the acidity of the moisture in clouds. When the moisture falls to Earth as acid rain, it can harm lakes and forests and all the living things in them (see Figure 20-5).

Figure 20-5. Pollutants in acid rain can harm lakes and forests.

Many industries and most forms of transportation produce carbon dioxide. Many scientists fear that a buildup of carbon dioxide in the atmosphere caused by human activities may lead to a rise in worldwide average temperatures, or ***global warming.*** This could have disastrous effects on many ecosystems by making climates warmer and drier.

Other examples of human activities that cause pollution and affect the environment are the burial and burning of garbage and toxic materials and the release of sewage into rivers and lakes. Today, many people are showing a concern for the environment and the ecosystems it supports.

2. *Growth and land exploitation.* As the human population increases on Earth, the need for land grows. Land is used to build homes, businesses, and roads, to provide water supplies, to provide recreation areas, to grow food, and to provide for waste disposal. When land is changed to meet human needs, the ecosystem is often disrupted or destroyed. How can the growing human population meet its needs without harming the environment? Humans can change land without completely destroying ecosystems. People can focus on living in harmony with nature by not completely

removing the ecosystem as they build. In some cases, people are revisiting damaged ecosystems and restoring them.

In 1941, the city of Los Angeles began diverting the streams flowing into Mono Lake to meet its water demands. This quickly lowered the water level of the lake and increased its salinity (saltiness). The rapid change in environment harmed the ecosystem. Algae in the water—the base of the food chain for all the living things in and around the lake—began to die. The brine shrimp population decreased. The lowering of the lake's water level produced land bridges to islands, which animals could cross to prey on the bird population. Stream ecosystems deteriorated, and air quality worsened due to organic decay around the lake.

In 1984, restoration processes began to reverse some of the stresses placed on Mono Lake and its ecosystems. Today, the living communities in and around the lake are slowly recovering, and humans are learning better ways to meet their needs while maintaining existing ecosystems.

3. *Introduction of new species.* Throughout history, humans have transported plants and animals as they traveled. When "foreign" organisms are introduced to a new location, they may upset the existing ecosystem. They often compete for food and space with the local or "native" species, and they may disrupt the entire ecosystem.

For example, in 1788, people introduced rabbits to Australia for the first time. The rabbits multiplied rapidly because they had few natural enemies in the ecosystem. By the mid-1800s, the grasslands were being stripped of vegetation by the rabbits. Cattle (also introduced by people) found it difficult to find food and were faced with starvation. Scientists introduced a disease-causing virus that killed many of the rabbits. The grasslands recovered and the cattle were saved.

EXERCISE 2

1. In 1980, after the Mount St. Helens eruption, volcanic ash covered the ground for thousands of square kilometers of eastern Washington. Although the ash was responsible for destroying much of the wheat and apple crop, it was also partially responsible for the higher than normal wheat and apple production later that summer. This was most likely due to the ash
   (1) covering the ground and holding moisture in the soil
   (2) introducing new plants into the area
   (3) introducing new animals into the area
   (4) replacing the soil

2. In the eastern United States, thousands of lakes are so acidic they cannot support fish life. This is probably due to
   (1) humans doing too much fishing
   (2) changing climate conditions
   (3) rain mixing with air pollutants to form acidic rain
   (4) decaying of fish in the lakes

Questions 3–5 refer to the following paragraph.

In the winter of 1949, Lake Michigan froze, and a pack of wolves made their way to Isle Royal National Park. A study was conducted to investigate the wolf-moose relationship and its effects on the ecosystem. During the 1970s, there were several severe winters that led to starvation conditions, causing malnutrition among the moose population. The wolves found the weak moose easy prey, and as the moose population decreased, the wolf population tripled.

3. The wolf-moose relationship at Isle Royal demonstrates the effects on the ecosystem caused by
   (1) human actions (2) weather events (3) geologic events (4) none of the above

4. If the moose population continues to decrease, the wolves' food supply will decrease. This will most likely cause the wolf population to
   (1) increase greatly (2) increase slightly (3) remain the same (4) decrease

5. Which graph best shows the wolf-moose relationship during the 1970s?

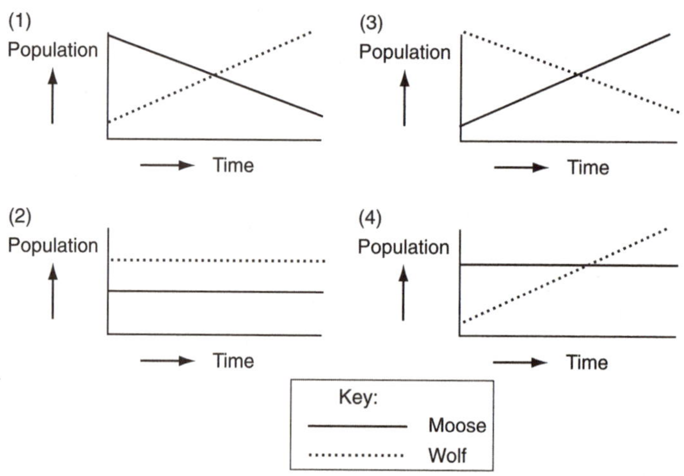

Questions 6 and 7 refer to the following paragraph.

The gypsy moth, which originally came from Europe and from Asia, was accidentally introduced to North America about 1869. After about 10 years in North America, there were major outbreaks of gypsy moths. Insecticide spraying tried to eliminate the moths. However, more damage was done to other plants and animals than to the gypsy moth. Attempts to destroy the gypsy moth failed, and since then its population has continued to spread. The gypsy moth caterpillar commonly feeds in oak and aspen trees. High densities of the caterpillars can *defoliate* trees (remove their leaves) over large areas, and if this continues for several years, many trees die.

6. The safest and most effective way to reduce the gypsy moth population may be to
   (1) introduce new birds to eat them
   (2) spray the forest with pesticides
   (3) kill the egg masses by hand
   (4) encourage natural diseases of the gypsy moth

7. Defoliation of trees in a forest area by the gypsy moth caterpillar may lead to
   (1) reduction of the squirrel population
   (2) reduction of the bird population
   (3) reduction of the insect population
   (4) all of the above

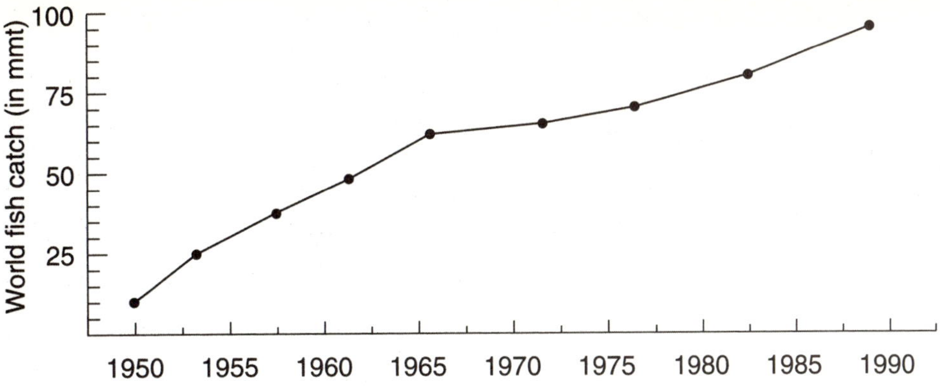

Questions 8 and 9 refer to the following paragraph and the above graph.

The world's fishing industry removed increasingly greater numbers of fish from the oceans between 1950 and 1990. The graph indicates the world's fish catch for this period of time.

8. From 1950 to 1990, the world's fish catch increased by about
   (1) three times (3) seven times
   (2) five times (4) nine times

9. Although most of the fish caught are used for food, worldwide fishing management appears unable to control harvest rates at levels that will maintain fish populations. Uncontrolled fishing practices are an example of
   (1) human exploitation of animal life
   (2) environmental factors affecting fish populations
   (3) geologic events decreasing fish populations
   (4) water pollution

Results of Environmental Change

We have already discussed how some environmental conditions can change an ecosystem. The effects on life can range from limiting diversity within an ecosystem, to endangering the species, and even to causing extinction, the ultimate destruction of life.

Biodiversity

Biodiversity describes the variety of life-forms that exist. There is great value to having a high number of species on Earth. Humans should preserve biodiversity because we depend on plants and animals for such things as our food supply, medicine, and much of our industry. Table 20-3 on page 260 indicates the biodiversity in the world's major coral reefs.

Table 20-3. Biodiversity in Coral Reefs

Coral Reef	*Coral Species*	*Fish Species*
Philippines	400	1,500
Great Barrier Reef	350	1,500
New Caledonia	300	1,000
French Polynesia	168	800
Aqaba	150	400

There are believed to be millions of living species on Earth that have not yet been identified. The potential usefulness of undiscovered plants and animals needs to be investigated. These living things could solve some of humanity's problems by providing important food supplies or cures for diseases. Failure to preserve the diversity of life on Earth could limit our ability to solve future problems and may limit our existence on Earth.

An endangered species is a plant or animal species that appears close to extinction. More and more species are being added to the endangered species list. Many of these plants and animals are endangered because of human activities. Organizations concerned with this problem are developing strategies to save species from extinction. Endangered species include the buffalo, blue whale, rhinoceros, tiger, snow leopard, gorilla, California condor, whooping crane, and manatee. (See Figure 20-6.) Some reasons why species decline and may become extinct are:

- loss of habitat
- introduced species
- overgrazing of lands
- developed areas
- low gene pool diversity
- predation
- vegetation changes
- competition among species
- forest clearing
- heavy equipment effects

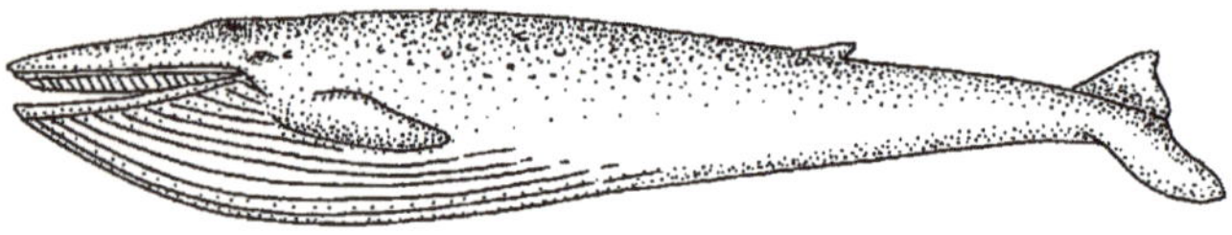

Figure 20-6. Blue whales are endangered as a result of overhunting by people.

In 1840, the buffalo (bison) population of North America was estimated to be between 30 and 40 million. As settlers moved across the plains, massive slaughter of these animals occurred. Professional hunters killed buffalo in great numbers for sport and hides. Often, the bodies were left to rot. In the early 1870s, it was estimated that 2.5 million buffalo were killed each year. The last large herd was slaughtered in 1883. By 1900, only 500 buffalo remained. Various organizations and parks have protected the buffalo since then, and today the number is back up to about 200,000.

Extinction

Extinction occurs when a species dies out. This usually happens when some essential part of the living or nonliving environment is removed. Common natural causes of extinction include climate changes and habitat invasion by a predator. Humans also cause extinction by excessive hunting and pollution of land, air, and water. Some animals known to have become extinct are dinosaurs, woolly mammoths, dodo birds, and passenger pigeons.

1. *Dinosaurs.* About 70 percent of all animal species, including all the dinosaurs, became extinct about 65 million years ago. Many theories have been proposed for this great extinction of life on Earth. However, scientists are still not exactly sure what caused it. Theories for the demise of the dinosaurs include climate change, impact of a large asteroid or comet, or massive volcanic eruptions. Certainly their extinction was natural and had nothing to do with human interference, since humans did not even exist yet. Some mammals existed on Earth during the heyday of the dinosaurs; but after their extinction, mammals diversified rapidly and soon became the dominant large land creatures. (See Figure 20-7.)

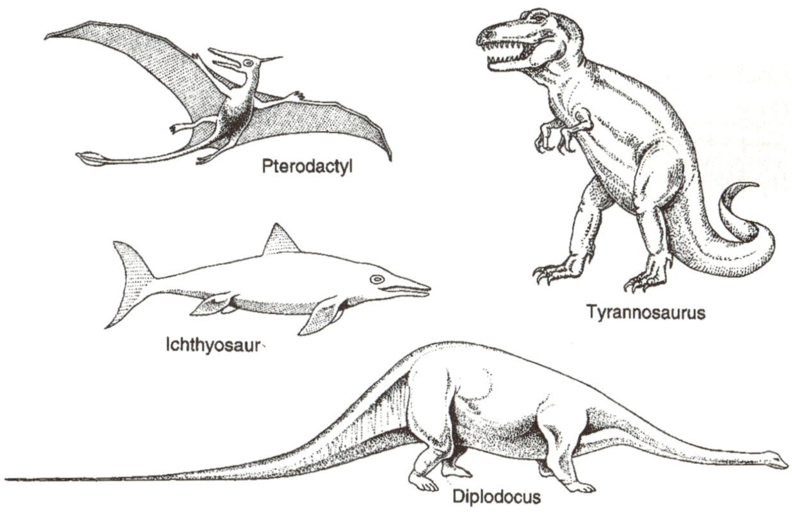

Figure 20-7. The extinction of dinosaurs was a result of natural causes.

2. *Passenger Pigeons.* Early in the 1800s, millions of passenger pigeons lived throughout eastern North America. They fed on acorns and beechnuts in forests. Passenger pigeons became a popular food of people and were killed in great numbers. By 1880, the huge flocks were gone, and the pigeon's numbers were decreasing rapidly. The last wild passenger pigeon was seen in 1889, and the last captive one died in the Cincinnati Zoo in 1914.

EXERCISE 3

Brief case histories of three species are presented below. Read each of the cases and answer questions 1–3.

Case 1: The dodo was discovered in 1598 by Portuguese sailors on the island of Mauritius in the Indian Ocean. Having no natural enemies, the birds lacked fear of the sailors and were easily killed. Those that survived were killed by the dogs and pigs introduced to the island by the sailors. By 1681, the dodo was completely eliminated.

Case 2: The elimination of the dodo disrupted the delicate balance of the environment on Mauritius. The calvaria tree stopped sprouting seedlings soon after the extinction of the dodo and it, too, seemed doomed. It appears that the calvaria seeds needed to be eaten, digested, and eliminated by the dodo. Fortunately, turkeys were found to do the same thing, and the calvaria tree was saved.

Case 3: The Iowa Pleistocene snail is a small land snail about .6 cm in diameter. The brown to greenish snail lives in cool leaf litter at about 30 sites in Iowa and Illinois. Fossil evidence indicates that the snails were much more widespread during the Ice Age. Although climate change appears to have been the most devastating cause of population decline, human building activities have also contributed to their loss.

1. Extinction occurs when a living species dies out. According to the case histories above, the organism that has become extinct is the
   (1) dodo
   (2) calvaria tree
   (3) Iowa Pleistocene snail
   (4) none of the above

2. An endangered species is a plant or animal that appears close to extinction. According to the case histories above, the organism that has become endangered is the
   (1) dodo
   (2) calvaria tree
   (3) Iowa Pleistocene snail
   (4) none of the above

3. The two organisms that most nearly perform the same role in their ecosystems are
   (1) humans and Iowa Pleistocene snails
   (2) calvaria trees and turkeys
   (3) calvaria trees and dodos
   (4) turkeys and dodos

4. Biodiversity refers to the variety of living species. It is important to have a great biodiversity because it provides humans with
   (1) food diversity
   (2) medicine diversity
   (3) industrial diversity
   (4) all of the above

5. Biodiversity would most likely be greatest in a
   (1) rainforest ecosystem
   (2) desert ecosystem
   (3) grassland ecosystem
   (4) tundra ecosystem

The four diagrams below represent a living community on an island over a four-year period. The symbols represent the total population of each species. Each large circle represents the total community. Study the diagrams carefully and answer questions 6–9.

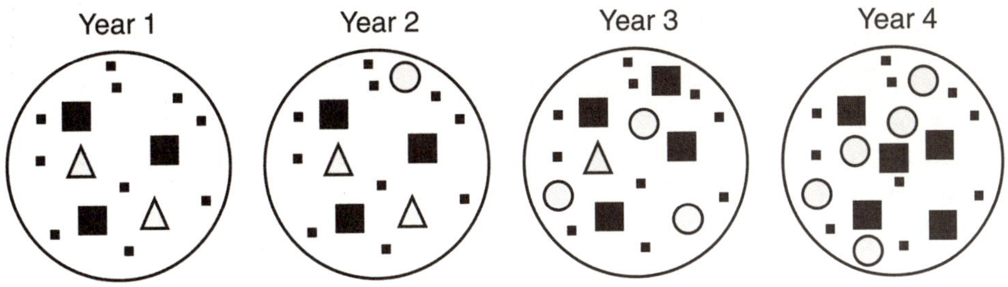

6. A new animal was introduced to the community in
   (1) Year 1 (2) Year 2 (3) Year 3 (4) Year 4

7. The animal that became extinct is represented by the
   (1) square (2) triangle (3) circle (4) dot

8. The animal that became an endangered species in Year 3 is represented by the
   (1) square (2) triangle (3) circle (4) dot

9. The species represented by the triangle was most likely the food source for the species represented by the
   (1) square (2) circle (3) dot (4) cannot determine

The graph shows the relationships among three species—frogs, grasshoppers, and snakes—in a community. Study the graph and answer questions 10–12.

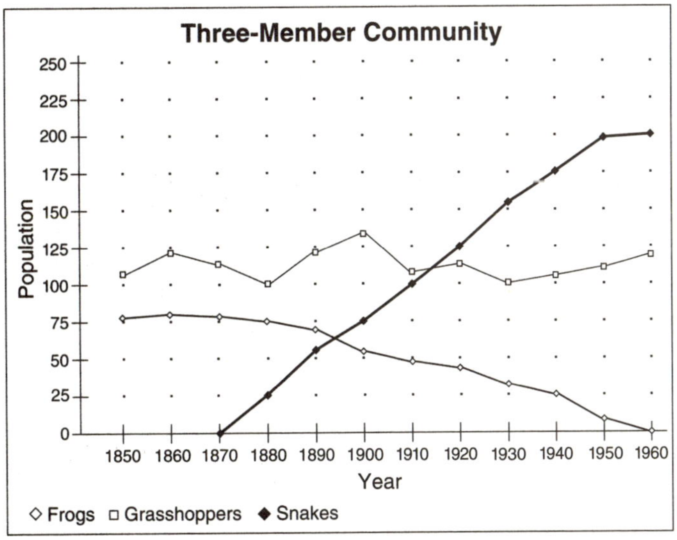

10. Which two species lived in a stable relationship in 1850?
    (1) frogs and snakes
    (2) frogs and grasshoppers
    (3) grasshoppers and snakes
    (4) none of the species listed

11. The decline of frogs appears to have been caused by
    (1) the decline in the number of snakes
    (2) the introduction of the snakes
    (3) the decline in the number of grasshoppers
    (4) climate change

12. Based on the graph, by 1980 the snake population would probably be
    (1) greater than in 1960
    (2) less than in 1960
    (3) the same as in 1960
    (4) the same as the grasshopper population

13. Scientists often say: "Extinction is forever." In one or two sentences, explain the meaning of this statement.

14. What are three natural causes of change to the environment? In what three ways do humans change the environment?

15. Why is habitat protection important for the survival of species?

Practice Test 1

Questions 1–6 refer to the situation below.

There are four blood types: A, B, AB, and O. Blood types are found by observing the reactions of blood with two antibodies called anti-A and anti-B. The reaction with anti-D tells if the blood type is positive or negative. The charts below show the blood types.

Reacts With:

Type	Anti-A	Anti-B
A	yes	no
B	no	yes
AB	yes	yes
O	no	no

Reaction With Anti-D

Positive	yes
Negative	no

1. A patient's blood reacts with anti-A and anti-D but not with anti-B. His blood type is
   a. A positive c. B positive
   b. A negative d. O positive

2. A patient at the hospital needs blood. Her body produces anti-A, anti-B, and anti-D. To keep her body from reacting with the new blood, the doctors should give her which type?
   a. A positive c. AB positive
   b. B negative d. O negative

3. A father has two genes for blood type A, and a mother's genes are both blood type B. According to this Punnett square, which percentage of their children will be type AB?

	A	A
B		
B		

   a. 0% b. 25% c. 50% d. 100%

4. The main function of blood in organisms is to
   a. help break down food to produce energy
   b. transport materials throughout the body
   c. carry messages to the individual cells
   d. produce hormones needed for growth

5. New technology allows blood to be stored for 35 days. If a person is planning on having surgery, he may donate several pints of blood for storage at the hospital. During the surgery, he can then receive his own blood. Which advantage would there be to a patient donating his own blood to himself?
   a. His body will have less blood to lose during the surgery.
   b. His body does not have to produce new blood cells for several days.
   c. There is no risk of his body rejecting the blood.
   d. This blood will mix better with the blood he has.

6. By spinning it in circles, blood can easily be separated into several parts, such as plasma, red blood cells, and platelets. This shows that blood could be classified as
   a. an atom
   b. an element
   c. a compound
   d. a mixture

Questions 7–11 refer to the diagram below.

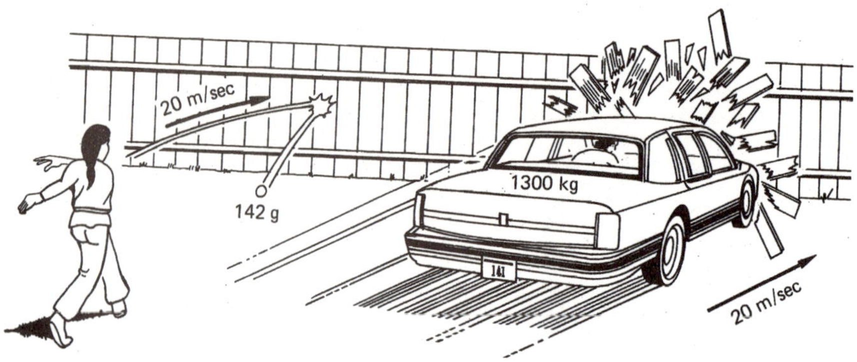

7. The car crashes through the fence because it
   a. is moving faster than the ball
   b. has more mass than the ball
   c. takes up more space than the ball
   d. is rectangular, while the ball is round

8. The car is powered by gasoline. Which type of energy transformation occurs that enables the car to move?
   a. chemical to heat
   b. electromagnetic to chemical
   c. mechanical to electromagnetic
   d. chemical to mechanical

9. When does the ball have the *least* kinetic energy?
   a. when it is released from the girl's hand
   b. when it is halfway to the fence
   c. when it has hit the fence but not bounced back yet
   d. just before it strikes the ground

10. The girl wants to break through the fence also. Assuming she can throw all the balls at the same speed, which type of ball would give her the best chance of breaking the fence?
    a. a smaller ball, like a superball
    b. a bigger ball, like a beach ball
    c. a lighter ball, like a whiffle ball
    d. a heavy ball, like a bowling ball

11. When the car crashes through the fence, the girl will
    a. see the crash first, then hear it
    b. hear the crash first, then see it
    c. see and hear the crash at the same time
    d. feel the force of the crash before she sees or hears it

12. The killdeer bird lays its eggs on the ground. To keep predators from finding her eggs, the bird distracts them by fluttering and hopping around as if it had a broken wing. The predator chases the bird, which leads it away from the nest. This action of the killdeer bird is an example of a
    a. mutation
    b. behavioral adaptation
    c. physical adaptation
    d. genetic trait

13. On the savannah, many different species of animals, such as giraffes, gazelles, and zebras, are often found herding together. Which is the reason for this behavior?
    a. The savannah does not cover a large area, and animals must share the limited space.
    b. Faster animals such as gazelles use sharp-eyed zebras to find food and then race over to eat it first.
    c. Single animals or small groups are more likely to be caught and eaten by predators.
    d. Some animals in the herd, such as gazelles, will steal and eat the offspring of the others.

Questions 14–18 refer to the food web diagram below.

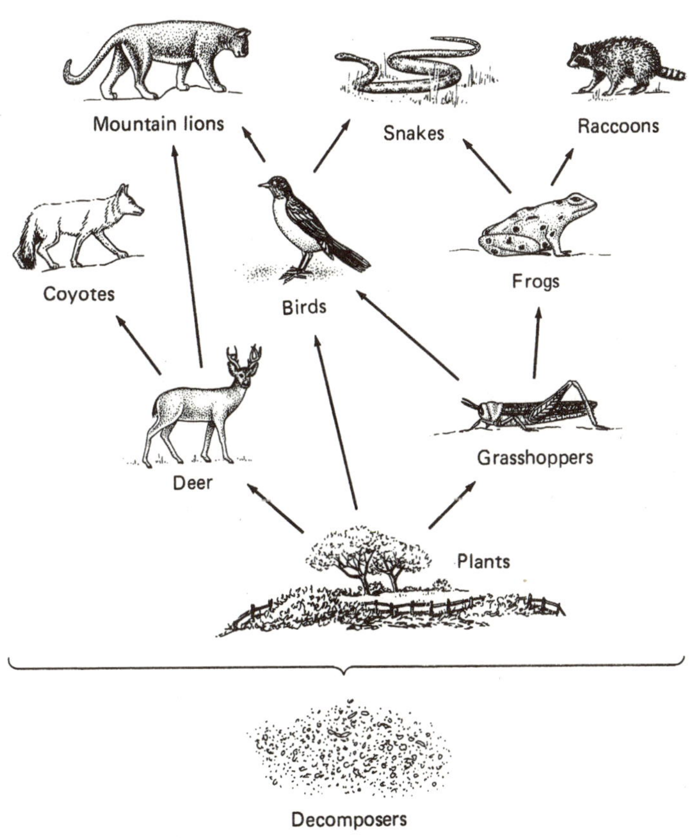

14. Which organism in the food web is a producer?
    a. bacteria
    b. deer
    c. grass
    d. raccoon

15. Which term best describes the relationship between the snake and the frog?
    a. competition
    b. predator/prey
    c. symbiosis
    d. mutualism

16. Over a lifetime, a single coyote eats many deer, and a single deer eats a lot of grass. This illustrates that energy
   a. is lost as it is transferred from one level of the food web to another
   b. is gained as it is transferred from one level of the food web to another
   c. can be created by each level of the food web
   d. does not change from one level of the food web to another

17. Which organism or organisms undergo the process of respiration?
   a. plants c. owl
   b. bacteria d. all of the above

18. A fire sweeps through this ecosystem, driving away all the animals and burning everything to the ground. In order for the ecosystem to reestablish itself, which member of the food web has to return first?
   a. mountain lion c. birds
   b. grasshoppers d. grass

Questions 19 and 20 are based on the diagram below.

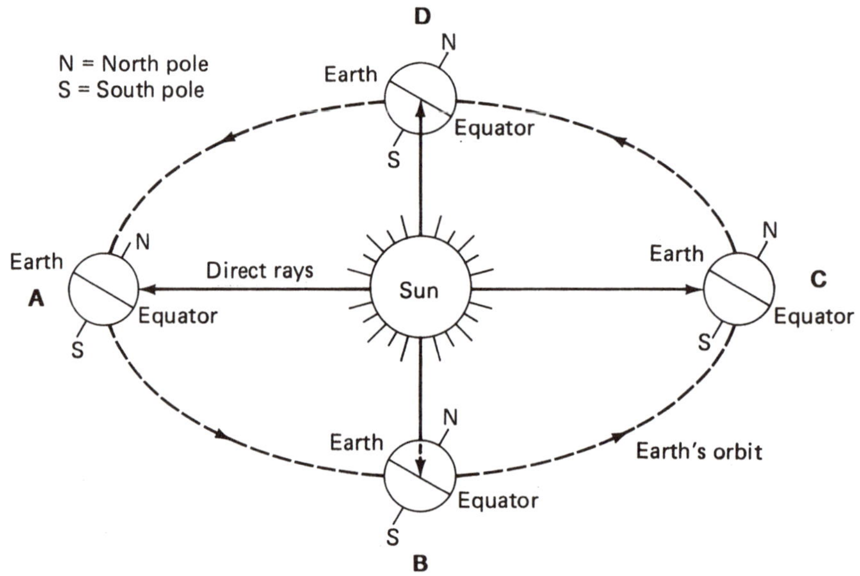

19. The diagram shows Earth at four positions in its orbit around the sun. When Earth is at position C, the season in the northern hemisphere is
   a. summer c. fall
   b. winter d. spring

20. The diagram shows that the seasons are caused by
   a. the tilt of Earth
   b. Earth's distance from the sun
   c. the rotation of Earth on its axis
   d. the orbital speed of Earth

Questions 21 and 22 are based on the diagram below.

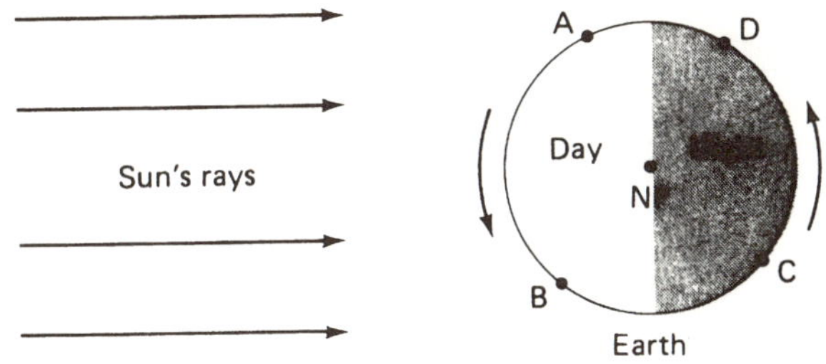

21. The diagram shows rotating Earth, viewed from above the North Pole (NP). At which point will nightfall take place next?
    a. A
    b. B
    c. C
    d. D

22. The diagram illustrates which property of light?
    a. Light travels in a straight line.
    b. Light bends as it enters a new medium.
    c. Light can be broken up into different colors.
    d. Light can be reflected, absorbed, or transmitted.

Questions 23–27 refer to the experiment below.

Erika wants to see how much water is in various fruits and vegetables. To do this, she finds the mass of five different foods and lets them sit under heat lamps for one week. She then finds the mass again and subtracts to find the mass of the water lost. Here are her results:

Food	*Part of Plant*	*Beginning Mass (grams)*	*Mass After 1 Week (grams)*	*Water Lost (grams)*	*Percentage of Water Lost*
Apple	Fruit	43	12	31	72
Potato	Root	54	36	18	33
Celery	Stem	25	6	22	76
Tomato	Fruit	62	20	42	68
Carrot	Root	22	5	19	77

23. Which food has the highest percentage of water content in its cells?
    a. apple b. celery c. tomato d. carrot

24. As the food dries out, it becomes smaller and harder. Is this an example of a chemical change or a physical change? In one or two sentences, explain your answer.

25. What process takes the water out of the food?
    a. condensation c. precipitation
    b. evaporation d. distillation

26. Koalas are found only in Australia, living in groves of eucalyptus trees. Since they can spend days in the tree branches, what keeps them from dehydrating?
   a. They absorb rainwater into the pores on their skins.
   b. They recycle the water within their bodies, so they have a new supply.
   c. The leaf of the tree contains enough water for the koala.
   d. Nothing stops dehydration; it is the top killer of koalas.

27. On her way home, Erika sees a cherry tree. She can infer that the least amount of water can be found in the leaves. Erika knows that water flows up from the roots through the stems to the leaves. Why is her inference correct? What happens to the water in the leaves?

Questions 28–31 refer to the diagram below.

28. The ramp is really another form of which simple machine?
   a. inclined plane
   b. pulley
   c. lever
   d. wedge

29. To make his work easier (by decreasing the effort needed to move the wheelbarrow), the man should
   a. use a longer ramp
   b. use a shorter ramp
   c. move his hands closer to the bricks
   d. move the bricks away from the wheel

30. The fulcrum is the fixed point around which a lever rotates. If the wheelbarrow is really a lever, where is the fulcrum?
   a. the man's hands
   b. the handles of the wheelbarrow
   c. the basket holding the bricks
   d. the wheel

31. This man is using the ramp to make it easier to get the bricks up on the wall. What is a disadvantage to using a ramp?
   a. The wheelbarrow will move faster.
   b. The wheelbarrow must be pushed a greater distance.
   c. More effort is required to overcome friction between the tire and the ramp.
   d. More effort is required to push the wheelbarrow up than just to lift the wheelbarrow by hand.

Questions 32–34 refer to the cell pictures below.

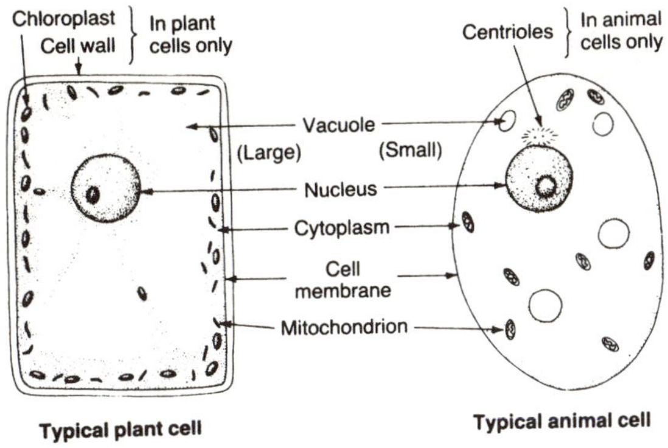

32. Which structure is found in plant cells but not in animal cells?
    a. chloroplast
    b. cell membrane
    c. centrioles
    d. nucleus

33. The cell wall provides structure and support to a plant. Animals do not need cell walls because
    a. animals have muscles
    b. animals have bones
    c. animals do not grow as tall as plants
    d. animals have specialized tissues for a variety of functions

34. Mitochondria release energy in the cell. Why do plants have fewer and smaller mitochondria than animals do?
    a. The vacuoles take up too much space, and they have no room to grow.
    b. Plant cells are not as advanced as animal cells.
    c. Plants need less energy because they do not move.
    d. Plants make energy through photosynthesis; they don't use any.

Question 35 refers to the diagram below.

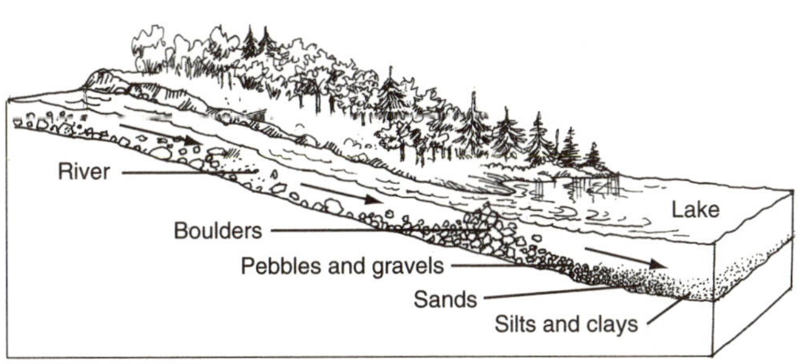

35. Running water is very efficient at separating rock particles. According to the diagram, how does moving water separate big and little rocks?
    a. Rocks of all sizes and shapes are heaped together, with little rocks filling in big spaces.
    b. The heavier rocks are carried farther than lighter ones because they have more momentum.
    c. The lighter rocks are carried farther than heavier ones because they stay suspended in water longer.
    d. Rounder rocks are smoother and likely to roll farther under water than square or rectangular rocks.

36. Which is an example of chemical weathering?
   a. Rock fragments are frozen to the base of a glacier and carried away.
   b. Dead plant material forms a weak acid that reacts with minerals in the rocks.
   c. Roots from a tree grow through cracks in a rock and split it apart.
   d. Sand grains in the wind chip away any surface they are blown against.

37. Scientists could infer that glaciers are responsible for
   a. sand along an ocean shore
   b. tiny rocks sitting along the bottom of streams
   c. big rocks sitting at the bottom of a mountain
   d. huge rocks sitting in an open, flat field

Questions 38 and 39 refer to the following passage.

When seat belts were first introduced in cars, they consisted of a single strap across the waist. However, this strap allowed a person's upper body to snap forward during an accident, sometimes causing head injuries. To remedy this, a strap was added that crosses the chest and goes over one shoulder. The shoulder strap prevents the upper body from flying forward in an accident.

38. Which of Newton's laws best explains why seat belts are needed in a car?
   a. An object in motion tends to stay in motion.
   b. The bigger it is, the more force needed to accelerate it.
   c. Every action has an equal and opposite reaction.
   d. What goes up, must come down.

   In one or two sentences, explain how this relates to the movement of a person in a car.

39. The introduction of the shoulder strap shows that
   a. science and technology help each other advance
   b. people's attitudes can affect acceptance and use of new technology
   c. a technological device may be modified to reduce or eliminate its disadvantages
   d. use of a technological device may be short-lived if its disadvantages outweigh its advantages

Questions 40–44 refer to the graph below.

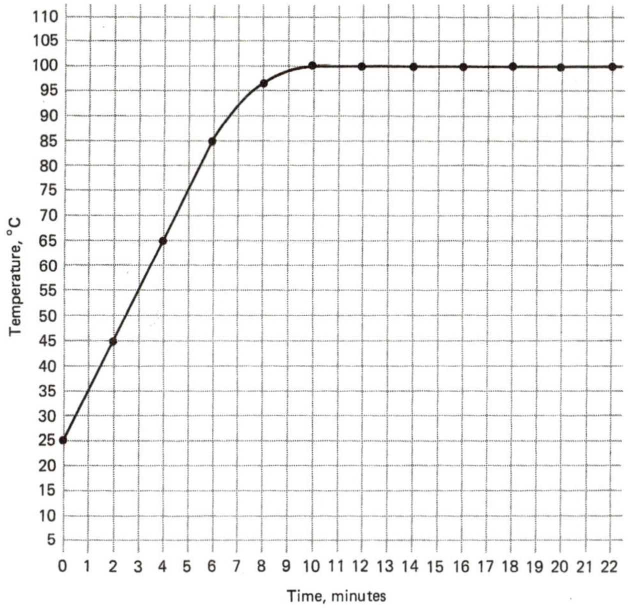

In the experiment above, Suki is given 50 mL of water, which she heats at a constant rate for 22 minutes. The graph shows the temperature of the water over time.

40. At which temperature is a phase change occurring?
    a. 45°C b. 65°C c. 85°C d. 100°C

41. In this experiment, how many minutes does it take the water to reach the boiling point?
    a. 2 minutes
    b. 6 minutes
    c. 10 minutes
    d. 22 minutes

42. Suki is using an alcohol burner and a very large test tube to heat the water. She should remember to
    a. make sure her sweater has long, baggy sleeves to protect her arms
    b. point her test tube away from others
    c. use a glass test tube to prevent the plastic ones from melting
    d. put a large cork in the test tube to hold in the heat

43. When the water is boiled, it turns into a gas. This is an example of a
    a. physical change
    b. chemical change
    c. reaction
    d. chemical property

44. At 15 minutes, why isn't the temperature rising?
    a. The phase change is using up all the heat energy.
    b. The heat source has been turned off.
    c. The heat energy is being absorbed and released by the liquid.
    d. The temperature *is* rising.

The diagram below shows the water cycle. Use it to answer questions 45–47.

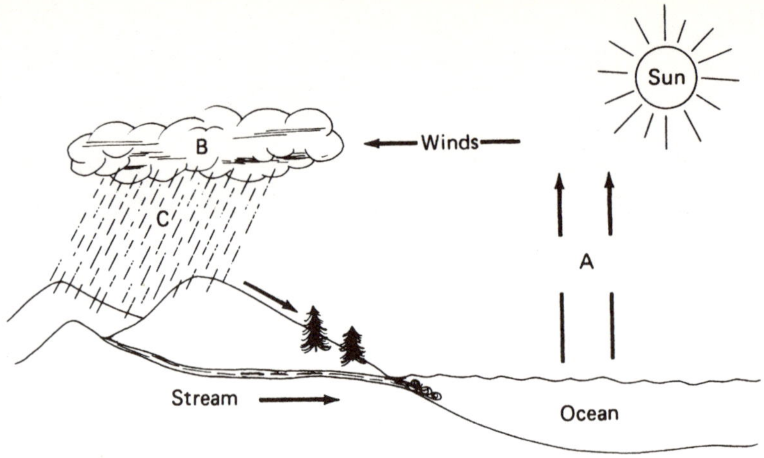

45. Which process is occurring at Point A?
    a. condensation
    b. evaporation
    c. precipitation
    d. respiration

46. This diagram shows that in nature, resources such as water are
    a. destroyed
    b. created when needed
    c. used up
    d. reused

47. As a resource, water would be considered
    a. renewable, because evaporated water returns to the earth as rain
    b. renewable, because it can be produced by our bodies
    c. nonrenewable, because once polluted it cannot be reused
    d. nonrenewable, because most of our water is used by animals and industries

Questions 48–50 refer to the following passage.

In ancient Greece, Democritus proposed that all matter was made up of different amounts of four elements: air, earth, fire, and water. The properties of an object depended on how much of each element was present.

48. Which observation would lead Democritus to believe that people have fire in their bodies?
    a. loud sneezing   c. blood flow
    b. warm skin   d. hard bones

49. If Democritus saw a piece of wood floating on water, he would probably say it floated because it contained lots of which element?
    a. air   b. earth   c. fire   d. water

50. Today, if we saw a piece of wood floating, we would infer that it
    a. weighs less than the water
    b. takes up less space than the surrounding water
    c. is less dense than the water
    d. is solid and contains no holes for water to leak in

Practice Test 2

A windlass is a device used to lift a bucket of water out of a well. Use the picture of the windlass below to answer questions 1 and 2.

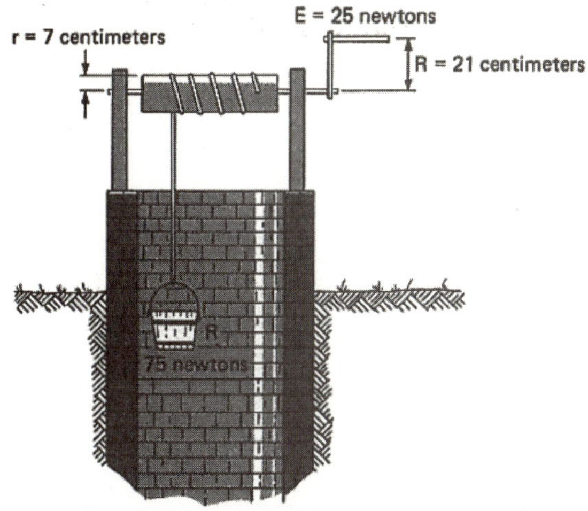

1. The windlass is which kind of simple machine?
   a. lever
   b. inclined plane
   c. wedge
   d. wheel and axle

2. To make it easier to lift the bucket, you should
   a. use a longer handle
   b. use a thicker rope
   c. use a metal bucket
   d. increase the speed at which you turn the handle

Use the pulley illustrations below to answer questions 3–5.

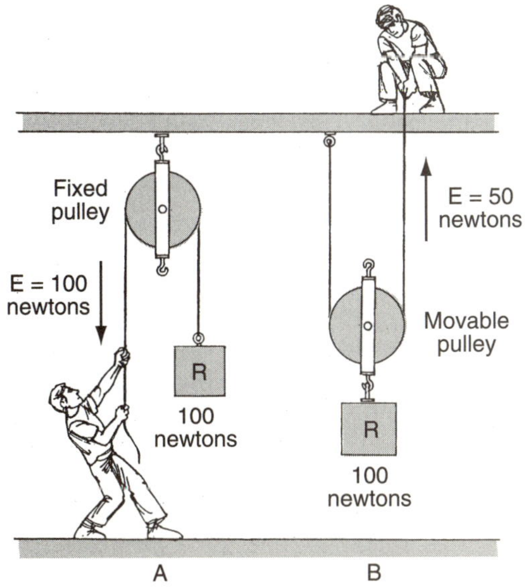

3. Pulleys can make work easier by changing the direction of the force or decreasing the effort needed to lift a weight. Which is a true statement concerning these pulleys?
   a. Pulley A reduces effort; Pulley B changes direction of force.
   b. Pulley A changes direction of force; Pulley B reduces effort.
   c. Both Pulley A and Pulley B reduce effort.
   d. Pulley B does not reduce effort or change direction of force.

4. A disadvantage to using Pulley B is that
   a. you will need more effort than with Pulley A
   b. you will need more resistance than with Pulley A
   c. you will need to pull the rope farther than with Pulley A
   d. you will have more friction than with Pulley A

5. Pulleys help make work easier and are used in many common devices. Which device uses a pulley to operate?
   a. window shade
   b. doorknob
   c. car jack
   d. pencil sharpener

Questions 6–8 are based on this weather map.

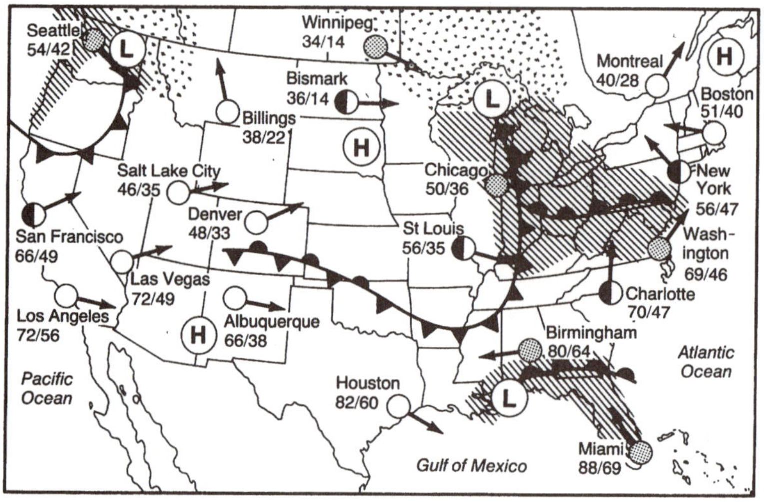

6. According to the map, the coldest temperatures are in which city?
   a. Boston
   b. San Francisco
   c. Miami
   d. Denver

7. In the next 12 hours, it will probably start raining in which city?
   a. Charlotte
   b. Denver
   c. New York
   d. St. Louis

8. Which city will *not* have a change in weather over the next few days?
   a. Birmingham
   b. Montreal
   c. Denver
   d. New York

Questions 9–12 refer to the following experiment.

Dexter wants to see if smaller particles dissolve faster than larger ones. He sets up four samples of salt and dissolves each sample in 50 mL of warm water. Here are his results:

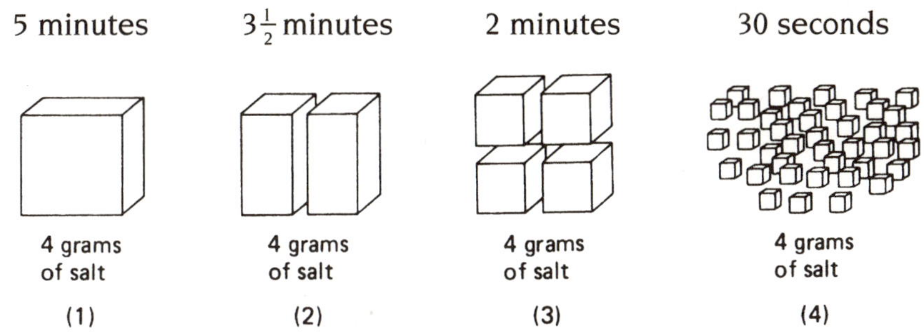

9. Which inference can Dexter make from his results?
   a. Smaller particles dissolve faster than large ones.
   b. Square particles dissolve faster than rectangular ones.
   c. Salt will dissolve faster in warmer water.
   d. Salt will dissolve faster than an equal amount of sugar.

10. If the water weighs 50 grams, and the salt weighs 4 grams, how much should one solution weigh after Dexter has totally dissolved the salt?
   (1) exactly 50 grams
   (2) exactly 54 grams
   (3) 51 to 53 grams
   (4) more than 54 grams

11. When dissolving, what happens to the salt particles?
   (1) They react with water.
   (2) They are destroyed.
   (3) They change in state.
   (4) They change in size.

12. While Dexter is dissolving salt in the lab, what safety rule is the most important for him to remember?
   a. Never smell fumes directly; fan them gently toward your nose.
   b. Never taste any substance in the laboratory.
   c. Tie back long hair and roll up sleeves when using a burner.
   d. Always add acid to water, never water to acid.

Questions 13–15 refer to the table below, which shows the high, low, and average monthly temperatures on the South Island of New Zealand for half a year.

Temperatures (°C)

Month	*High*	*Low*	*Average*
Jan.	29	5	17
Feb.	28	5	16.6
Mar.	27	2.7	15
Apr.	24	.5	12
May	20	–1.6	9.4
June	16	–3.3	6.6

13. On South Island, the coldest average temperatures occur in which month?
    a. January
    b. March
    c. May
    d. June

14. Based on this trend, which season does South Island have in January?
    a. summer
    b. spring
    c. winter
    d. fall

15. Which statement best explains why South Island is so cold in June?
    a. Earth is much farther away from the sun.
    b. The southern hemisphere is tilted away from the sun.
    c. The sun's rays are weakened by increased smog and pollution.
    d. It is too far from the equator to be warm.

Questions 16–20 refer to the graphs below.

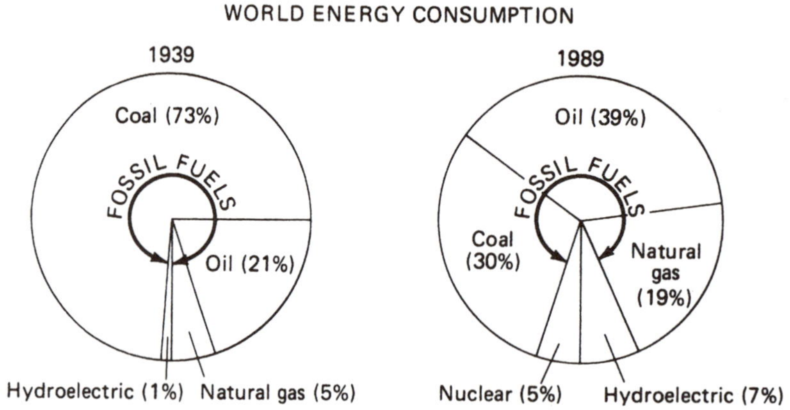

16. How did our energy usage change between 1939 and 1989?
    a. Use of coal increased.
    b. Use of natural gas decreased.
    c. Use of hydroelectric power did not change.
    d. Overall use of fossil fuels decreased.

17. Technological advances help meet a need in society, such as the need for new energy sources. Which energy source was available in 1989 that was not available in 1939?
    a. natural gas
    b. oil
    c. nuclear
    d. hydroelectric

18. Renewable resources can be replaced by nature within one's lifetime. Which is considered a renewable resource?
    a. oil
    b. hydroelectric
    c. coal
    d. natural gas

19. When burning coal in a stove, which kind of energy transformation is taking place?
    a. mechanical to nuclear
    b. electromagnetic to chemical
    c. chemical to heat
    d. kinetic to potential

20. Based on your observations of how energy consumption has changed from 1939 to 1989, how do you predict a graph would look in 2039?
    a. Use of fossil fuels will decrease, and use of renewable resources will increase.
    b. More coal will be used because it is so plentiful.
    c. New sources such as hydroelectric and nuclear will decrease due to lack of raw materials.
    d. Natural gas will continue to increase because it is a renewable.

Questions 21–22 refer to the diagram below.

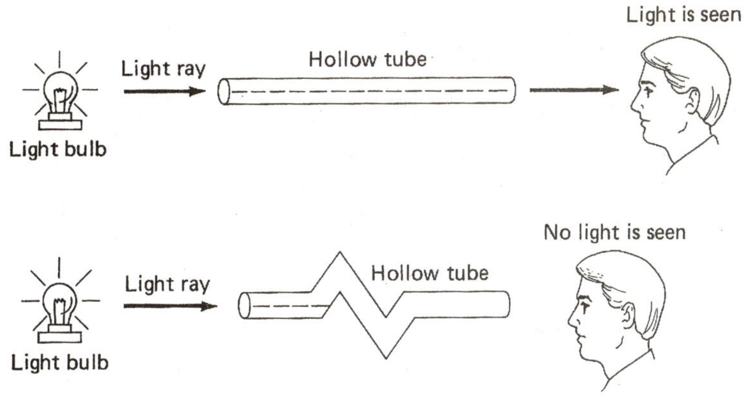

21. The diagram best illustrates which property of light?
    a. Light travels faster than sound.
    b. Light travels in straight paths.
    c. Light can be absorbed, transmitted, or reflected.
    d. Light can travel through a vacuum.

22. A lightbulb would be much brighter if much of the bulb's energy were not wasted as
    a. heat
    b. friction
    c. sound
    d. X rays

---

23. The Greek philosopher Empedocles believed that our vision is a result of light rays coming from our eyes, and objects become visible when touched by these rays. Which observation best *disproves* his theory?
    a. When you look into a mirror, you can see your reflection in it.
    b. When you close your eyes, it appears to be dark.
    c. Deep in a mine shaft, you cannot see even with your eyes open.
    d. Convex and concave lenses can magnify the appearance of objects.

24. Now we know that our vision depends on our ability to receive light that interacts with objects. If an object did not reflect or absorb any light, it would appear
    a. dark
    b. bright
    c. white
    d. invisible

Use the diagram below to answer questions 25–28.

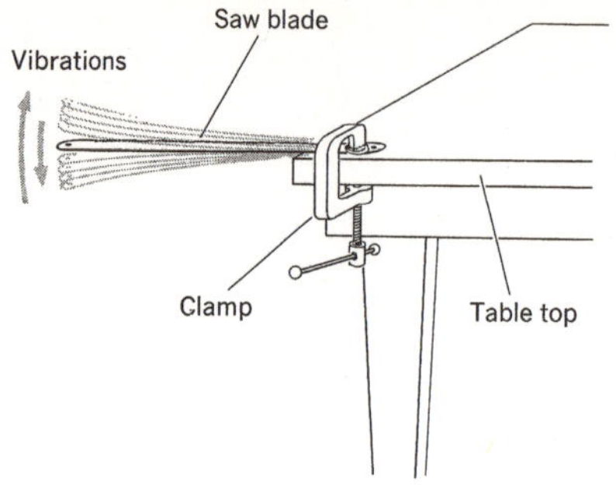

The saw blade has 30 centimeters hanging off the table. When pushed down and suddenly released, it produces a sound.

25. Which type of energy is being converted into sound by the vibrating blade?
    a. nuclear
    b. electromagnetic
    c. heat
    d. mechanical

26. If the blade were shortened so that only 10 centimeters were hanging off the end of the table, and struck in the exact same way, the pitch of the sound would
    a. be higher
    b. be lower
    c. be the same
    d. start high and get lower

27. Joe hits the blade too hard, and it breaks in half. This is a
    a. chemical change
    b. physical change
    c. change in state
    d. property

28. The sound of the blade would travel fastest through which medium—solid, gas, or liquid? In one or two sentences, explain the reason for your answer.

Questions 29 and 30 refer to the information below.

A boy from the city gets a summer job on a farm. He is amazed at how many different buildings there are on the farm and keeps going into the wrong buildings. So the farmer makes him this handy key to find the tools and animals he needs.

1a. If it is a living thing, choose 2a *or* 2b.
1b. If it is not a living thing, choose 3a *or* 3b.

2a. If it has two legs, choose 4a *or* 4b.
2b. If it has more than two legs, choose 5a *or* 5b.

3a. If it has an engine, go to Building A.
3b. If it does not have an engine, go to Building B.

4a. If it lays eggs, go to Building C.
4b. If it does not lay eggs, go to Building D.

5a. If it can be milked, go to Building E.
5b. If it cannot be milked, go to Building F.

29. First, the boy needs to collect eggs from the chickens. Which building should he go to?
   a. Building A c. Building D
   b. Building C d. Building F

30. Next, he needs to hook the plow up to a mule to work a very small field. Where should he go to find the plow and mule?
   a. plow, Building A; mule, Building F
   b. plow, Building C; mule, Building E
   c. plow, Building A; mule, Building B
   d. plow, Building B; mule, Building F

31. In the barnyard, there are chickens eating small rocks on the ground. They do this to supply their bodies with
   a. food c. oxygen
   b. minerals d. water

32. The farmer's wife has a garden with pea plants. Although most of the pea pods are smooth, a few plants have wrinkled pods. Smooth pods (S) is a dominant trait, and wrinkled pods (s) is a recessive trait. The plants with wrinkled pods probably have what genotype?
   a. Ss b. SS c. ss d. sS

33. If a pure smooth-pod plant were crossed with a pure wrinkled-pod plant, which would happen? Explain your answer.
   a. All of the offspring would appear smooth.
   b. Some of the offspring would appear smooth.
   c. All of the offspring would appear wrinkled.
   d. Some of the offspring would appear wrinkled.

Questions 34–37 refer to the experiment and diagram below.

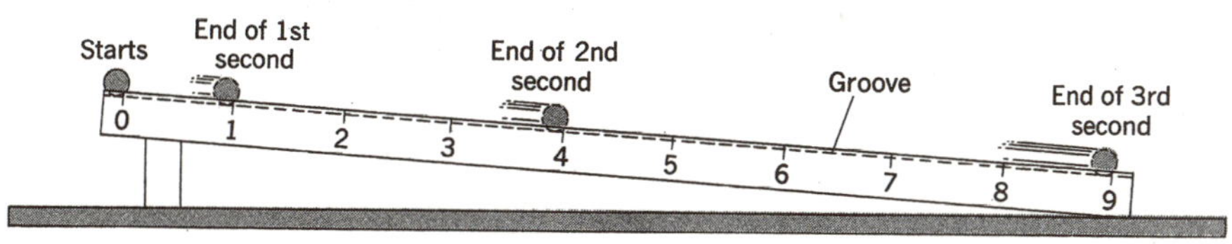

In 1590, Galileo experimented with moving objects. The diagram shows a modern version of one of his experiments.

34. Which property of motion is being demonstrated by the diagram?
   a. acceleration
   b. average speed
   c. inertia
   d. terminal velocity

35. This picture shows the ball as it accelerates for 3 seconds. If the ramp were longer and the acceleration stayed constant, how far would the ball travel in the fourth second?
   a. 1 cm
   b. 3 cm
   c. 5 cm
   d. 7 cm

36. One way to make the ball accelerate faster would be to
   a. use a lighter ball that is the same size
   b. use a bigger ball that is the same weight
   c. make the ramp wider
   d. make the ramp steeper

37. Aristotle believed that a heavy object would roll faster than a light one. Almost 2,000 years later, Galileo showed that both heavy and light objects roll at the same speed, unless there is
   a. air resistance c. inertia
   b. gravity d. momentum

Use the diagram of the food web below to answer questions 38–40.

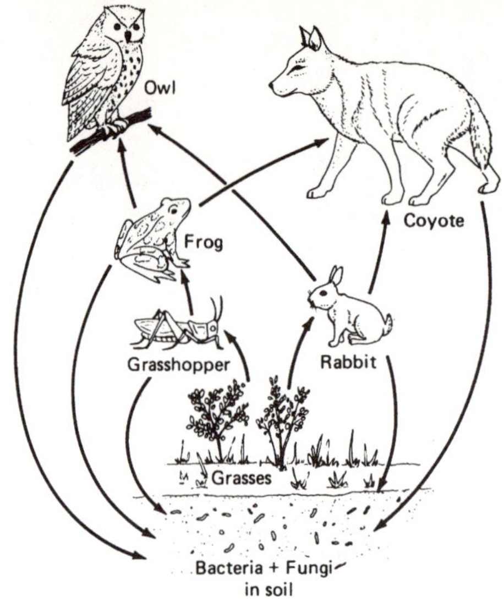

38. Which organism in the food web is a decomposer?
   a. grass c. bacteria
   b. owl d. rabbit

39. Which term best describes the relationship between the coyote and the owl?
   a. mutualism
   b. competition
   c. symbiosis
   d. predator/prey

40. If the frog population died out, what might happen in this ecosystem?
   a. The owls would adapt by eating grasshoppers.
   b. The rabbit population would increase due to less competition for food.
   c. The coyotes would eat more rabbits, causing the rabbit population to decline.
   d. There would be no change, as all organisms have other food to eat.

---

41. To find food, all of the animals must be able to move around. Most animals can walk, fly, swim, or crawl. Plants can also "move" in many different ways. Which is *not* a way in which plants can move?
   a. They grow in different directions to obtain the light.
   b. Their seeds are scattered to new locations by insects and winds.
   c. They can grow new branches and shoots to spread themselves out.
   d. They can apply pressure to the ground, causing them to shift sideways.

42. All giraffes have long necks, which enable them to reach leaves in tall trees. Why might there be no short-necked giraffes?
   a. Short-necked giraffes could not see as far and were easier prey for predators.
   b. Long-necked giraffes eat better, so they have more energy to grow longer necks.
   c. Short-necked giraffes could not compete with long-necked giraffes to reach leaves in tall trees.
   d. Female ancestral giraffes only mated with long-necked male giraffes.

Use the information and graph below to answer questions 43–46.

Tyrone wanted to see how watering plants would affect their growth. He put six plants into two groups. The first group (group A) was watered four times a week; the second group (group B) was watered two times a week. Here is a graph of his results.

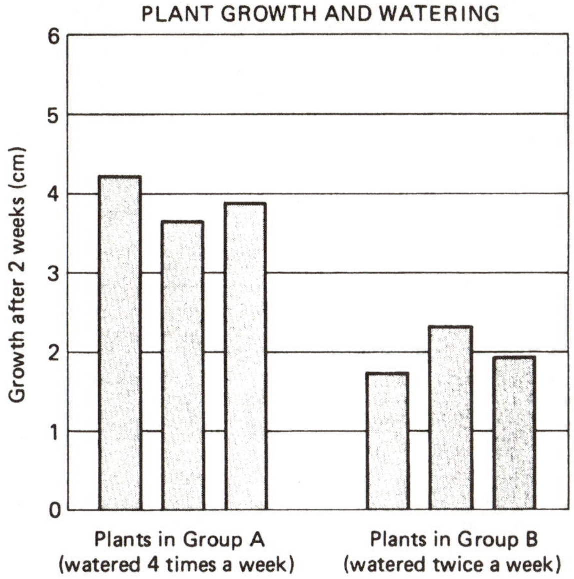

43. This graph shows that the plants in group A grew
    a. four times as much as the plants in group B
    b. twice as much as the plants in group B
    c. the same amount as the plants in group B
    d. half as much as the plants in group B

44. From this experiment, one can infer that
    a. living things come from other living things
    b. living things need oxygen to grow
    c. living things depend on materials from the environment
    d. living things undergo chemical changes to keep them alive

45. Suppose that these plants were growing in a field that suddenly flooded, and they were covered with water for a week. How would the plants adapt to the changed conditions?
    a. Plant growth would increase because these plants like water.
    b. Plant growth would stay the same because they take only the water they need.
    c. Plant growth would decrease because too much water can be harmful.
    d. Only growth of leaves and roots would increase to help them move the extra water.

46. Another student tried the same experiment with different plants and found the plants in group B grew better than in group A. You can infer that she probably used what type of plant?
    a. sunflower
    b. cactus
    c. corn
    d. pea

Use the diagram below to answer questions 47 and 48.

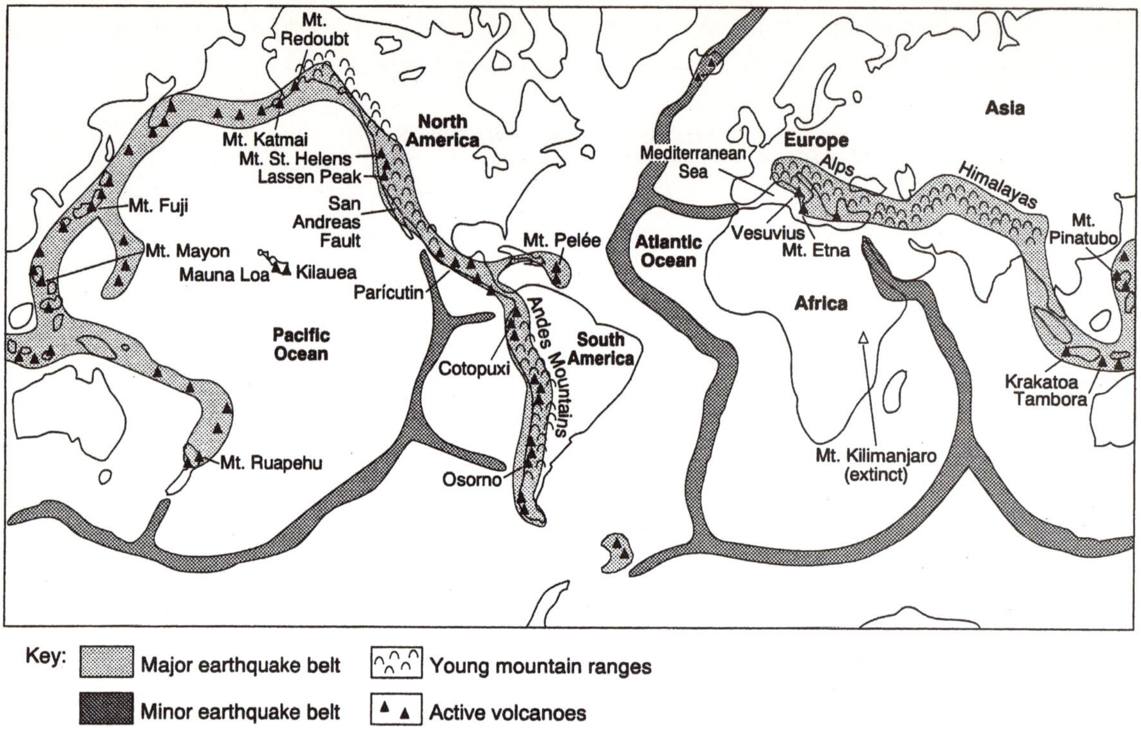

47. According to the diagram, you can infer that most active volcanoes are likely to be found in an earthquake belt and in young mountain ranges. In one or two sentences, explain why this is true.

48. What is happening in Earth's crust to cause the mountains, earthquakes, and volcanoes in these specific areas?
    a. Different sections (plates) of the crust are pushing against each other.
    b. Erosion due to weathering has caused weaknesses in the crust.
    c. The pressure of Earth's molten core is causing the crust to buckle.
    d. The gravitational pull from the moon is causing ruptures in the crust.

---

49. In nature, many rocks end up at the bottom of the oceans. Which process is responsible for moving rock material from the continents to the ocean basins?
    a. volcanism
    b. evaporation
    c. faulting
    d. erosion

50. The planet Venus is not significantly tilted on its axis. This means that Venus probably has no
    a. day and night
    b. seasons
    c. years
    d. tides

Practice Test 3

Use the following information to answer questions 1 and 2.

The ability to roll your tongue into a U-shape is a dominant trait. Suppose a father has one dominant gene for rolling (R) and one gene for nonrolling (r), and a mother also has one dominant gene for rolling and one gene for nonrolling. Both parents would then have an Rr genotype (combination of genes).

		DAD R	DAD r
MOM	R	RR	Rr
MOM	r	Rr	rr

1. Which percentage of their children will likely have an RR genotype?
   a. 25%
   b. 50%
   c. 75%
   d. 100%

2. As long as a child has at least one dominant R trait, he or she will be able to roll his or her tongue. Which percentage of children are likely to be able to roll their tongue?
   a. 25% b. 50% c. 75% d. 100%

Use the diagram below to answer questions 3 and 4.

3. The screwdriver being used to remove the lid from a can of paint is acting as
   a. a wedge
   b. a lever
   c. an inclined plane
   d. a screw

4. Compared to the amount of work put into a machine, the amount of work put out by a machine is
   a. always greater
   b. always less
   c. always the same
   d. sometimes greater, sometimes less

Questions 5–8 refer to the paragraph and the graph below.

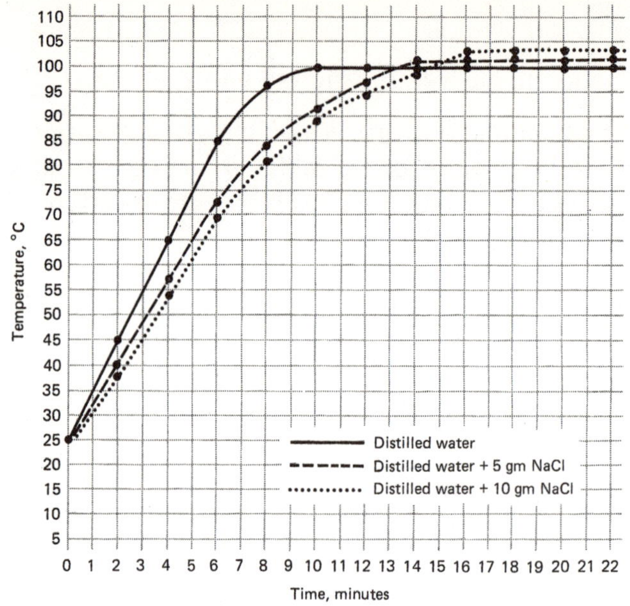

Luna performs an experiment to see how salt (NaCl) affects the boiling point of water. She boils three solutions of distilled water: one with no salt, one with 5 g of salt, and one with 10 g of salt. Her data are in the graph.

5. Which is the temperature of pure distilled water after 4 minutes?
   a. 53°C
   b. 57°C
   c. 65°C
   d. 95°C

6. Which solution has the highest boiling point?
   a. distilled water with no salt
   b. distilled water with 5 grams of NaCl
   c. distilled water with 10 grams of NaCl
   d. all water has the same boiling point

7. When water reaches its boiling point and heat continues to be applied, the heat energy
   a. raises the water temperature
   b. lowers the water temperature
   c. changes water to steam
   d. is not used by the water

8. To prevent cars from overheating, a mixture of antifreeze and water is placed in the coolant system. Car engines can get hotter than 105°C without the water boiling away. You could infer that the antifreeze has what effect on the boiling point of the water? Explain.

Questions 9 and 10 refer to the paragraph below.

A polar bear's fur is actually clear, not white. Some light rays bounce off, which makes the fur look white. However, most of the sun's rays reach the black skin underneath the fur, which keeps the polar bear warm.

9. How does the black skin of a polar bear help keep it warm?
   a. It reflects the sun's rays.
   b. It bends the sun's rays.
   c. It transmits the sun's rays.
   d. It absorbs the sun's rays.

10. The fur and the skin of the polar bear are both examples of
    a. an adaptation to the environment
    b. a mutation in the genes
    c. a dominant and recessive trait
    d. an instinct for survival

Questions 11 and 12 refer to the drawings below.

Figure 1

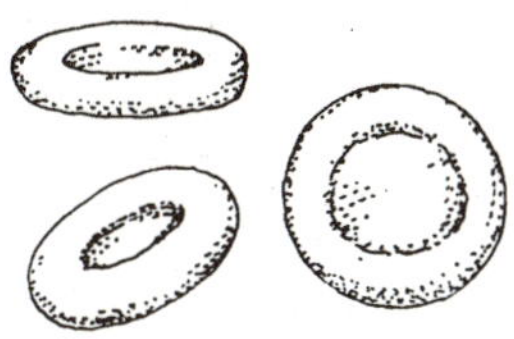

Figure 2

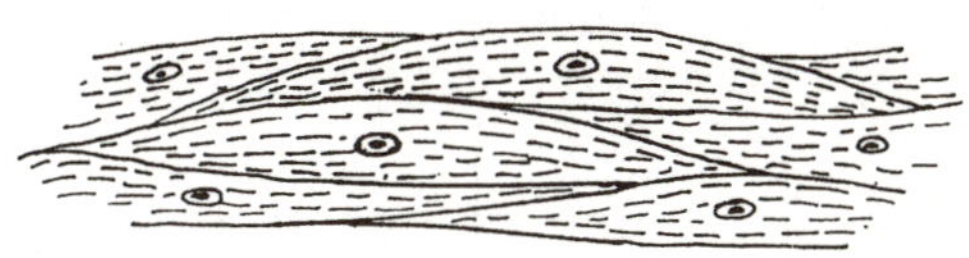

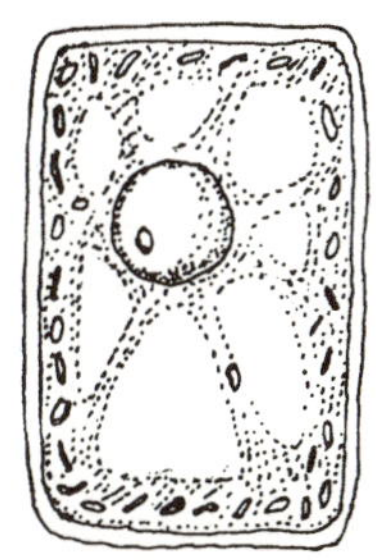

Figure 3

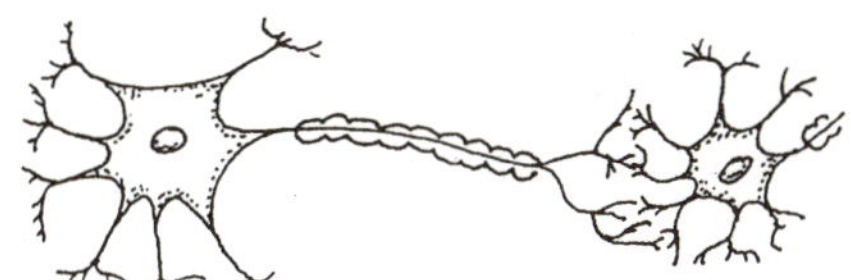

Figure 4

11. Which is a plant cell?
    a. Figure 1
    b. Figure 2
    c. Figure 3
    d. Figure 4

12. Which cell is specialized to carry messages throughout the human body?
    a. Figure 1
    b. Figure 2
    c. Figure 3
    d. Figure 4

Questions 13–17 refer to the diagram below.

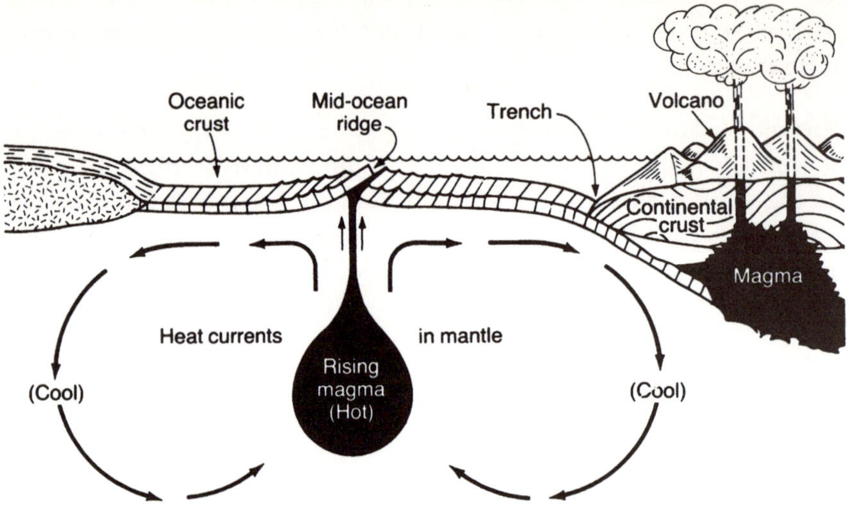

13. In the diagram, where would scientists find the newest rocks?
    a. right next to the mid-ocean ridge
    b. halfway between the ridge and the trench
    c. down at the bottom of the trench
    d. at the edge of the continent

14. In the diagram, where is an earthquake most likely to occur?
    a. in the mid-ocean ridge
    b. between layers of rock along the ocean floor
    c. where the edge of the ocean plate melts and becomes magma
    d. where the continental plate and the ocean plate meet

15. What happens to the oldest rocks on the ocean floor to continue the rock cycle?
    a. They are melted under a continent and become magma again.
    b. They push onto the continent and become dry land.
    c. They dissolve in the ocean water and are redeposited as sediment.
    d. The rock molecules are destroyed by intense heat and pressure.

16. Why does the hot magma rise to the ocean floor instead of sinking?
    a. Its molecules are moving more slowly than cool magma.
    b. It is less dense than cool magma.
    c. It is being forced up by cool magma trying to escape.
    d. Its molecules are packed closely together.

17. To find a mountain range built by volcanoes, you would look
    a. under the water, where Earth's crust is thinnest
    b. in the center of a continent, where two continental plates have pushed together
    c. along a coastline, where the ocean floor has gone beneath the continent
    d. in the tropics, where temperatures are much hotter

Use the table below to answer questions 18–22.

Commonly Eaten Birds

Bird	*Size*	*Diet*	*Habitat*	*Reproduction*	*Ability to Fly*
Pheasant	Medium	Seeds and berries	Meadows	Eggs	Limited
Quail	Small	Seeds and insects	Meadows	Eggs	Excellent
Turkey	Large	Seeds and insects	Woods	Eggs	Limited
Squab	Small	Grains	Trees	Eggs	Excellent

18. Which characteristic could not be used to distinguish these birds?
    a. size
    b. diet
    c. ability to fly
    d. reproduction

19. The pheasant and the turkey could be placed in the same group based on their
    a. size
    b. diet
    c. habitat
    d. ability to fly

20. Which physical adaptation do all of these birds probably have in common?
    a. talons for catching and carrying prey
    b. sharp beaks for breaking through shells and nuts
    c. strong wings to catch their prey
    d. webbed feet for swimming

21. A farmer uses a pesticide on his crops to kill insects. One year, park rangers find that the pheasant population in a nearby meadow has greatly decreased, and the remaining birds have large amounts of this pesticide stored in their body tissue. How did the pesticide most likely reach the meadow?
    a. It was carried on the legs of insects as they moved from plant to plant pollinating flowers.
    b. During rainstorms, it was dissolved in water and carried there by streams and ditches.
    c. As it was sprayed onto the crops, high winds and air currents carried it to the meadow.
    d. It gradually was passed along from one soil molecule to another by diffusion.

22. A coyote also lives in the meadow, and he has eaten several pheasants. Compared to a single pheasant, he probably
    a. has significantly less pesticide in his body
    b. has the same amount of pesticide in his body
    c. has accumulated more pesticide in his body
    d. is too large to be affected by the pesticide

Use the diagrams below to answer questions 23–27.

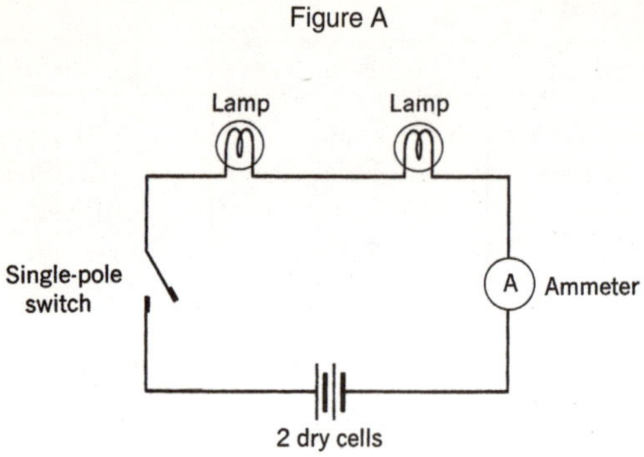

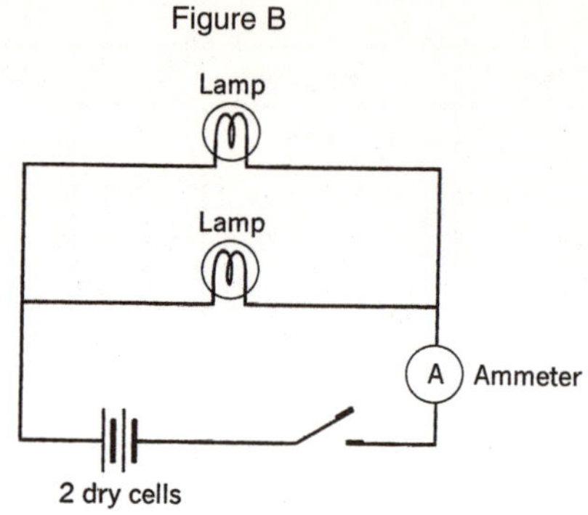

23. If one bulb burns out in each figure, what will happen to the other bulb?
    a. Figure A's bulb will go out; Figure B's bulb will stay lit.
    b. Figure A's bulb will stay lit; Figure B's bulb will go out.
    c. Both Figure A's and Figure B's bulbs will stay lit.
    d. Neither Figure A's nor Figure B's bulbs will light at all.

24. What could you do to make the bulbs in Figure B brighter?
    a. Add a bulb.
    b. Remove a bulb.
    c. Add a battery.
    d. Remove a wire.

25. In Figure A, about 90 percent of the energy from the battery is used to light the bulbs. What happens to the rest of the energy?
    a. It is transformed into heat and makes the bulbs hot.
    b. It takes a different path through the circuit.
    c. It is used up by pushing electrons through the wire.
    d. It is destroyed inside the circuit.

26. To light the bulbs, what energy transformation is taking place?
    a. chemical to potential
    b. electromagnetic to mechanical
    c. mechanical to heat
    d. chemical to electromagnetic

27. Which is a chemical change?
    a. energy being produced in the battery
    b. the wires being attached together
    c. a bulb being broken by a hammer
    d. the switch being opened

Use the planetary data table below to answer questions 28–32.

Planetary Data

Planet (in order from sun)	*Distance from sun in Earth-sun units†*	*Time to revolve once around sun*	*Time to rotate once on axis*	*Number of satellites (moons)*
Inner planets:				
Mercury	0.4	88 days	59 days	0
Venus	0.7	225 days	243 days	0
Earth	1.0	365.25 days	24 hours	1
Mars	1.5	1.88 years	24.6 hours	2
		Asteroid belt		
Outer plants:				
Jupiter	5.2	11.86 years	9.9 hours	16*
Saturn	9.5	29.63 years	10.6 hours	18*
Uranus	19.2	83.97 years	17 hours	15*
Neptune	30.1	165 years	16 hours	8*
Pluto	39.5	248 years	6.4 days	1*

†An Earth-sun unit is the average distance from Earth to the sun (149,600,000 kilometers).
*Number of known moons; the actual number may be higher.

28. Which planet has the longest year?
   a. Venus
   b. Earth
   c. Neptune
   d. Pluto

29. Which factor causes each of these planets to have a different length of year?
   a. size of the planet
   b. distance from the sun
   c. length of the day (speed of rotation)
   d. number of moons

30. If you were born on a certain planet, the day would be longer than the year. You could actually have more birthdays than you'd see sunrises. Which planet is this?
   a. Mercury
   b. Venus
   c. Saturn
   d. Pluto

31. The largest planets in the solar system are Jupiter, Saturn, Uranus, and Neptune. This may help explain why
   a. their days are longer than the other planets' days
   b. their years are longer than the other planets' years
   c. they are farther from the sun than any other planet
   d. they have more moons than any other planet

32. In 1543, Copernicus published a book that explained planetary motion. His explanation had to assume that Earth and the other planets all traveled around the sun. He was the first to claim that
   a. daylight occurs when the sun rises and sets
   b. Earth is round, not flat
   c. Earth cannot be at the center of the solar system
   d. seasons are caused by the weakening of the sun as its fuel runs low

Use the description of the lab to answer questions 33–37.

One day in the science lab, Omar decides to test five substances to see if they are living organisms. He puts a small sample of each into five different test tubes, along with 3 mL of bromthymol blue. Test tube 6 is his control; it contains only the bromthymol blue. Omar knows that bromthymol blue will turn yellow in the presence of carbon dioxide gas. He corks all the tubes and waits 24 hours. Here is a table of his results.

Test Tube	*Substance*	*Color of Bromthymol Blue*
1	Yogurt	Dark yellow
2	Salt	Dark blue
3	Yeast	Yellow-orange
4	Bean seeds	Dark yellow
5	Sand	Dark blue
6	—	Dark blue

33. According to this table, Omar should conclude that the following substances are living.
    a. yogurt, salt, and yeast
    b. yogurt, bean seeds, and sand
    c. salt and sand only
    d. yogurt, yeast, and bean seeds

34. To come to this conclusion, what inference must Omar make about living things?
    a. They give off carbon dioxide from respiration.
    b. They have the element carbon in their cells.
    c. They are able to reproduce in the bromthymol blue.
    d. They absorb the blue nutrients from the liquid, leaving yellow.

35. If Omar put his experiment in a dark closet for a week, which item would grow?
    a. sand
    b. yeast
    c. bean seeds
    d. salt

36. How do these living cells get energy for growth?
    a. They all absorb it from the sun.
    b. They all absorb it from the bromthymol blue.
    c. They all chemically change food into energy.
    d. They don't have to; it doesn't take any energy to grow.

37. While cleaning up at the sink, Omar breaks one of his test tubes. Which action should he take?
    a. quickly pick up the larger pieces with his hands
    b. use a broom to sweep all the pieces into the trash can
    c. break it again until all the pieces are small enough to be rinsed down the sink
    d. not touch any pieces but tell the teacher about it

The diagram below shows a card held against the teeth of a spinning wheel. This makes the card vibrate, which makes a sound. Use the diagram to answer questions 38–40.

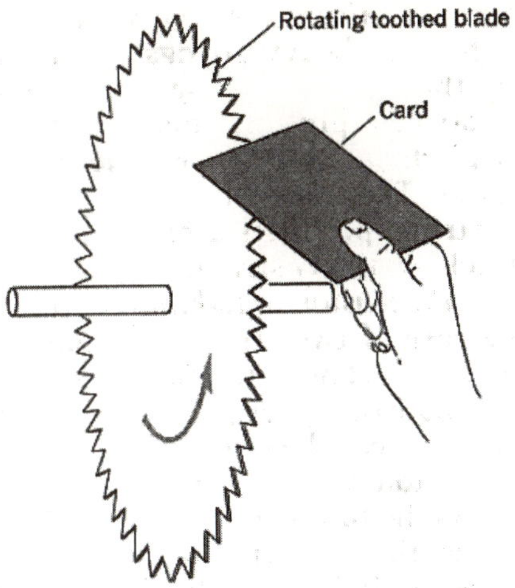

38. Which force is present as the card rubs against the edge of the wheel?
    a. friction
    b. gravity
    c. inertia
    d. weight

39. When the wheel is spun faster, what will happen to the sound from the card?
    a. It will have a lower pitch.
    b. It will have the same pitch.
    c. It will have a higher pitch.
    d. It will not make a sound anymore.

40. An astronaut floating in outer space spins the wheel and holds the card against it. How will the sound be different than if it were made on Earth?
    a. The sound will be quieter.
    b. The sound will have a higher pitch.
    c. The sound will last longer.
    d. There will be no sound produced.

    In one or two sentences, explain the reason for your answer.

Use the picture below to answer questions 41–43.

41. The type of energy shown by the windmills could be described as
  a. renewable
  b. nonrenewable
  c. electromagnetic
  d. potential

42. Which type of simple machine is most similar to the windmill?
  a. wheel and axle
  b. lever
  c. wedge
  d. inclined plane

43. Unlike many other sources of energy, the windmills do not give off any pollution and are environmentally safe. Which is a disadvantage to using windmills?
  a. They are expensive and complicated to build.
  b. They cannot change their energy into electricity.
  c. They function only when it is windy.
  d. They would put the oil companies out of business.

---

44. In 1626, Francis Bacon stuffed a dead chicken with snow to see if it would decay less rapidly than one kept at room temperature. It did, which shows that
  a. chicken meat should not be cooked, as the heat causes faster decay
  b. bacteria and other decomposers are more active when it is warm
  c. snow contains natural preservatives that slow decay
  d. a chicken's feathers help protect it from the cold

45. Which invention may have been developed from an experiment such as this?
  a. oven
  b. microwave
  c. penicillin
  d. refrigerator

Use the information in the passage below to answer questions 46–48.

In the African savannah, the rhinoceros population is usually accompanied by a large population of dung beetles. The dung beetles live in and eat the rhinos' manure. This helps fertilize the ground and allows more plants to grow to feed the rhinoceroses.

46. The relationship between the dung beetle and the rhino is an example of
  a. competition
  b. mutualism
  c. symbiosis
  d. predator/prey

47. The dung beetle's eating habit best shows that
  a. all living things have their own job to do in an ecosystem
  b. all animals should get along and not fight or eat each other
  c. if a creature such as a beetle dies out, the food web will not be affected
  d. these beetles could probably survive without the rhinos by eating plants directly

48. Tree shrews eat dung beetles. If a large group of tree shrews were introduced into the ecosystem, what would probably happen?
  a. The rhinos would not be affected because they don't eat beetles.
  b. The rhino population would increase because they might begin eating tree shrews.
  c. The rhino population would decrease because the plants wouldn't grow as well without beetles.
  d. The rhinos would become extinct because the tree shrew would eat all the plants.

Use the picture below to answer questions 49 and 50.

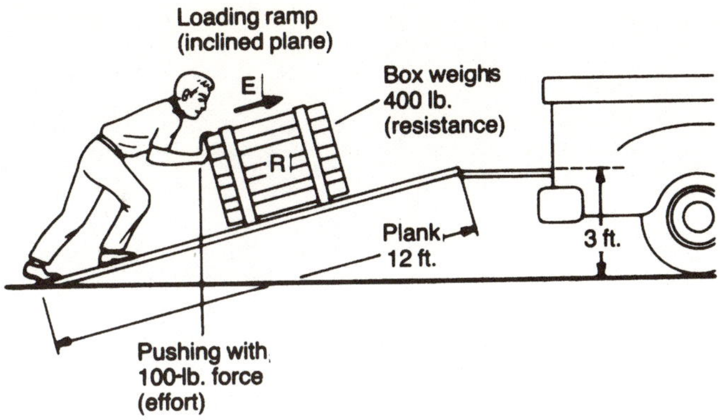

49. Which of Newton's laws of motion is best demonstrated by the man exerting a force to push the box? Explain your answer.

50. If the box weighed 200 lb, how much force would be needed to move it up the same inclined plane?
    a. 0 lb
    b. 50 lb
    c. 100 lb
    d. 200 lb

GLOSSARY

Acid rain Rain that has been made more acidic than normal by pollutants in the atmosphere.

Adaptation A characteristic that helps an organism survive in its habitat.

Air mass A large body of air that has roughly uniform temperature and humidity throughout.

Air pressure The force with which air presses down on Earth's surface.

Air temperature A measurement of the amount of heat energy in the atmosphere.

Altitude The height above sea level of a place.

Arteries Blood vessels that carry blood away from the heart.

Asexual reproduction Reproduction that involves only one parent, producing offspring that are genetically identical to the parent.

Atom The smallest particle of an element that has the properties of that element.

Axis of rotation An imaginary line through an object, around which the object spins.

Bedrock The unbroken, solid rock portion of Earth's crust.

Benefit An advantage to a person or society of a technological device or process.

Biodiversity Describes the great variety of life-forms on Earth or within a habitat.

Blood A liquid tissue that contains red and white blood cells, and platelets, and also carries dissolved gases, nutrients, hormones, and wastes.

Blood vessels Tubes through which the blood flows.

Boiling The rapid change in phase from liquid to gas, during which bubbles of gas form within the liquid.

Boiling point The temperature at which a substance changes rapidly from a liquid to a gas.

Brain The organ, located within the skull, that controls thinking and body activities.

Bronchi The two tubes that branch off from the lower end of the trachea, connecting it to the lungs.

Burden A disadvantage to a person or society of a technological device or process.

Calorie A unit used to measure and compare the amount of heat energy contained in substances such as foods and fuels.

Capillaries Tiny blood vessels, connecting arteries to veins, through which materials are exchanged between the blood and the body's cells.

Cartilage A flexible tissue that acts as a cushion between bones and provides flexibility at the ends of bones.

Cell The basic unit of all living things.

Cell division The process by which cells reproduce, wherein a parent cell splits into two new daughter cells.

Cell membrane The outer covering, or "skin," of a cell, which controls the flow of materials into and out of the cell.

Cellular respiration A life process that occurs in all cells, in which nutrients from digested food are combined with oxygen to release energy and produce the wastes carbon dioxide and water.

Chemical bond The link that joins one atom to another in a molecule.

Chemical change A change that results in the formation of one or more new substances; a chemical reaction.

Chemical property A characteristic that a substance displays when it undergoes a change to a new substance or substances.

Circuit breaker A device that prevents overloading of an electric circuit by interrupting the flow of electricity when it reaches a dangerous level.

Climate The general character of the weather in an area over many years.

Cloud A mass of tiny water droplets or ice crystals suspended high in the atmosphere.

Coal A black rock formed from the remains of ancient swamp plants; it is a fossil fuel.

Cold front The boundary formed when a cool air mass pushes into and under a warm air mass.

Community All the different organisms that live within a habitat.

Condensation The changing of water vapor into droplets of liquid water; more generally, the change in phase from gas to liquid.

Conductor A material through which electricity can flow.

Conservation The saving of natural resources through wise use.

Consumer An organism that obtains nutrients by eating other organisms.

Crust The outermost rock layer of Earth, which contains all of Earth's surface features.

Cytoplasm The watery substance that fills the cell, where most life processes occur.

Decomposer An organism that breaks down the remains and wastes of other living things.

Disinfection The destruction of all or most of the harmful microorganisms in an area by using chemicals called germicides.

Dormancy A state in which an organism is inactive while it awaits more favorable conditions in its environment.

Earthquake A shaking or vibrating of Earth's crust, usually caused by the sudden movement of rocks sliding along a fault.

Ecological succession The natural process by which one community of living things is replaced by another community, until a stable climax community appears.

Ecosystem The living members of a community, plus the nonliving elements of their environment.

Electric circuit A complete path for the flow of electricity.

Electricity A form of energy produced by the flow of electrons from one point to another point.

Electromagnetic waves Energy waves that travel at the speed of light and can move through a vacuum; they include radio waves, microwaves, infrared waves, visible light, ultraviolet waves, X rays, and gamma rays.

Element One of the 109 known basic substances that form the building blocks of matter.

Energy The ability to do work.

Environment The surroundings in which an organism lives, including both living and nonliving things.

Erosion The process whereby rock material at Earth's surface is removed and carried away.

Evaporation The changing of liquid water into water vapor (gaseous water); more generally, the change in phase from liquid to gas.

Faulting The process in which internal forces cause Earth's crust to break and slide along a fracture called a fault.

Fertilization The joining of an egg cell and a sperm cell, during sexual reproduction, to begin the development of a new individual.

Flammable Capable of catching fire and burning easily.

Folding The process whereby rock layers in Earth's crust are squeezed into wavelike patterns called folds.

Food chain A sequence of organisms through which nutrients are passed along in an ecosystem.

Food web A number of interconnected food chains.

Fossil The remains or traces of an ancient organism.

Fossil fuel A fuel that was formed from the remains of ancient plants or animals; examples include oil, coal, and natural gas.

Freezing (1) The change in phase from liquid to solid. (2) The storing of food at temperatures below 0°C (32°F), to slow the growth of microorganisms that can spoil food.

Freezing point The temperature at which a substance changes from a liquid to a solid.

Front The boundary between two different air masses.

Full moon The phase of the moon that occurs when Earth is between the sun and the moon, so that all of the moon's lighted side can be seen from Earth.

Fuse A device, used in an electric circuit, containing a thin metal strip that melts to interrupt the flow of electricity when the circuit becomes overheated.

Gland An organ that makes and secretes (releases) chemicals called hormones.

Greenhouse effect The trapping of heat in Earth's atmosphere by carbon dioxide; can lead to global warming.

Grounding A safety feature of an electrical device or circuit in which an extra wire attached to the device or circuit conducts any excess charge to the earth, where it is absorbed.

Habitat The particular environment in which an organism lives.

Heart An organ, made mostly of muscle, that contracts (beats) regularly to pump blood throughout the body.

Heat energy The energy of motion of the vibrating particles that make up matter.

Hibernate To enter a sleeplike state of reduced body activity; how some animals survive the winter.

High-pressure system A large area where air is sinking, causing high surface air pressure; also called a high.

Hormone A chemical "messenger" secreted by a gland into the bloodstream, which carries the hormone to an organ that responds in some way.

Humidity The amount of moisture (water vapor) present in the atmosphere.

Hurricane A huge, rotating storm that forms over the ocean in the tropics, with strong winds and heavy rains.

Hydroelectric energy Electricity produced by using the energy of flowing water to turn the turbines of a generator.

Igneous rock A rock formed by the cooling and hardening of hot, liquid rock material.

Inclined plane A simple machine that consists of a flat surface with one end higher than the other, such as a loading ramp.

Infectious disease A disease caused by microorganisms that can be transmitted from one individual to another.

Insulation Material used to reduce or slow the flow of heat from one area to another.

Insulator A material through which electricity cannot flow.

Involuntary muscles Muscles that we do not consciously control.

Joint A place where one bone is connected to another bone.

Kidneys A pair of organs that filter wastes from the blood and help control the water and mineral balance of the body.

Kinetic energy Energy that an object has because of its motion.

Latitude Distance from the equator, measured in degrees.

Lens A piece of transparent glass or plastic with curved surfaces that bend light rays to form an image.

Lever A simple machine consisting of a bar or rod that can turn around a point called the fulcrum.

Life cycle The changes that an organism undergoes as it develops and produces offspring.

Light A visible form of energy.

Light-year The distance that light travels in one year, about 9.46 trillion kilometers.

Liver An organ that produces urea from excess amino acids, removes harmful substances from the blood, and secretes bile, a digestive juice.

Locomotion The movement of the body from place to place.

Low-pressure system A large area where air is rising, causing low surface air pressure; also called a low.

Lungs A pair of organs, located in the chest, that contain millions of tiny air sacs, in which the exchange of respiratory gases between the blood and the environment takes place.

Lymph A fluid that bathes all body cells and acts as a go-between in the exchange of materials between the blood and the cells.

Lymph vessels Tubes in which waste-laden lymph is collected and returned to the bloodstream.

Machine A device that transfers mechanical energy from one object to another object.

Mammary glands The female breasts, which produce milk to nourish newborn offspring.

Mass The amount of matter in an object.

Matter Anything that has mass and takes up space.

Melting The change in phase from solid to liquid.

Melting point The temperature at which a substance changes from a solid to a liquid.

Metamorphic rock A rock produced when existing igneous or sedimentary rock undergoes a change in form caused by great heat, pressure, or both.

Meteor A rock fragment traveling through space that enters Earth's atmosphere and burns up, producing a bright streak of light.

Microorganism An organism that is very small, usually too small to be seen with the unaided eye.

Migrate To move from one environment to another, where conditions are more favorable; how some animals survive the change in seasons.

Mineral A naturally occurring, solid inorganic substance with characteristic physical and chemical properties.

Mountain A feature on Earth's surface that rises relatively high above the surrounding landscape.

Muscles Masses of tissue that contract to move bones or organs.

Natural gas A gaseous fossil fuel found trapped deep underground, often with oil deposits.

Nerves Thin strands of tissue, composed of neurons, that carry impulses throughout the body.

Neurons Cells that make up the nervous system, which receive and transmit information in the form of impulses.

New moon The phase of the moon that occurs when the moon is between Earth and the sun, so that the moon cannot be seen from Earth.

Noninfectious disease A disease that cannot be transmitted from one individual to another.

Nonrenewable resource A resource that is not replenished by nature within the time span of human history.

Nuclear energy The energy stored within the nucleus of an atom, used by nuclear power plants to produce electricity.

Nuclear waste The poisonous, radioactive remains of the materials used to fuel nuclear power plants.

Nucleus (1) The structure within the cell that controls cell activities and contains genetic material. (2) The center of an atom.

Nutrients Food substances that supply an organism with energy and with materials for growth and repair.

Oil A thick, black, liquid fossil fuel, found trapped underground; also called petroleum.

Orbit The path of an object in space that is revolving around another object.

Organ A group of tissues that act together to perform a function.

Organism A living thing.

Organ system A group of organs that act together to carry out a life process.

Ovaries The female reproductive organs that produce egg cells.

Oviducts Tubes that connect the ovaries to the uterus.

Pasteurization The process of heating milk, and some other foods, to kill bacteria that cause spoilage and disease.

Phases (1) The changing apparent shape of the moon, as seen from Earth. (2) The three forms, or states, of matter—solid, liquid, and gas.

Photosynthesis The process by which green plants produce food, using sunlight, carbon dioxide, and water; oxygen is given off as a by-product.

Physical change A change in the appearance of a substance that does not alter the chemical makeup of the substance.

Physical property A characteristic of a substance that can be determined without changing the identity of the substance.

Plain A broad, flat landscape region at a low elevation, usually made of layered sedimentary rocks.

Plateau A large area of Earth's surface made up of horizontally layered rocks, found at a relatively high elevation.

Plate tectonics The theory that Earth's crust is broken up into a number of large pieces, or plates, that move and interact, producing many of Earth's surface features.

Pollutants Harmful substances that contaminate the environment, often produced by human activities.

Potential energy Stored energy that an object has because of its position or chemical makeup.

Precipitation Water, in the form of rain, snow, sleet, or hail, falling from clouds in the sky.

Prevailing winds The winds that commonly blow in the same direction at a given latitude.

Producer An organism that makes its own food. Most producers are green plants.

Refrigeration The storing of food at cold temperatures to slow the growth of harmful bacteria.

Renewable resource A resource that is replenished by nature within a relatively short time span.

Reproduction The life process by which organisms produce new individuals, or offspring.

Respiration (1) The process of taking in oxygen from the environment and releasing carbon dioxide and water vapor. (2) See also cellular respiration.

Response The reaction of a living thing to a change in its environment.

Revolution The movement of an object in space around another object, such as the revolution of the moon around Earth.

Rock A natural, stony material composed of one or more minerals.

Rotation The spinning of an object around its axis.

Science The study of the natural world.

Sedimentary rock A rock formed from layers of particles, called sediments, that are cemented together under pressure.

Sense organs Organs that receive information from the environment. The sense organs include the eyes, ears, nose, tongue, and skin.

Sexual reproduction Reproduction that involves two parents, producing offspring that are not identical with either parent.

Skin The organ that covers and protects the body, and excretes wastes by perspiring.

Smog A haze in the atmosphere produced by the reaction of sunlight with pollutants from cars and factories.

Soil A mixture of small rock fragments and decayed organic material that covers much of Earth's land surface.

Solar energy Energy from the sun.

Solar system The sun and all the objects that revolve around it, including the planets and their moons, asteroids, comets, and meteors.

Sound A form of energy produced by a vibrating object.

Sound waves Alternating layers of compressed and expanded air particles that spread out in all directions from a vibrating object.

Species A group of organisms of the same kind that can produce fertile offspring.

Sperm ducts Tubes through which sperm cells pass upon leaving the testes.

Spinal cord The thick cord of nerve tissue that extends from the brain down through the spinal column.

Sterilization The killing of all micro-organisms in an area, usually by heating.

Stimulus A change in the environment that causes an organism to react in some way.

Storm A natural disturbance in the atmosphere that involves low air pressure, clouds, precipitation, and strong winds.

Succession See ecological succession.

System A group of related elements or parts that work together for a common purpose.

Technology The application of scientific knowledge and other resources to develop new products and processes.

Testes The male reproductive organs that produce sperm cells.

Thermal pollution An increase in the temperature of a body of water, caused by human activities, that may be harmful to living things in that environment.

Thunderstorm A brief, intense rainstorm that affects a small area and is accompanied by thunder and lightning.

Tides The rise and fall in the level of the ocean's waters that take place twice each day.

Tissue A group of similar cells that act together to perform a function.

Tornado A violent whirling wind, sometimes visible as a funnel-shaped cloud.

Toxic Poisonous.

Trachea The tube that connects the nose and mouth to the bronchi, which lead to the lungs; also called the windpipe.

Uranium A radioactive element found in certain rocks and used as a fuel for nuclear power plants.

Uterus The organ of the female reproductive system within which the offspring develop; also called the womb.

Vapors Fumes or gases given off by a substance.

Veins Blood vessels that return blood to the heart.

Volcano (1) An opening in Earth's surface through which hot, liquid rock flows from deep underground. (2) A mountain formed by a series of volcanic eruptions.

Volume The amount of space an object occupies.

Voluntary muscles Muscles that we consciously control.

Warm front The boundary formed when a warm air mass slides up and over a cool air mass.

Water cycle The process in which water moves back and forth between Earth's surface and the atmosphere by means of evaporation, condensation, and precipitation.

Watt A unit that measures the rate at which electrical energy is used.

Weather The changing condition of the atmosphere, with respect to heat, cold, sunshine, rain, snow, clouds, and wind.

Weathering The breaking down of rocks into smaller pieces.

Wind The movement of air over Earth's surface.

Wind direction The direction from which the wind is blowing.

Work The moving of an object over a distance by a force.

INDEX

Note: "t" and "f" denote table and figure references.

Absorbed light, 137
Acceleration, 112
Acid rain, 91, 223, 242, 256
Adaptations, 178
  of organisms, 178t
Adrenaline, 195
Air masses, 83-84
  types of, 86t
Air pressure, 76
Air pressure systems, 86t
Air temperature, 76
Altitude, climate and, 79
Ammonia, handling, 22t
Analyzing data
  in tables, 174
  using math in, 116
Animal cells, 172
  plant cells compared with, 172f
Animal wastes, energy from, 229-230
Appliances, high-efficiency, 225-226
Aristotle, 210
Arm muscular system, 193f
Arteries, 197-198
Artificial heart, 236t
Asexual reproduction, 179
Atmosphere, 72, 75-80
  composition of, 73
  interaction of humans and, 89-92
  interactive cycles in, 73-74
Atomic number, 145
Atomic structure, 145
Atomic theory, 213
Atoms, 145
Averaged results, 161
Axis of rotation, 61

Balanced forces, 112t
Balance of nature
  biological, 250
  upsetting, 166
Bar graphs, 17-18
Bedrock, 43
Binary fission, 179f
Biodiversity, 259-260
Biological balance in nature, 250
Bird species, 174t
Black lung disease, 223
Bladder, 202
Blizzards, 91
Blood, 197
Blood vessels, 197-198
Body systems. *See* Human body systems
Bohr, Niels, 213
Boiling, 148
Boiling point, 149, 149t
Bones, 191-192
Brain, 194
Breastbone, 191-192
Bronchi, 200
Budding, 179f
Buildings, insulated, 226
Butterfly, life cycle of, 181f

Calculations, performing
  with density, 36t
  with measurements, 34
Calendar year, 65
Calorie, 217
Capillaries, 197-198
Carbon dioxide, 162f, 200
Cartilage, 191-192
Caustic substances, 21
Cell division, 182
Cell membrane, 171
Cells, 171-172, 190
Cell structures, functions of, 172t
Cellular reproduction, 182
Cellular respiration, 200
Celsius thermometer, 38-39
Centrioles, 172
Chemical bonds, 146
Chemical changes
  energy and, 154-155
  examples of, 152t
  in matter, 151-153
Chemical energy, 123
Chemical equations, 152
Chemical formulas, 146t
Chemical properties of minerals, 44
Chemical reaction, 146
Chemicals
  household, 21-22
  working with, 22-23
Chemical weathering, 51
Chlorine bleach, handling, 22t
Chlorophyll, 162
Chloroplasts, 171
Chromosomes, 182, 183
Circulatory system, human, 197-198
Class, 3
Classification, 2
  of living things, 2-4
  of rocks, 4-5
Classifying
  data, 4
  observing and, 108-109
Cleavage of minerals, 44
Climates, 79-80
  adapting to changing, 175
  changes in, 254
Climax community, 250, 251f
Clouds, 76
Coal, 218
  advantages and disadvantages of, 229t
  problems with, 223
Cold fronts, 85f
Color of minerals, 44
Commensalism, 166t
Communications satellites, 243
Communities, biological, 160, 250
Competition for resources, 251

Compound machines, 103-104
Compounds, 145-146
Concave lens, 138f
Condensation, 73, 148
Conglomerate, 47t
Conservation, 225
  of energy, 225
  of Energy, Law of, 125-126
  of matter, 153
Constellations, 68
Consumers, animals as, 163-164
Consumption of energy. *See* Energy consumption
Continental drift, 55, 209-210
Continental shelves, 57
Continental slopes, 57
Controls, 13, 207
Convection, 77
Convex lens, 138f
Copernicus, Nicholas, 210
Corrosive substances, 21
Crustal plates, 54-55
Crust of Earth. *See* Earth's crust
Cubic centimeters, 32
Cytoplasm, 171

Daily time, 62-63
Dalton, John, 213
Data
  graphing, 156
  interpreting, 149
    and classifying, 4
  in tables
    analyzing, 174
    interpreting, 45, 56, 92-93, 174
    putting, 10
  using math in analyzing, 116
Decision making, technology and, 246
Decomposers, organisms as, 163-164
Density, 35-38
  calculations with, 36t
Density currents, 96-97
Dental X rays, 246f
Designing experiments, 78
Diagrams, interpreting, 201
Digestive juices, 196t
Digestive system, human, 196-197
Dinosaur extinction, 261
Discoveries, major scientific, 208-213
Diseases, 10t
Diversity of life, 178-179
Dodo extinction, 262
Doppler effect, 131
Droughts, 254
Drugs, 21

Ear, human, 132
Earlobes, 18f
Earth, 61
Earthquakes, 54
Earth science, 235
Earth's crust, 43
  elements in, 145t
Earth's mantle, 55
Earth's oceans, 95-97
Earth's revolution, 64-65
  proof of, 68
  seasonal changes and, 66-67
  yearly seasonal time and, 65-66
Earth's rotation, 61-62
  daily time and, 62-63
Earth's spheres, 72
  energy exchange between, 72t
  interaction of, 72-74
Earth's surface, 43-48
  external forces on, 50-51
  forces that change, 50-57
  internal forces on, 53-55
Earth's wind belts, 79f
Ecosystems, 160, 162
  causes of change in, 250t
  changing, 250-251
  environment threats upsetting, 253-257
  human interferences in, 256-257
Efficiency of machines, 104
Effort, 101
Egg cells, 203
Electrical devices, 217t
Electrical energy, 123
Electricity, 218
  safe use of, 24-25
Electromagnetic spectrum, 138-139
Electromagnetic waves, 138-139
Electrons, 145
Elements, 5-6, 144-145
  in Earth's crust, 145t
  number discovered, by decade, 214t
  periodic table of, 5-6
Endocrine system, human, 194, 195f
Energy, 122
  chemical changes and, 154-155
  forms of, 123
  for future, 225-231
  Law of Conservation of, 125-126
  light, 136-137
  measuring, 217
  recycling and, 227-228
  sound, 130
  states of, 122
  from sun, 76-77
  unwanted, 125-126
Energy conservation, 225
Energy consumption, 217-218
  reducing, 225-226
Energy efficiency rating label, 226f
Energy exchange between Earth's spheres, 72t
Energy resources, 228t
Energy sources, 217-221
  advantages and disadvantages of, 229t
  problems with, 223-224
Energy transformation, nature of, 125-126
Environment, 160
  organisms and, 160-166
  results of change in, 259-261
  technology and, 238t
  threats to, upsetting ecosystems, 253-257
Enzymes, 197
Erosion, 51, 152-153
Evaporation, 73, 148
Evolution, 187
Excretory system, human, 202
Experiments, scientific, 207f
  designing, 78
  interpreting, 180-181
    results of, 186
  predicting results in, 52
Extinction, 261
Extrapolation, 16
Eye, human, 139-140

Fahrenheit thermometer, 38-39
Families of elements, 7
Family, 3
Faulting in crust, 53
Feedback system, 237f
Female reproductive system, 203f
Fertilization, 179, 182-183
Fires, forest, 255
First-class lever, 108
First law of motion, 114-115
Flammable substances, 21
Fleming, Alexander, 13
Floating objects, 37-38
Floods, 254
Flower, reproductive parts of, 173f
Folding of land, 53
Food chains, 162-164
Food pyramids, 165

Food webs, 164-165
Force, 101, 111-112
Forecasting weather, 87-88
Forest fires, 255
Fossil fuels, 218
  problems with, 223
Freezing, 148
Freezing point, 148
Frequency, wave, 131
Frog, life cycle of, 181f
Fronts, 84-85
Fuels, fossil. *see* Fossil fuels

Galaxy, 211
Galilei, Galileo, 211
Gallbladder, 196
Gases, 147-149
Gasoline, 218
  handling, 22t
Genes, 183
Genetic information, 182
Genetics
  Mendelian, 184
  traits studied in, 184t
Genus, 3
Geologic events, 253
Germ theory, 212-213
Glacial episode, 254
Glaciers, 51
Glands, 195
Global effects of technology, 242-243
Global warming, 256
Glossary, 297-302
Graduated cylinder, 33
Graphing data, 156
Graphs, interpreting, 133, 161, 219
Greenhouse effect, 91
Groundwater, 51
Groups of elements, 7

Habitat, 160, 178
Hair spray, handling, 22t
Hardness of minerals, 44
Harmful sound, 134
Hazards, weather, 89-91
Heart, human, 197f
  artificial, 236f
Heat, working with, 23-24
Heat energy, 123
Heliocentric solar system, 210-211
Herschel, William, 211
Hess, Harry, 209
Hibernation, 175
High-efficiency appliances, 225-226
High-pressure systems, 84

Home, safety at, 21-22
Home-heating system, 237f
Hormones, 195
Household chemicals, 21-22
Hubble telescope, 211f
Human body systems
  interdependent, 190
  levels of organization in, 190-191
  organization, support, and movement of, 190-193
  regulation, digestion, and circulation in, 194-198
  regulation of, 194-195
  respiration, excretion, and reproduction in, 200-203
Human circulatory system, 197-198
Human digestive system, 196-197
Human ear, 132
Human endocrine system, 195f
Human excretory system, 202
Human eye, 139-140
Human heart, 197f
Human interactions with atmosphere, 89-92
Human interferences in ecosystems, 256-257
Human muscular system, 193
Human nervous system, 194
Human organ systems, 191
Human reproductive system, 202-203
Human respiratory system, 200f
Human skeletal system, 191-192
Human skeleton, 192f
Human teeth, 173f
Human urinary system, 202f
Humidity, 76
Hurricanes, 90, 255
Hydroelectric dam effect, 224f
Hydroelectric energy, 220-221
  advantages and disadvantages of, 229t
  problems with, 223
Hydrosphere, 72, 95-97

Igneous rocks, 4, 45-46
Inclined plane, 103
Inertia, 114
Inferences, 12
  making valid, 13
Information, organizing, into tables, 86
Inheritance of traits, 183
Inner ear, human, 132
Insulated buildings, 226

Interpreting
  data, 4, 149
    in tables, 45, 56, 92-93, 174
  diagrams, 201
  experiments, 180-181
    results of, 186
  graphs, 133, 161, 219
  tables, 7, 161
Involuntary muscles, 193

Joints, 192

Kepler, Johannes, 211
Kidneys, 202
Kilograms, 31
Kinetic energy, 122
Kingdom, 3

Laboratory, safety in, 22-25
Laboratory equipment, using safely, 26-27
Land exploitation, 256-257
Laser light, 139
Latitude, 79
Lava, 4, 45-46
Law of Conservation
  of Energy, 125-126
  of Matter, 153
Laws of motion, 114-116
Leap year, 65
Lenard, Philip, 213
Lenses, 138
Lever, 102, 108-109
Life, diversity of, 178-179
Life cycles, 180-181
Life processes, functions of, 171t
Life science, 235
Ligaments, 192
Light, 123, 136-140
  laser, 139
  reflected, absorbed, or transmitted, 137
  speed of, 137
  white, 139
Light energy, 136-137
Lightning, 90
Limbs, 191-192
Limestone, 48t
Line graphs, 15-16
Linnaeus, Carolus, 3
Liquids, 147-149
Liter, 32
Liver, 196, 202
Living things
  characteristics of, 171
  classification of, 2-4
Locomotion, 173, 193
Low-pressure systems, 84

Lungs, 202
Lye, handling, 22t
Lymph, 198
Lymph vessels, 198

Machines, 101
compound, 103-104
efficiency of, 104
simple, 102-103
work and, 101-102
Magma, 4, 45-46
Making predictions, 156
Male reproductive system, 203f
Mammary glands, 203
Mantle, Earth's, 55
Mass, 31, 112, 144
Materials, reusing, 227
Math, using, in analyzing data, 116
Matter, 144-149
changes in, 151-155
chemical changes in, 151-153
conservation of, 153
defining, 144
phases of, 147-149
physical changes in, 151
properties of, 152-153
Matthews, Drummond, 209
Measurements
making, 29-39
tools for, 29-33
performing calculations with, 34
precise, making, 30
Mechanical energy, 123
Melting, 148
Melting point, 148, 149t
Mendeleev, Dimitri, 5
Mendelian genetics, 184
Meniscus, 33
Metamorphic rocks, 5, 48
Meter, 29
Methane, 230
Metric system, 29
Middle ear, human, 132
Mid-ocean ridge, 55
Migration, animal, 175
Milky Way Galaxy, 211
Mineral identification chart, 45t
Minerals, 43-44
Mining, 223
Mitosis, 182
Mixtures, 146-147
Molecules, 146
Motion, laws of, 114-116
Motor neurons, 195
Mountains, 53
Mount St. Helens, 253
Muscular system, human, 193
Music, noise versus, 134
Mutations, 186-187
Mutualism, 166t

Natural gas, 218
advantages and disadvantages of, 229t
problems with, 223
Natural selection, 187
Nature, balance of. *See* Balance of nature
Nerve impulses, 194
Nerves, 194-195
Nervous system, human, 194
Neurons, 195
Neutrons, 145
Newton's laws of motion, 114-116
Nitrogen cycle, 74
Noise, music versus, 134
Nonrenewable resources, 228
Noon, 67
Nuclear energy, 123, 221
advantages and disadvantages of, 229t
problems with, 223-234
Nuclear power plant, 221f
Nuclear waste, 224
Nucleus
of atom, 145
of cell, 171
Nutrition
of organisms, 162
of plants and animals, 172-173

Observations, 12
organizing, 14-15
Observing, classifying and, 108-109
Ocean floor features, 55-57
Oceans of Earth, 95-97
Ocean water, motions of, 95-97
Oil, 218
advantages and disadvantages of, 229t
daily uses of, 219
problems with, 223
Orbit, 64
Order, 3
Organic matter, 43
Organisms, 160, 171
adaptations of, 178t
environment and, 160-166
nutrition of, 162
Organizing information into tables, 86
Organs, human, 191
functions of, 191t
Organ systems, human, 191
functions of, 191t
Outcrop, 43
Outer ear, human, 132
Ovaries, 203
Oviducts, 203
Oxygen, 162f, 200
Oxygen-carbon dioxide cycle, 74
Ozone layer, 242

Pancreas, 196
Paper, recycling, cases for and against, 228t
Parasitism, 165-166
Passenger pigeon extinction, 261
Pasteur, Louis, 212
Penis, 202-203
Periodic table of elements, 5-6
Perspiration, 202
Phases of matter, 147-149
Photosynthesis, 162
Phylum, 3
Physical changes, 151
in matter, 151
Physical properties, 147
of minerals, 43-44
Physical science, 235
Pictographs, 219
Plains, 54
Planets, 9t
Plant cells, 171-172
animal cells compared with, 172f
Plant matter, energy from, 229-230
Plants, sexual reproduction in, 173, 175
Plateaus, 54
Plate tectonics, 54-55
history of theory of, 209-210
Pollutants, 91, 256
Pollution, 91-92, 256
thermal, 224
Potential energy, 122
Precipitation, 76
Predicting
experimental results, 52
results, 78, 154
Predictions, making, 156
Prefixes used in metric system, 29t
Pressure, air, 76
Pressurized containers, 21
Prevailing winds, 79
Producers, plants as, 163-164
Properties of matter, 152-153
Protons, 145
Ptolemy, 210
Pulley, 102

Pulling forces, 111t
Punnett squares, 184-185
Pushing forces, 111t

Quantitative relationships
determining, 156

Radiation, 242
Ramsay, Sir William, 214
Rate of reaction, 15
Rayleigh, Lord, 214
Reactants, 152
Reactions
rate of, 155
reversible, 155
Recycling, energy and, 227-228
Recycling paper, cases for and against, 228t
Redi, Francesco, 180
Reflected light, 137
Regeneration, 179f
Regulation of human body systems, 194-195
Renewable resources, 228-230
Reproduction, 179
cellular, 182
sexual. *See* Sexual reproduction
Reproductive system, human, 202-203
Resistance to effort, 101
Respiration, 200
Respiratory system, human, 200f
Results
averaged, 161
interpreting, of experiments, 186
predicting, 78, 154
Reusing materials, 227
Reversible reactions, 155
Revolution, 64
of Earth. *See* Earth's revolution
Ribs, 191-192
Rock cycle, 48
Rocks, 44-48
classification of, 4-5
Rotation, 61
of Earth. *See* Earth's rotation
Rush, Benjamin, 212
Rutherford, Ernest, 213

Safety
at home, 21-22
in laboratory, 22-25
Safety equipment, 22
Safety goggles, 24
Samples, 13
San Andreas Fault, 58
Sandstone, 47t, 48t
Satellites, communications, 243
Scales, 31-32
Schematic diagrams, 201
Science, 2, 235
defined, 206
effects of, on society, 240-241
effects of society on, 241-242
interaction of technology and society and, 240-243
major fields of, 235
relationship of technology and, 235-238
Scientific discoveries, major, 208-213
Scientific experiments, 207f
Scientific method, 206-207
Scientific procedures, 13
Screw, 103
Seafloor spreading, 55, 209
Seamounts, 57
Seasonal changes, 66-67
Seasonal time, yearly, 65-66
Second-class lever, 108
Second law of motion, 115, 116
Sedimentary rocks, 5, 47
Sediments, 5, 47
Seed dispersal, 175
Semen, 202-203
Sensation
of sight, 139-140
of sound, 132
Sense organs, 194-195
Sensory neurons, 195
Sexual reproduction, 179
in organisms, 182-183
in plants, 173, 175
Shale, 47t, 48t
Sharp tools, working with, 25
Sight, sensation of, 139-140
Siltstone, 47t
Simple machines, 102-103
Skeletal system, human, 191-192
Skeleton, human, 192f
Skin, 202
Skull, 191-192
Smog, 91
Snow line, 59t
Society
effects of, on science and technology, 241-242
effects of science and technology on, 240-241
interaction of science and technology and, 240-243
Soil, 43
Solar cell, 230
Solar collector, 230f
Solar energy, 230-231
advantages and disadvantages of, 229t
Solar system, 61f
heliocentric, 210-211
Solids, 147-148
Sound, 130-134
harmful, 134
sensation of, 132
speed of, 131-132
temperature and, 133f
Sound energy, 130
Sound waves, 130-131
Sources of energy. *See* Energy sources
Species, 3, 179
declining, 260
new, introduction of, 257
Speed
of light, 137
of sound, 131-132
temperature and, 133f
Sperm ducts, 202-203
Spinal column, 191-192
Spinal cord, 194
Spontaneous generation, 180
Stars, 68
Storms, 90-91, 255
number of, per month, 92t
Strip mining, 223
Subatomic particles, properties of, 145t
Succession, 250, 251f
Sun, 61
angle of rays from, 77f
energy from, 76-77
Surface currents, 96
Surface of Earth. *See* Earth's surface
Sweat glands, 202f
Symbiosis, 165-166
Systems, 236-237

Tables
data in
analyzing, 174
interpreting, 45, 56, 92-93, 174
interpreting, 7, 161
organizing information into, 86
Technology, 235
assessing, 245
benefits and burdens of, 246t
decision making and, 246
effects of, on society, 240-241
effects of society on, 241-242
environment and, 238t
global effects of, 242-243
increasing choices with, 245

interaction of science and society and, 240-243
making decisions about, 245-246
relationship of science and, 235-238
as system, 236-237
in use, 237-238
Teeth, human, 173f
Temperature
air, 76
measuring, 38-39
sound speed and, 133f
Tendons, 192
Testes, 202-203
Test tube holders, 23-24
Thermal pollution, 224
Thermostat, 237
Third-class lever, 108
Third law of motion, 115-116
Thomson, Joseph J., 213
Thunderhead, 90f
Thunderstorms, 90, 255
Tides, 96
Time
daily, 62-63
yearly seasonal, 65-66
Tissues, 190
Tongs, 23-24
Tools, sharp, working with, 25
Tornadoes, 90, 91f, 255
Toxic substances, 21
Trachea, 200
Traits
inheritance of, 183
studied in genetics, 184t
Transformation, energy, nature of, 125-126
Transmitted, 137
Trenches, oceanic, 56
Triple-beam balance, 31
Tsunami waves, 97
Turbine, 220

Unbalanced forces, 112t
Unwanted energy, 125-126
Uranium, 221
Urea, 202
Urinary system, human, 202f
Urine, 202
Uterus, 203

Vagina, 203
Vapors, 21
Variables, 13
Veins, 197-198
Vine, Fred, 209
Volcanic rocks, 45-46
Volcano, 54
Volume, 32-33
Voluntary muscles, 193

Warm fronts, 85f
Warming, global, 256
Waste, nuclear, 224
Water cycle, 73-74
Watt, 217
Wave frequency, 131
Waves, winds and, 96
Weather, 75
defining, 75
Weather elements, 76
Weather forecasting, 87-88
Weather hazards, 89-91
Weathering, 50
Weather map, 87f
Weather systems, large-scale, 83-88
Wedge, 103
Wegener, Alfred, 209-210
Weight, 31
Wheel and axle, 102
White light, 139
Wildfires, causes of, 255t
Wind belts, Earth's, 79f
Wind energy, advantages and disadvantages of, 229t
Winds, 51, 76
prevailing, 79
waves and, 96
Wind turbine, 229, 230f
Winter storms, 91
Work, 101, 217
machines and, 101-102

X rays, dental, 246f

Year, 65
Yearly seasonal time, 65-66